Progress in Systems and Control Theory
Volume 6

Diederich Hinrichsen Bengt Mårtensson
Editors

Control of Uncertain Systems

Proceedings of an International Workshop, Bremen, West Germany, June 1989

With 48 Illustrations

Springer Science+Business Media, LLC 1990

Diederich Hinrichsen
Institut für Dynamische Systeme
Universität Bremen
Postfach 330 440
D-2800 Bremen 33
Federal Republic of Germany

Bengt Mårtensson
Institut für Dynamische Systeme
Universität Bremen
Postfach 330 440
D-2800 Bremen 33
Federal Republic of Germany

Library of Congress Cataloging-in-Publication Data
Control of uncertain systems : proceedings of an international workshop held at Bremen, West Germany, June 1989 / Diederich Hinrichsen, Bengt Mårtensson.
p. cm. — (Progress in systems and control theory ; v. 6)
Includes bibliographical references.
ISBN 978-1-4757-2110-2 ISBN 978-1-4757-2108-9 (eBook)
DOI 10.1007/978-1-4757-2108-9

1. Control theory—Congresses. 2. Stochastic processes—Congresses. I. Hinrichsen, Diederich, 1939– . II. Mårtensson, Bengt. III. Series.
QA402.3.C637 1990
629.8'312—dc20 90-1004

Printed on acid-free paper.

ISBN 978-1-4757-2110-2

Camera-ready copy provided by the editors.

9 8 7 6 5 4 3 2 1

Preface

Invoking the machinery of Control Theory to control a real-world plant means, on the basis of available a priori knowledge of the plant, designing/selecting a feasible controller accomplishing the control objective. A priori knowledge about the plant is normally represented as a *mathematical model*, given by physical laws or measurement data. Such a model is inevitably *uncertain*, due to measurement errors, simplified models of natural laws, neglected dynamics, and the desire to get a model tractable for further computations. *Control of Uncertain Systems* is therefore an important issue, both practically and philosophically. Basically, two different approaches exist: Adaptive Control and Robust Control.

On four beautiful summer days in June 1989, researchers from 12 countries gathered together in Bremen, West Germany, for a workshop devoted to these issues. 49 talks, invited and contributed, were presented. The present work collects 18 of these talks. (Program and original abstracts from the workshop are available as Report 209, Institute for Dynamical System.) The papers represent a broad scope of current trends in robust and adaptive control. Topics covered include: New directions in adaptive control, stability analysis of uncertain systems, robust and adaptive stabilization, numerical methods, and topological aspects of robust control. The papers were carefully refereed and we would like to thank the referees for their efforts, which have resulted in substantial improvements.

The scientific committee consisted of R. F. Curtain (Groningen), D. Hinrichsen (Bremen), and A. J. Pritchard (Warwick). The workshop was sponsored by a grant from the Commission of the European Communities to the Universities of Bremen, Groningen and Warwick. We gratefully acknowledge their support.

Bremen, April 19, 1990

DIEDERICH HINRICHSEN
BENGT MÅRTENSSON

Contents

Contributors

Eyad H. Abed, Department of Electrical Engineering and Systems Research Center, University of Maryland, College Park, MD 20742, U.S.A.

Jan Bontsema, Department of Agricultural, Engineering and Physics, University of Wageningen, Duivendaal 1, 6701 AP Wageningen, The Netherlands

Fritz Colonius, Institut für Mathematik, Universität Augsburg, Universitätsstraße 2, D-8900 Augsburg, F.R.G.

Ruth F. Curtain, Mathematics Department, University of Groningen, P.O. Box 800, 9700 AV Groningen, The Netherlands

E. J. Davison, Department of Electrical Engineering, University of Toronto, Toronto M5S 1A4, Canada

Filippo De Mari, Institut für Dynamische Systeme, Universität Bremen, D-2800 Bremen 33, F.R.G.

T. T. Georgiou, Department of Electrical Engineering, University of Minnesota, Minneapolis, MN 55455, U.S.A.

G. C. Goodwin, Department of Electrical Engineering and Computer Science, University of Newcastle, NSW 2308, Australia

U. Helmke, Naturwissenschaftliche Fakultät I — Mathematik — , Universität Regensburg, F.R.G.

D. Hinrichsen, Institut für Dynamische Systeme, Universität Bremen, D-2800 Bremen 33, F.R.G.

K. H. Kienitz, Institut für Automatik und Industrielle Elektronik, Eidgenössische Technische Hochschule Zürich, ETH-Zentrum / ETL, CH-8092 Zürich, Switzerland

Wolfgang Kliemann, Department of Mathematics, 400 Carver Hall, Iowa State University, Ames, Iowa 50011, U.S.A.

B. A. León de la Barra, Department of Electrical Engineering and Computer Science, University of Newcastle, NSW 2308, Australia

Izchak Lewkowicz, Department of Electrical Engineering, Technion — Israel Institute of Technology, Haifa 32000, Israel

R. J. Mazzaferri, Department of Electrical Engineering and Computer Science, University of Newcastle, NSW 2308, Australia

J. V. Medanić, University of Illinois at Urbana-Champaign, Coordinated Science Laboratory, Urbana, IL, U.S.A.

M. Milanese, Dipartimento di Automatica e Informatica, Politecnico di Torino, Corso Duca degli Abruzzi, 24 - 10129 Torino, Italy

D. E. Miller, Engineering Department, Cambridge University, Trumpington Street, Cambridge CB2 1PZ, GB

Raimund Ober, Engineering Department, Cambridge University, Trumpington Street, Cambridge CB2 1PZ, GB

W. R. Perkins, University of Illinois at Urbana-Champaign, Coordinated Science Laboratory, Urbana, IL, U.S.A.

V. Peterka, Institute of Information Theory and Automation, Pod vodárenskou věží 4, CS 182 08 Prague, Czechoslovakia

D. Prätzel-Wolters, Fachbereich Mathematik, Universität Kaiserslautern, F.R.G.

A. J. Pritchard, Control Theory Centre, University of Warwick, Coventry CV4 7AL, UK

E. P. Ryan, School of Mathematical Sciences, University of Bath, Claverton Down, Bath BA2 7AY, UK

Lahcen Saydy, Faculté des Sciences, Université Cadi Ayyad, Marrakech, Morocco

S. Schmid, Fachbereich Mathematik, Universität Kaiserslautern, F.R.G.

Raphael Sivan, Department of Electrical Engineering, Technion — Israel Institute of Technology, Haifa 32000, Israel

M. C. Smith, Department of Electrical Engineering, The Ohio State University, Columbus, OH 43210, U.S.A.

G. Sonnevend, (on leave from Inst. Math. Univ. Budapest, Hungary), Institut für Angewandte Mathematik, Universität Würzburg, Am Hubland, D-8700 Würzburg, F.R.G.

Eduardo D. Sontag, SYCON — Rutgers Center for Systems and Control, Department of Mathematics, Rutgers University, New Brunswick, NJ 08903, U.S.A.

André L. Tits, Department of Electrical Engineering and Systems Research Center, University of Maryland, College Park, MD 20742, U.S.A.

H. Unbehauen, Ruhr-University Bochum, Dept. of Electrical Engineering, D-4630 Bochum 1, F.R.G.

R. J. Veillette, University of Illinois at Urbana-Champaign, Coordinated Science Laboratory, Urbana, IL, U.S.A.

A. Vicino, Dipartimento di Sistemi e Informatica, Universitá di Firenze, Via di Santa Marta, 3 - 50139 Firenze, Italy

Robust Stabilization of a Flexible Beam Model Using a Normalized Coprime Factorization Approach

Jan Bontsema Ruth F. Curtain

The problem of robustly stabilizing a linear system subject to H_∞-bounded perturbations in the numerator and the denominator of its normalized left coprime factorizations is considered for a class of infinite-dimensional systems. This class has possibly unbounded, finite-rank input and output operators which includes many delay and distributed systems. The optimal stability margin is expressed in terms of the solutions of the control and filter algebraic Riccati equations. The applicability of this theory is demonstrated by a controller design for a flexible beam with uncertain parameters.

1 Introduction

This paper extends the results of Glover and McFarlane [17], [19] on robust stabilizability of normalized coprime factors to a large class of infinite-dimensional systems, the Pritchard-Salamon class. The problem of robust stability of closed-loop systems has received much attention in the literature and for a discussion of and references to the finite-dimensional literature we refer the reader to [17], [18], [19]. In section 2 the Pritchard-Salamon class of systems is introduced and its special properties are listed and discussed. It is a state-space class of systems and its relationships to the Callier-Desoer class of transfer matrices is discussed in section 3, where concepts and properties relevant to control synthesis are detailed. The main result of section 4 is that for the class of Pritchard-Salamon systems, (C, A, B), such that (A, B) is exponentially stabilizable and (C, A) is exponentially detectable it is possible to extend the arguments in [17], [18] in a natural way. The details of the arguments may be found in [11]. The exponentially stabilizable and detectable Pritchard-Salamon class is not the most general class of infinite-dimensional linear systems (see [29]), but it does include many delay and p.d.e. systems with unbounded inputs and outputs (see [24], [10]). In [15] it was shown that the problem of robustness optimization for normalized coprime factor perturbations is equivalent to robust optimization in the gap metric. This covered both finite- and infinite-dimensional systems.

Of course these theories only consider unstructured perturbations, whereas in flexible structures (and other applications [20]) one usually has to take structured perturbations into account. Here we consider a prototype example of a p.d.e. model of a damped beam in which we suppose that the damping and the stiffness coefficients are unknown. This p.d.e. model retains some essential characteristics typical of large flexible structures, such as uncertain damping, and point actuators and sensors, while at the same time it is possible to obtain a rigorous mathematical formulation in both time and frequency domain [1]. In particular, it belongs to the class of infinite-dimensional systems discussed in section 2 and 3 (see [1]). In the absence of a theory for robustness optimization under structured perturbations for infinite-dimensional systems, we decided to investigate how unstructured theories of robustness optimization would work on this prototype model. In [1] the theory of robustness optimization under additive perturbations from [12] was applied and here we apply the theory of robustness optimization with respect to additive stable perturbations of a normalized left-coprime factorization. Although there are countless numerical examples in the literature demonstrating that various controller theories produce good controllers for flexible systems [18], they do not provide much insight into the effect of uncertainty in the damping or other parameters. It is hoped that this parameter study will help in this direction.

The first part of investigation was the dependence of the maximal robustness margin on the damping and stiffness parameters. Taking a fixed pair of parameter values to specify the nominal model we designed the controller which (nearly) achieves the maximal robustness margin. Then we mapped the parameter region this controller actually stabilized which was larger than the region guaranteed by the theory. The region guaranteed by the theory can be calculated in terms of the T-gap between the nominal plant and the perturbed plant [15]. In section 5 we summarize the relevant results on the relationships to the T-gap metric from [15].

An important question is how one can best apply the infinite-dimensional robustness theories to this p.d.e. model and this is discussed in section 6 together with various approximation questions. In section 7 we give our numerical results and in section 8 we conclude with some remarks.

2 The Pritchard Salamon Class (C, A, B)

Definition 2.1. We suppose that there exist three separable Hilbert spaces V, X and W with continuous, dense injections satisfying

$$\mathcal{Z} = D_V(A) \hookrightarrow W \hookrightarrow X \hookrightarrow V \tag{2.1}$$

and we suppose that A generates a C_0-semigroup on V, X and W, which we denote by the same symbol, $S(t)$. The input and output maps are $B \in \mathcal{L}(U, V), C \in \mathcal{L}(W, Y)$, where $U = \mathbb{R}^q$ and $Y = \mathbb{R}^p$ are finite-dimensional spaces. Furthermore we suppose that B induces a smooth reachability map with respect to W on $[0, t_1]$ for all finite t_i, i.e. for some $\beta > 0$ and all $u \in L_2[0, t_1]$

$$\left\| \int_0^{t_1} S(t_1 - s)Bu(s)\, ds \right\|_W \leq \beta \|u\|_{L_2[0,t_1]} \tag{2.2}$$

and C induces a smooth observability map which extends to V on $[0, t_1]$ for all finite t_1, i.e. for some $\gamma > 0$ and all $x \in W$

$$\|CS(\cdot)x\|_{L_2[0,t]} \leq \gamma \|x\|_V. \tag{2.3}$$

Under the above assumptions we call (C, A, B) *a Pritchard-Salamon system.*

Remarks

R1: In fact in Weiss [27] and [28] it is shown that without loss of generality we may always take $\mathcal{Z} = D_V(A) = W$ and $V = D(A^*)^*$ and so (2.1)(i) and (2.1)(ii) are automatically satisfied. However, sometimes it proves convenient to allow the choice of V and W depend on the example in question, as for example in [10]. □

R2: The role of X is to some extent artificial and its only purpose is to define a dual system (B^*, A^*, C^*). To do this we identify the duals of U, Y and X. Using X as a pivot space we take V^* to be the space of all linear functionals on V which are continuous with respect to the topology on X. W^* is defined analogously and so we have $V^* \hookrightarrow X \hookrightarrow W^*$. Then $S^*(t) \in \mathcal{L}(V^*) \cap \mathcal{L}(W^*)$ generates a C_0-semigroup on both V^* and W^* and $B^* \in \mathcal{L}(V^*, U)$ satisfies a (2.3) condition with respect to S^* if and only if B satisfies a (2.2) condition with respect to S. This and a similar statement can be found in [24]. It is possible to choose $X = V$ or W, thus eliminating X, but this formulation is not symmetric. □

R3: B is an admissible control operator for A under the weaker assumption that the controllability map be bounded on X ([27]) and so (2.2) requires a much smoother reachability map with its range in the smaller space, W. □

R4: C is an admissible observation operator for A under the weaker assumption that the observability map be bounded on X ([28]) and so (2.3) requires a smoother observability map defined on the larger space, V. □

R5: These assumptions still allow for large classes of p.d.e. and delay systems with unbounded B and C operators and these technical assumptions can be readily verified (see [24], [10]). □

Important properties of Pritchard Salamon class

P1: The Pritchard-Salamon class is closed under perturbations. If $F \in \mathcal{L}(V,U)$, then $A+BF$ generates a C_0-semigroup on W, X and V and if $H \in \mathcal{L}(Y,W)$, $A+HC$ generates a C_0-semigroup on W, X and V. ([24]). In addition, if (C,A,B) is a Pritchard-Salamon system, F and H as above, then $(C,A+BF,B)$ and $(C,A+HC,B)$ are also Pritchard-Salamon systems with respect to the same triple $W \hookrightarrow X \hookrightarrow V$, ([24]). □

P2: The Pritchard-Salamon class has important frequency domain properties which were established in [10]. For all s in $\mathrm{Re}\, s > w$, the growth bound of $S(t)$, $(sI-A)^{-1}B \in \mathcal{L}(U,V) \cap \mathcal{L}(U,W)$ and $C(sI-A)^{-1} \in \mathcal{L}(W,Y)$ has a bounded extension to $\mathcal{L}(V,Y)$. So the *transfer function* $C(sI-A)^{-1}B$ is a well-defined analytic function in $\mathrm{Re}\, s > w$ and it is norm-bounded there. □

P3: We say that (A,B) is *exponentially stabilizable* if there exists an $F \in \mathcal{L}(V,U)$ such that $A+BF$ generates an exponentially stable C_0-semigroup (on V, X and W according to P1). In this case $g(s) = K(sI-A-BF)^{-1}B$ is in the Hardy space H_2 and $|g(s)| \to 0$ as $|s| \to \infty$. (i.e. as $\rho \to \infty$ in $\{s \in \mathbb{C}^+ : |s| \geq \rho\}$). □

P4: We say that (C,A) is *exponentially detectable* if there exists an $H \in \mathcal{L}(Y,W)$ such that $A+HC$ generates an exponentially stable C_0-semigroup on V, X and W. In this case $h(s) = C(sI-A-HC)^{-1}L$ for any $L \in \mathcal{L}(U,V)$ and $|h(s)| \to 0$ as $|s| \to \infty$. □

P5: If (C,A) is exponentially detectable and (A,B) is exponentially stabilizable, then the *Control Algebraic Riccati Equation* (CARE) (2.4) has a unique non-negative definite self-adjoint solution $Q \in \mathcal{L}(V,V^*)$, and the *Filter Algebraic Riccati Equation* (FARE) (2.5) has a unique non-negative definite self-adjoint solution $P \in \mathcal{L}(W^*,W)$ such that

$$A^*Qz + QAz - QBB^*Qz + C^*Cz = 0 \quad \text{for } z \in \mathcal{Z} \tag{2.4}$$

$$APz + PA^*z - PC^*CPz + BB^*z = 0 \quad \text{for } z \in \mathcal{Z}_2 = D_{W^*}(A^*) \tag{2.5}$$

Furthermore both $A - BB^*Q$ and $A - PC^*C$ generate exponentially stable semigroups on V, X and W. $PQ \in \mathcal{L}(V,W) \cap \mathcal{L}(W) \cap \mathcal{L}(V)$ and $QP \in \mathcal{L}(W^*,V^*) \cap \mathcal{L}(V^*) \cap \mathcal{L}(W^*)$. □

3 The Callier-Desoer Class of Transfer Functions

It is known that controller synthesis can be generally formulated over the Callier-Desoer class [5,6,7] of transfer functions (see section 8.2 in [25]).

Definition 3.1. Let $\mathcal{A}$ comprise all distributions $f(\cdot)$ with support in $(0,\infty)$ of the form

$$f(t) = f_a(t) + \sum_{i=1}^{\infty} f_i \delta(t - t_i) \tag{3.1}$$

where $\delta(\cdot)$ is the delta distribution, $t_0 = 0$ and t_i are positive real numbers; $t_0 \leq t_1 \leq t_2 \leq, \ldots$; f_i are real numbers and f_a is a measurable function in $L_1(0,\infty)$ and $\sum_{i=0}^{\infty} |f_i| < \infty$.

We say that $f \in \mathcal{A}_-(\alpha)$, if and only if for some $\alpha_1 < \alpha$, f has the decomposition (3.1) where $e^{-\alpha_1 \cdot} f_a(\cdot) \in L_1(0,\infty)$, and $\sum_{i=0}^{\infty} e^{-\alpha_1 t_i}|f_i| < \infty$.

$\hat{\mathcal{A}}_-(\alpha)$ denotes the set of Laplace transforms of elements in $\mathcal{A}_-(\alpha)$.

$\hat{\mathcal{A}}_-^\infty(\alpha)$ denotes the subset of $\hat{\mathcal{A}}_-(\alpha)$ consisting of those $\hat{f}$ which are bounded away from zero at infinity in $\mathbb{C}_\alpha = \{\operatorname{Re} s \geq \alpha\}$.

$\hat{\mathcal{B}}(\alpha)$ is the following commutative algebra of fractions: $\hat{\mathcal{B}}(\alpha) := [\hat{\mathcal{A}}_-(\alpha)][\hat{\mathcal{A}}_-^\infty(\alpha)]^{-1}$.

$\mathcal{M}(\hat{\mathcal{B}}(\alpha))$ and $\mathcal{M}(\hat{\mathcal{A}}_-(\alpha))$ will denote the class of matrix-valued transfer functions whose elements are in $\hat{\mathcal{B}}(\alpha)$ and $\hat{\mathcal{A}}_-(\alpha)$ respectively. We do not distinguish between the sizes of the matrices.

Properties of the Callier-Desoer class

CD1: Elements $G \in \mathcal{M}(\hat{\mathcal{B}}(\alpha))$ can be decomposed: $G = G_a + G_f$ where $G_a \in \mathcal{M}(\hat{\mathcal{A}}_-(\alpha))$ and G_f is a rational transfer function with all its poles in $\operatorname{Re} s \geq \alpha$. □

CD2: $G \in \mathcal{M}(\hat{\mathcal{B}}(\alpha))$ always has a *left-coprime* factorization over $\mathcal{M}(\hat{\mathcal{A}}_-(\alpha))$: $G = \tilde{M}^{-1}\tilde{N}$, where $\tilde{M}, \tilde{N} \in \mathcal{M}(\hat{\mathcal{A}}_-(\alpha))$, $\det(\tilde{M}) \in \mathcal{M}(\hat{\mathcal{A}}_-(\alpha))$ and there exist $X, Y \in \mathcal{M}(\hat{\mathcal{A}}_-(\alpha))$ such that $\tilde{N}Y - \tilde{M}X = I$. (i.e. $\tilde{M}$ and $\tilde{N}$ are *left coprime* over $\mathcal{M}(\hat{\mathcal{A}}_-(\alpha))$. An analogous statement holds for *right-coprime*. (In the sequel we shall only consider coprime factorizations over $\mathcal{M}(\hat{\mathcal{A}}_-(0))$ and so we shall omit the qualification "over $\mathcal{M}(\hat{\mathcal{A}}_-(0)$"). □

CD3: If the Pritchard-Salamon system (C, A, B) is such that (A, B) is exponentially stabilizable and (C, A) is exponentially detectable, then its transfer function $G(s) = C(sI - A)^{-1}B$ is in $\mathcal{M}(\hat{\mathcal{B}}(0))$. [10]. □

Extra properties of the Pritchard-Salamon class

P6: A *doubly coprime factorization* is a pair of left- and right-coprime factorizations $G = NM^{-1} = \tilde{M}^{-1}\tilde{N}$, where

$$\begin{bmatrix} \tilde{X} & -\tilde{Y} \\ -\tilde{N} & \tilde{M} \end{bmatrix}\begin{bmatrix} M & Y \\ N & X \end{bmatrix} = \begin{bmatrix} I & 0 \\ 0 & I \end{bmatrix} \quad \text{and } X, Y, \tilde{X}, \tilde{Y} \in \mathcal{M}(\hat{\mathcal{A}}_-(0)). \tag{3.2}$$

If (C, A, B) is a Pritchard-Salamon system such that (A, B) is exponentially stabilizable and exponentially detectable, then $G(s) = D + C(sI - A)^{-1}B$ with $D \in \mathcal{L}(U, Y)$ has the doubly coprime factorization given by

$$\begin{cases} M(s) = I + F(sI - A_F)^{-1}B; & \tilde{M}(s) = I + C(sI - A_H)^{-1}H \\ N(s) = D + C_F(sI - A_F)^{-1}B; & \tilde{N}(s) = D + C(sI - A_H)^{-1}B_H \\ X(s) = I - C_F(sI - A_F)^{-1}H; & \tilde{X}(s) = I - F(sI - A_H)^{-1}B_H \\ Y(s) = -F(sI - A_F)^{-1}H; & \tilde{Y}(s) = -F(sI - A_H)^{-1}H \end{cases} \tag{3.3}$$

where $F \in \mathcal{L}(V, U)$ is chosen such that $A_F = A + BF$ generates an exponentially stable semigroup and $H \in \mathcal{L}(Y, W)$ is chosen such that $A_H = A + HC$ generates an exponentially stable semigroup; $C_F = C + DF$; $B_H = B + HD$. □

P7: If (C, A, B) is a Pritchard-Salamon system with (A, B) exponentially stabilizable and (C, A) exponentially detectable then $G(s) = C(sI - A)^{-1}B$ has *normalized right- and left-coprime factorizations:* $G = MN^{-1} = \tilde{M}^{-1}\tilde{N}$ such that

$$\tilde{N}(j\omega)\tilde{N}^t(-j\omega) + \tilde{M}(j\omega)\tilde{M}^t(-j\omega) = I \tag{3.4}$$

$$N^t(-j\omega)N(j\omega) + M^t(-j\omega)M(j\omega) = I. \tag{3.5}$$

These normalized factorizations may be obtained by choosing $F = -B^*Q$ and $H = -PC^*$ in (3.3) where Q and P are the solutions of the CARE, (2.4) and the FARE, (2.5). □

Remarks

R6: CD1 and CD2 are well-known properties which can be found in [5,6,7]. CD3 was proved for bounded B and C in [22] and for the Pritchard-Salamon class in [10]. P6 was proved for general rings in [21] and for bounded B and C in [23] and for the Pritchard-Salamon class in [11]. P7 was proved for bounded B and C in [9] and for the Pritchard-Salamon class in [11]. □

R7: Notice that the extension of all these formulas to the Pritchard-Salamon class depends essentially on the special properties P1–P5. This allows us to manipulate expressions as in (3.3) just as if they were all bounded operators (or even matrices). This is of course not true for more general infinite-dimensional systems for which the expressions in (3.3) may not be well-defined, (see [29]). □

4 Robust Stabilization for Normalized Coprime Factors

This is a brief outline of the main ideas from [11] underlying the extension of the finite-dimensional theory of robust stabilization for normalized coprime factors in [17,19] to infinite-dimensional plants $G(s) = C(sI-A)^{-1}B$, where (C, A, B) is a Pritchard-Salamon system such that (A, B) is exponentially stabilizable and (C, A) is exponentially detectable. The robust stabilization problem is that of finding a feedback controller $K \in \mathcal{M}(\hat{\mathcal{B}}(0))$ which stabilizes not only the nominal plant G, but also the family of perturbed plants defined by

$$\mathcal{G}_\varepsilon = \{G_\Delta = (\tilde{M} + \Delta_M)^{-1}(\tilde{N} + \Delta_N) \text{ such that } \Delta_M, \Delta_N \in \mathcal{M}(\hat{\mathcal{A}}_-(0)) \text{ and } \|[\Delta_M, \Delta_N]\|_\infty < \varepsilon\} \tag{4.1}$$

where $\tilde{M}^{-1}\tilde{N}$ is a normalized left-coprime factorization of $G \in \mathcal{M}(\hat{\mathcal{B}}(0))$.

By *stability* of the feedback system (G, K) of figure 4.1 with $G, K \in \mathcal{M}(\hat{\mathcal{B}}(0))$ we mean that $\det(I - GK) \in \mathcal{M}(\hat{\mathcal{A}}_-^\infty(0))$ and $S = (I - GK)^{-1}$, KS, SG, $I - KSG \in \mathcal{M}(\hat{\mathcal{A}}_-(0))$.

We say that $(\tilde{M}, \tilde{N}, K, \varepsilon)$ is ***robustly stable*** if (G_Δ, K) is stable for all $G_\Delta \in \hat{\mathcal{G}}_\varepsilon$. Given $\varepsilon > 0$, if there exists a K such that $(\tilde{M}, \tilde{N}, K, \varepsilon)$ is robustly stable we say that (G, ε) is ***robustly stabilizable with robustness margin*** ε.

Using the properties of Pritchard-Salamon systems stated in section 2 and 3 we can extend all the finite-dimensional arguments in [17,19] in a natural way to reduce our problem to the following minimum distance one:

$$\inf_{J \in \mathcal{M}(\hat{\mathcal{A}}_-(0))} \left\| \begin{bmatrix} R + J \\ I \end{bmatrix} \right\|_\infty \tag{4.2}$$

where R is antistable and $R^+(s) := R^t(-s) \in \mathcal{M}(\hat{\mathcal{A}}_-(0))$ has the realisation

$$R^+(s) = C(I + PQ)(sI - A + BB^*Q)^{-1}B \tag{4.3}$$

where P and Q are the solution of the FARE, (2.5), and CARE, (2.4). This realization is a Pritchard-Salamon system and it is exponentially stabilizable and detectable. Furthermore the controllability and observability gramians for this realization of R^+ are given by

$$\begin{cases} Q_0 = Q + QPQ \in \mathcal{L}(V, V^*) \\ P_0 = (I + PQ)^{-1}P \in \mathcal{L}(W^*, W) \end{cases} \tag{4.4}$$

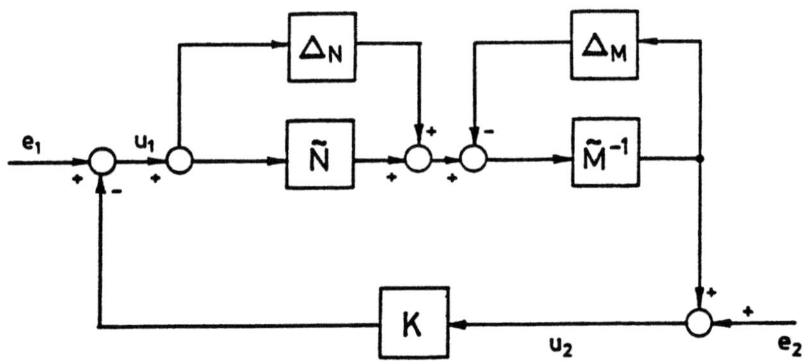

Figure 1. Left comprime factor perturbations

$$PQ = P_0Q_0 \in \mathcal{L}(V, W) \cap \mathcal{L}(W) \cap \mathcal{L}(V) \tag{4.5}$$

and the singular values of R^+ equal the square roots of the non-zero eigenvalues of $PQ = P_0Q_0$.

At this stage we are confronted with a difficulty which is peculiar to the infinite-dimensional case. While the minimization problem (4.2) always has a solution over the larger space, H_∞, ([3]), namely

$$\inf_{J \in H_\infty} \left\| \begin{bmatrix} R+J \\ I \end{bmatrix} \right\|_\infty = \sqrt{1 + \lambda_{\max}(PQ)} \tag{4.6}$$

it is not known if the minimizing J is in the smaller space $\mathcal{M}(\hat{\mathcal{A}}_-(0))$ as required by (4.2). The recent results in [13] show that the *relaxed distance problem* does have a solution over $\mathcal{M}(\hat{\mathcal{A}}_-(0))$:

$$\inf_{J \in \mathcal{M}(\hat{\mathcal{A}}_-(0))} \left\| \begin{bmatrix} R+J \\ I \end{bmatrix} \right\|_\infty \le \sigma, \quad \text{where } \sigma > \sqrt{1 + \lambda_{\max}(PQ)} \tag{4.7}$$

and it seems likely that (4.2) will too (cf. [4]). However, at present our conclusions are a little weaker than those in the finite-dimensional case, namely:

(G, ε) is robustly stabilizable with robustness margin ε if $\varepsilon < [1 + \lambda_{\max}(PQ)]^{-\frac{1}{2}} := \varepsilon_{\max}$ and it will not be robustly stabilizable if $\varepsilon > \varepsilon_{\max}$.

Finally we remark that it is possible to give explicit formulas for the robust controllers analogous to [17,19] using the parametrization in [13].

These are in general infinite-dimensional controllers which depend on the solutions of infinite-dimensional Riccati equations. So we have obtained a nice generalization of the robustness theory of [17,19] to the exponentially stabilizable and detectable Pritchard Salamon class of infinite-dimensional linear systems.

5 Optimal Robustness and the Gap Metric

Here we summarize results from [15] which we need in the sequel. For simplicity we suppose that P_1 and P_2 are in the Pritchard-Salamon class defined in section 2, although the results in [15] apply to more general plants. The directed T-gap is defined by

$$\vec{\delta}_T(P_1, P_2) = \inf_{Q \in H_\infty} \|[\tilde{M}_1, \tilde{N}_1] - Q[\tilde{M}_2, \tilde{N}_2]\|_\infty \tag{5.1}$$

and the T-gap by

$$\delta_T(P_1, P_2) = \max\{\vec{\delta}_T(P_1, P_2), \vec{\delta}_T(P_2, P_1)\} \tag{5.2}$$

where $P_i = \tilde{M}_i^{-1}\tilde{N}_i$ is a normalized left-coprime factorization of P_i, $i = 1, 2$.

In general $\vec{\delta}_T(P_1, P_2) \neq \vec{\delta}_T(P_2, P_1)$, but if $\delta_T(P_1, P_2) < 1$, then they are both equal.

There exists the following relationship between the Glover-McFarlane class of perturbations $\mathcal{G}_\varepsilon$ of (4.1) and the directed T-gap ball:

$$\begin{aligned} B_T(P, \varepsilon) = \{P_1 : \vec{\delta}_T(P, P_1) < \varepsilon\} = \{&P_1 = (\tilde{M} + \Delta_M)^{-1}(\tilde{N} + \Delta_N) : \\ &P = \tilde{M}^{-1}\tilde{N} \text{ is a normalized coprime factorization, } \Delta_M, \Delta_N \in H_\infty \\ &\text{and } \|[\Delta_M, \Delta_N]\|_\infty < \varepsilon\}. \end{aligned} \tag{5.3}$$

Notice that the set in (5.3) is larger than the set $\mathcal{G}_\varepsilon$ in (4.1), where Δ_m, Δ_n are restricted to the smaller class $\mathcal{M}(\hat{\mathcal{A}}_-(0))$. The main result in [15] is that a controller $K \in H_\infty$ stabilizes all P_1 with $\vec{\delta}_T(P, P_1) < \varepsilon$, if and only if K stabilizes all $P_1 = (\tilde{M} + \Delta_M)^{-1}(\tilde{N} + \Delta_N)$ where $\Delta_M, \Delta_N \in H_\infty$ satisfy $\|[\Delta_m, \Delta_n]\|_\infty < \varepsilon$. The advantage of the T-gap metric is that it can be calculated (at least for finite-dimensional plants) as a 2-block H_∞-optimization problem [14], [8].

6 Flexible Beam Model and Approximations

Consider the following model for a flexible Euler-Bernoulli beam with viscous damping: ([1, p.68])

$$w_{tt}(x, t) + \alpha_1 A w_t(x, t) + \alpha_2 A w(x, t) = \frac{1}{\rho a}(\delta(0)u_1(t) - \delta_x(0)u_2(t)) \tag{6.1a}$$

$$y_1(t) = w(0, t); \qquad y_2(t) = w_x(0, t) \tag{6.1b}$$

where $w(x,t)$ is the deflection of the beam, $-1 \leq x \leq 1$, α_1 and α_2 are the damping and stiffness coefficients, ρa is the mass per unit length and A is the stiffness operator $A = \frac{d^4}{dx^4}$ with domain

$$D(A) = \{f \in H^4(-1,1) | w_{xx}(\pm 1, t) = 0; w_{xxx}(\pm 1, t) = 0\}$$

It is easy to show that the transfer matrix for this model is given by

$$G(s) = \frac{1}{\rho a}\begin{bmatrix} \frac{1}{2s^2} & 0 \\ 0 & \frac{3}{2s^2} \end{bmatrix} + \frac{1}{\rho a}\sum_{i=3}^{\infty} \frac{1}{s^2 + \alpha_1 \lambda_i^4 s + \alpha_2 \lambda_i^4} \begin{bmatrix} v_i^2(0) & 0 \\ 0 & v_{ix}^2(0) \end{bmatrix} \quad (6.1c)$$

where λ_i and $v_i(x)$ are the eigenvalues and eigenfunctions of the stiffness operator A. For more details see [1] or [2].

It has been shown in [1] that this beam model has a Pritchard-Salamon state-space realization, and so we may apply the theory of section 4 directly. However, this would involve solving infinite-dimensional Riccati equations, which is very time consuming at best. In addition there are no error estimates available, which motivates us to consider a priori approximation.

Our approximation approach is based on the known properties of the transfer function (6.1c). In [1, Sec. 3.4] it is shown that G is the sum of an finite-dimensional part G_f, which contains the unstable modes and an infinite-dimensional stable part G_s :

$$G = G_f + G_s \quad (6.2)$$

where $G_s \in \mathcal{M}(\hat{\mathcal{A}}_-(0))$. Furthermore G_s is nuclear, which means that it is easy to approximate by a finite-dimensional system and a bound on the error can be calculated ([16]). For our investigations we take the 8th order approximation

$$G(s) = \frac{1}{\rho a}\begin{bmatrix} \frac{1}{2s^2} & 0 \\ 0 & \frac{3}{2s^2} \end{bmatrix} + \frac{1}{\rho a}\sum_{i=3}^{4} \frac{1}{s^2 + \alpha_1 \lambda_i^4 s + \alpha_2 \lambda_i^4} \begin{bmatrix} v_i^2(0) & 0 \\ 0 & v_{ix}^2(0) \end{bmatrix} \quad (6.3)$$

and for the nominal values of $\rho a = 42.7$, $\alpha_1 = 3.89 \cdot 10^4$ and $\alpha_2 = 1.129$ we have for the error:

$$\|G_s\|_\infty = \mu \leq 0.02 \quad (6.4)$$

For other values of α_1 and α_2 in the neighbourhood of the nominal values the error is of the same order of magnitude. G_f is rational and we can apply the theory of section 5 or [15] to G_f allowing for infinite-dimensional perturbations. If G_f has the normalized coprime factorization

$$G_f = \tilde{M}_f^{-1}\tilde{N}_f \quad (6.5)$$

then

$$G = G_f + G_s = \tilde{M}_f^{-1}(\tilde{N}_f + \Delta_N^f) \tag{6.6}$$

where

$$\|\Delta_N^f\|_\infty = \|\tilde{M}_f G_s\|_\infty \leq \|G_s\|_\infty \leq \mu. \tag{6.7}$$

From (6.7), and (5.1) with $Q = I$ we may conclude that

$$\vec{\delta}_T(G_f, G) \leq \|\Delta_N^f\|_\infty \leq \mu. \tag{6.8}$$

Reversing the roles of G and G_f, we can write

$$G_f = G - G_s = \tilde{M}^{-1}(\tilde{N} + \Delta_N) \tag{6.9}$$

where

$$\|\Delta_N\|_\infty = \|\tilde{M} G_s\|_\infty \leq \|G_s\|_\infty \leq \mu \tag{6.10}$$

and so we can apply the same argument above to conclude that $\vec{\delta}_T(G, G_f) \leq \mu < 1$ and hence they are equal

$$\vec{\delta}_T(G, G_f) = \vec{\delta}_T(G_f, G). \tag{6.11}$$

Suppose now that we apply the finite-dimensional theory of [18] on G_f to obtain a controller K_f with robustness margin ε. Then from the results quoted in section 5, we see that K_f is a robust controller for G with a robustness margin of at least $\varepsilon - \mu$. In other words, replacing G by G_f in our calculations incurs an error of at most μ and we have chosen μ to be negligible compared to the robustness margins of G_f for our range of parameter values in Table 7.1 below. This justifies using G_f in our calculations.

7 Numerical Results

The dependence of the maximal robustness margin $\varepsilon_{\max}$ (see [18]) on the parameters α_1 and α_2 is shown in table 7.1. For the system of order 8 we choose as nominal values $\alpha_1 = 3.89 \cdot 10^{-4}$ and $\alpha_2 = 1.129$. Then α_1 and α_2 are varied and we calculate the distance between the perturbed plant and the nominal one. If the distance (the directed T-gap, see section 5) between the nominal and perturbed plant is smaller than $\varepsilon_{\max}$ then both plants are guaranteed to be stabilized by a maximally robust controller (see [18]). In order to calculate the directed T-gap (eqn. 5.1) we have to solve a 2-block H_∞-problem which can be a numerically hard problem. For this reason we used the following estimates for the directed T-gap: Let G_1 and G_2 have normalized left and right coprime factorizations: $G_1 = \tilde{M}_1^{-1}\tilde{N}_1 = N_1 M_1^{-1}$,

α_1 / α_2	$3.89 \cdot 10^{-2}$	$3.89 \cdot 10^{-3}$	$3.89 \cdot 10^{-4}$	$3.89 \cdot 10^{-5}$	$3.89 \cdot 10^{-6}$
112.9	0.3827	0.3827	0.3827	0.3827	0.3827
11.29	0.3827	0.3827	0.3827	0.3827	0.3827
1.129	0.3828	0.3828	0.3828	0.3828	0.3828
0.1129	0.3838	0.3838	0.3837	0.3836	0.3836
0.01129	0.3929	0.3935	0.3912	0.3896	0.3894

Table 1. The dependence of $\varepsilon_{\max}$ on α_1 and α_2.

α_1 / α_2	$3.89 \cdot 10^{-2}$	$3.89 \cdot 10^{-3}$	$3.89 \cdot 10^{-4}$	$3.89 \cdot 10^{-5}$	$3.89 \cdot 10^{-6}$
112.9	0.2115	0.2115	0.2115	0.2113	0.9098
	0.2115	0.2115	0.2115	0.2116	0.9549
11.29	0.2114	0.2113	0.2114	0.5649	0.9895
	0.2114	0.2114	0.2114	0.5716	1.0781
1.129	0.2092	0.1901	0.0	0.7985	0.967
	0.2092	0.1901	0.0	0.8249	1.0405
0.1129	0.2114	0.2114	0.5644	0.9892	1.0004
	0.2114	0.2114	0.5711	1.0778	1.1148
0.01129	0.2114	0.2124	0.9073	0.9990	0.9999
	0.2117	0.2132	0.9525	1.1070	1.1171

Table 2. The upper and lower bounds for the T-gap between nominal and perturbed plant.

$G_2 = \tilde{M}_2^{-1}\tilde{N}_2 = N_2 M_2^{-1}$. Defining $\tilde{R}_1 = \tilde{M}_1\tilde{M}_2^* + \tilde{N}_1\tilde{N}_2^*$, $R_2 = \tilde{M}_1 N_2 - \tilde{N}_1 M_2$ we have from [8] that:

$$\max\left(\|R_2\|_\infty, \tilde{H}_{R1}\right) \leq \vec{\delta}_T(G_1, G_2) \leq \sqrt{\|R_2\|_\infty^2 + \tilde{H}_{R1}^2} \qquad (7.1)$$

where $H_{R1} = \inf_{Q \in H_\infty} \|\tilde{R}_1 - Q\|_\infty$. Both the upper and lower bounds for different values of α_1 and α_2 are given in table 7.2. So we can find the region in (α_1, α_2) plane where a maximally robust controller is guaranteed to stabilize the perturbed plant. In order to see how conservative the guaranteed region is we also calculated the actual region where the central maximally robust controller ($K_{\max}$) stabilizes the perturbed plant. The same was done for a suboptimal controller (K_{subopt}), (see [18]) with robustness margin 0.33 . The results are shown in fig. 7.1. In [1] robust stabilization w.r.t. additive perturbations on the transfer function was considered and there the objective was to design a controller with an overall decay rate of 0.0065 (i.e. real parts of the closed loop poles left of -0.0065). It turned out that the guaranteed stability region in the (α_1, α_2) plane was very small and that the prediction was rather conservative. If the method

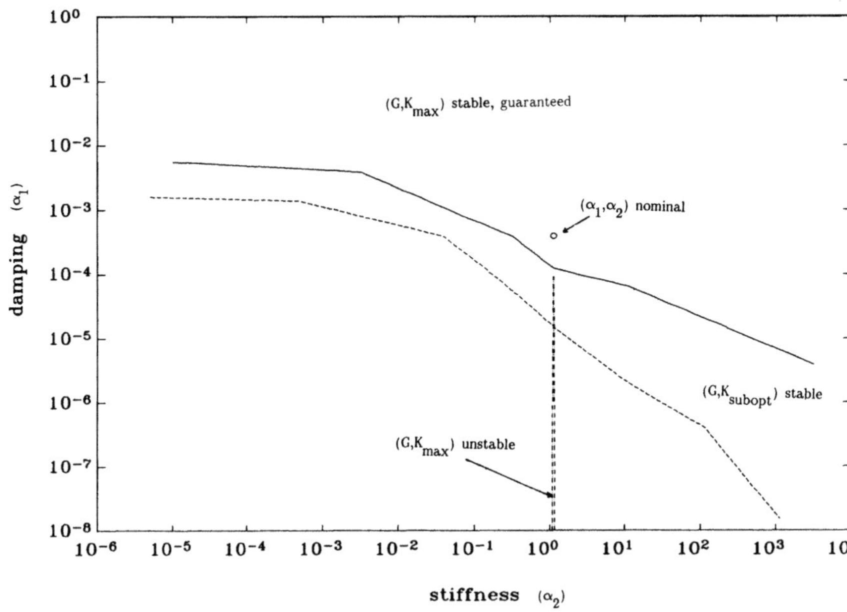

Figure 2. A priori guaranteed and actual stability regions in the (α_1, α_2)-plane.

of [18] is applied to the transfer function $G^{\beta}(s) = G(s - \beta)$ then the controller $K(s) = K^{\beta}(s + \beta)$ will move all the closed loop poles left of $-\beta$. In fig. 7.2 the guaranteed stability region (with $\beta = 0.0065$) is shown near the nominal values of α_1 and α_2. The actual stability region is much larger.

8 Conclusions

The theory of robust stabilization with respect to additive, stable perturbations of a normalized coprime factorization proposed in [17,18,19] seems to be a useful method for designing controllers even when the perturbations are structured. Here we have applied it to stabilize a flexible structure model with parameter uncertainty and it gave better results than the robustness theory with respect to additive perturbations proposed in [12]. For both methods the robustness margin decreases as the desired overall decay rate increases. Furthermore for our example it turns out that the maximal robustness margin is relatively insensitive to the order of the approximation of the transfer functions and to variations in the parameters. It seems that

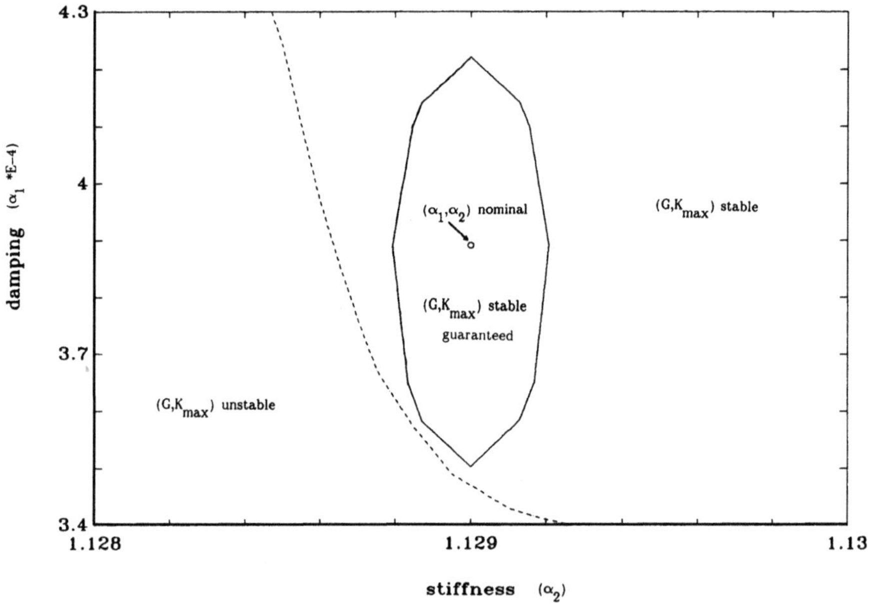

Figure 3. A priori guaranteed and actual stability region, with $\beta =$ 0.0065.

for this example the maximal robustness margin is determined by the rigid body modes.

REFERENCES

[1] J. Bontsema, *Dynamic stabilization of large flexible space structures*, Ph.D. thesis University of Groningen, 1989.

[2] J. Bontsema, R.F. Curtain, and J.M. Schumacher, *Robust control of flexible systems: a case study*, Automatica, vol.24, p.177–186, 1988.

[3] J.A. Ball, and J.W. Helton, *A Beurling-Lax theorem for the Lie group $U(m, n)$ which contains most classical interpolation theory*, J. Operator Theory, **8**, p. 107–142, 1983.

[4] J.A. Ball, and A. Ran, *Optimal Hankel Norm Model Reductions and Wiener-Hopf Factorization II: the Non-Canonical Case*, Integral Equations and Operator Theory (1987), **10**, p.416–436.

[5] F.M. Callier, and C.A. Desoer, *Simplifications and New Connections on an Algebra of Transfer Functions of Distributed Linear*, Time-Invariant Systems", IEEE Trans. Circuits of Systems, CAS-27, p. 320–323, 1980.

[6] F.M. Callier, and C.A. Desoer, *An Algebra of Transfer Functions for Distributed linear Time-Invariant Systems*, IEEE Trans. Circuits and Systems **25**, p. 651–663, 1978.

[7] F.M. Callier and C.A. Desoer, *Stabilization, Tracking and Distributed Rejection in Multivariable Convolution Systems*. Ann. Soc. Sci. Bruxelles, **94**, p. 7–51, 1980.

[8] C.C. Chu, J.C. Doyle and E.B. Lee, *The general distance in H_∞-optimal control theory*, Int. J. Control, vol.44, p.565–596, 1986.

[9] F.M. Callier and J. Winkin, *On Spectral Factorization and LQ-optimal Regulation of Multivariable Distributed Systems*, Int. J. Control (to appear).

[10] R.F. Curtain, *Equivalence of Input-Output Stability and Exponential Stability for Infinite-Dimensional Systems*, Math. System Theory, 1988, **21**, p.19–48.

[11] R.F. Curtain, *Robust Stabilizability of Normalized Coprime Factors: The Infinite-Dimensional Case*, 1990, International Journal of Control (to appear).

[12] R.F. Curtain and K. Glover, *Robust stabilization of infinite-dimensional systems by finite-dimensional controllers*, Systems and Control Letters, vol.7, pp.41–47, 1986.

[13] R.F. Curtain and A. Ran, *Explicit Formulas for Hankel Norm Approximations of Infinite-dimensional Systems*, Integral Equations & Operator Theory, 1989, **12**, p.455–469.

[14] T.T. Georgiou, *On the computation of the gap metric*, System and Control Letters, vol.11, pp.253–257, 1988.

[15] T.T. Georgiou, and M.C. Smith, *Optimal robustness in the gap metric*, 1990, IEEE Trans. Autom. Control (to appear).

[16] K. Glover, R.F. Curtain, and J.R. Partington, *Realization and approximation of linear infinite-dimensional systems with error bounds*, SIAM J. Control and Optim., vol. 26, pp. 863–899, 1989.

[17] K. Glover and D. McFarlane, *Robust stabilization of normalized coprime factors: An explicit H_∞-solution*, Proc. of the ACC, pp.842–847, 1988.

[18] D. McFarlane, and K. Glover, *Robust controller design using normalized coprime factor plant descriptions*, 1990, Springer Verlag LNCIS, no 138, Berlin.

[19] K. Glover and D. McFarlane, *Robust Stabilization of Normalized Coprime Factor Plant Descriptions with H_∞-bounded Uncertainty*, submitted to IEEE Trans. on Automatic Control, 1988.

[20] K.Glover and J.R. Partington, *Robust Stabilization of Delay Systems*, Proc. MTNS, 1989 (to appear).

[21] P.P. Khargonekar and E. Sontag, *On the Relation between Stable Matrix Fraction Factorizations and Regulable Realizations of Linear Systems over Rings*, IEEE Trans. AC-**27**, 1982, p.627–638.

[22] C.N. Nett, C.A. Jacobson and M.J. Balas, *Fractional Representation Theory: Robustness with Applications to Finite-Dimensional Control of a Class of Linear Distributed Systems*, Proc. IEEE Conf. on Decision and Control, p. 269–280, 1983.

[23] C.N. Nett, C.A. Jacobson and M.J. Balas, *A Connection Between State Space and Doubly Co-prime Fractional Representations*, Trans. IEEE, Vol. AC-21, 9, p. 831–832, 1984.

[24] A.J. Pritchard and D. Salamon, *The Linear Quadratic Optimal Control Problem for Infinite Dimensional Systems with Unbounded Input and Output Operators*, SIAM J. Control and Optimiz, **25**, p. 121–144, 1987.

[25] M. Vidyasagar, *Control System Synthesis: A Coprime Factorization Approach*, MIT Press, 1985.

[26] M. Vidyasagar and H. Kimura, *Robust controllers for uncertain linear multivariable systems*, Automatica, vol.22, pp.85–94, 1986.

[27] G. Weiss, *Admissibility of Unbounded Control Operator*, SIAM J. Control and Optim. **27**, p.527–545, 1989.

[28] G. Weiss, *Admissible Observation Operators for Linear Semigroups*, Israel J. Math., **65**, p. 17–43, 1989.

[29] G. Weiss, *The Representation of Regular Linear Systems on Hilbert Spaces*, Proceedings of the Conference on Distributed Parameter Systems, Vorau, Austria, July 1988, to appear.

Ruth F. Curtain
Mathematics Department
University of Groningen
P.O. Box 800
9700 AV Groningen
The Netherlands

Jan Bontsema
Department of Agricultural
Engineering and Physics
University of Wageningen
Duivendaal 1
6701 AP Wageningen
The Netherlands

Stability Radii and Lyapunov Exponents*

Fritz Colonius Wolfgang Kliemann

In the state space approach to stability of uncertain systems the concept of stability radius plays a central role. In this paper we use the classical concept of Lyapunov exponents, which describe the exponential growth behavior, in order to define a variety of stability and instability radii for families of linear systems $\dot{x} = [A + u(t)]x$, $u(t) \in U_\rho$, $\rho \geq 0$. Here $\{U_\rho, \rho \geq 0\}$ can denote sets of real or complex matrices, and the perturbation $u(t)$ can be deterministic or belong to different classes of stochastic processes. We analyze the stability radii and their relations, using Lyapunov exponents of associated bilinear control systems. The well developed theory of Lyapunov exponents for stochastic systems provides the background for the discussion of stability radii of systems excited by random noise. The example of the linear oscillator with uncertain restoring force is discussed in detail.

1 Introduction

It is the purpose of this paper to show that the concept of Lyapunov exponents can be used to define and to analyze various stability radii of matrices with respect to deterministic or stochastic (time-varying) perturbations. We hope to demonstrate that the well-developed theory of Lyapunov exponents for stochastic linear, parameter-excited systems [2, 3, 4] and also very recent results on Lyapunov exponents of bilinear control systems [8, 9, 10] (both theories are closely related) are of interest for researchers studying robustness properties of linear systems.

*Research supported in part by NSF grant No. DMS 8813976

Consider the family of linear systems in $\mathbb{R}^d$

$$\dot{x}(t) = [A + u(t)]x(t), \quad u(t) \in U_\rho, \rho > 0 \tag{1.1}$$

where $\{U_\rho; \rho \geq 0\}$ is an increasing family of sets of real $d \times d$-matrices. The admissible perturbations (presenting uncertainties) for each ρ with values in U_ρ are denoted by $\mathcal{U}_\rho$. For a stable matrix A, we define a (Lyapunov) stability radius r_L with respect to $\{U_\rho : \rho \geq 0\}$ as the lower bound of those ρ for which there exists $u \in \mathcal{U}_\rho$ such that the system (1.1) has a nonnegative exponential growth rate, i.e. a nonnegative Lyapunov exponent.

For deterministic perturbations $u(\cdot)$ the approach of this paper is an alternative to the one in [13, 16], where instead of Lyapunov exponents, Bohl exponents are used in order to define a stability radius. For the theory of Lyapunov and Bohl exponents of linear time varying differential equations we refer to [11] and [12]. The Bohl exponent is negative iff uniform asymptotic stability holds while negativity of the largest Lyapunov exponent is equivalent to asymptotic stability. Hence, for a fixed time-varying linear differential equation, the Bohl exponent is — in general — larger than the largest Lyapunov exponent. This implies that for a system (1.1) the corresponding Bohl stability radius is not greater than the corresponding Lyapunov stability radius. In fact, Theorem 5, below, states that for real perturbations the Bohl and the Lyapunov stability radii coincide. The (Bohl) stability radius studied in [16, 13] is based on complex perturbations (for special families $\{U_\rho; \rho \geq 0\}$). Hence it presents a more conservative robustness estimate than the radius for real perturbations. It is shown in [16], that the complex (Bohl-) radius coincides with the stability radius $r_{\mathbb{C}}$, which only takes into account constant complex perturbations. Below we will discuss in more detail the relation of our results to those in the literature. Here we remark only that (i) the Lyapunov exponents approach allows for a very general class of real perturbations and (ii) that stochastic perturbations can be analyzed within the same framework. The price for this is that contrary to [13, 15, 16], we do not obtain algebraic expressions, like associated Riccati equations, or explicit formulas, except in very special cases, (see e.g. Theorem 7, (2.1.2) and (2.2), Corollary 9 or the example of the linear oscillator, Example 9). For general systems with many degrees of freedom, reliable numerical methods are required (compare [14] and [18] for numerics in the case of time invariant uncertainties).

Section 2 introduces concepts of stability radii via Lyapunov exponents. In Section 3 a basic idea for the description of Lyapunov exponents of (bi-)linear systems is presented: The separation of the radial and the angular part of the solution, allowing the characterization of Lyapunov exponents on the sphere (resp. the projective space), a compact manifold.

Section 4 presents some basic facts about the maximal and minimal spectral values of *bilinear* control systems, which are crucial for the characterization of Lyapunov (in-)stability radii of *linear* systems. This is discussed in Section 5. Two examples of 2-dimensional systems are presented in Section 6, in particular the linear oscillator with uncertain restoring force: It is a simple, but prototypical example exhibiting much of the (surprising) behaviour of bilinear systems, and thus of the robustness properties of linear systems.

In Section 7 stability radii for stochastic excitations are defined for two classes of stochastic processes: stationary ones and stationary, ergodic, nondegenerate diffusions. While the theory for the first class parallels the results for the deterministic r_L, Markovian uncertainties yield a considerably more complicated behavior: Their radius of stability for all moments agrees with r_L, while the almost sure radius is in general greater. The difference is explained via the theory of large deviations. Finally, Section 8 discusses robust design, based on different information about the uncertainties.

2 Definitions of Deterministic Stability Radii

In this section we define several stability and instability radii using the concept of Lyapunov exponents.

Denote by $M(n,m;\mathbb{K})$ the set of $n \times m$ matrices over a field $\mathbb{K}$, $\mathbb{K} = \mathbb{R}$ or $\mathbb{C}$. Let $A \in M(d,d;\mathbb{R})$, let $\{U_\rho, \rho \geq 0\}$ be an increasing family of subsets in $M(d,d;\mathbb{K})$ and denote

$$\mathcal{U}_\rho := \{u : [0,\infty) \to U_\rho; u \text{ measurable and locally integrable}\} \tag{2.1}$$

Define for $u \in \mathcal{U}_\rho$ and $0 \neq x_0 \in \mathbb{R}^d$ the Lyapunov exponent

$$\lambda(x_0, u) = \limsup_{t\to\infty} \frac{1}{t} \log |\varphi(t, x_0, u)| \tag{2.2}$$

where $\varphi(\cdot, x_0, u)$ solves

$$\dot{x}(t) = [A + u(t)]x(t),\ t \geq 0, \quad x(0) = x_0. \tag*{$(2.3)_\rho$}$$

Note that for fixed u, there are at most d different $\lambda(x_0, u)$, $x_0 \neq 0$. Let for $\rho \in \mathbb{R}_+ = [0,\infty)$

$$\kappa_\rho(A) := \sup\{\lambda(x_0, u);\ 0 \neq x_0 \in \mathbb{R}^d,\ u \in \mathcal{U}_\rho\}. \tag{2.4}$$

Thus κ_ρ denotes the supremal Lyapunov exponent of the control system (2.3) with controls u in $\mathcal{U}_\rho$. A corresponding Lyapunov stability radius of $A \in M(d,d;\mathbb{R})$ is given by

$$r_L(A) = \inf\{\rho \geq 0; \kappa_\rho(A) \geq 0\}. \tag{2.5}$$

Naturally, $r_L(A)$ depends crucially on the family $\{U_\rho; \rho \geq 0\}$. For simplicity we suppress this dependence in our notation. For a stable matrix A (i.e. $\operatorname{Re}\lambda < 0$ for all λ in the spectrum $\sigma(A)$), $r_L(A)$ is the lower bound of those indices ρ, for which there exists $u \in \mathcal{U}_\rho$ such that the corresponding system $\dot{x} = [A + u]x$ is not asymptotically stable.

Similarly, instability radii can be defined via

$$\kappa_\rho^*(A) = \inf\{\lambda(x_0, u);\ 0 \neq x_0 \in \mathbb{R}^d, u \in \mathcal{U}_\rho\} \tag{2.6}$$

$$\kappa_\rho^*(A, x_0) = \inf\{\lambda(x_0, u);\ u \in \mathcal{U}_\rho\} \tag{2.7}$$

and

$$r_L^*(A) = \inf\{\rho \geq 0;\ \kappa_\rho^*(A) \leq 0\} \tag{2.8}$$

$$\tilde{r}_L(A) = \inf\{\rho \geq 0;\ \kappa_\rho^*(A, x_0) \leq 0 \text{ for all } x_0 \neq 0\} \tag{2.9}$$

The radius $r_L^*(A)$ makes sense for totally unstable matrices A (i.e. $\operatorname{Re}\lambda > 0$ for all $\lambda \in \sigma(A)$) and describes the lower bound of those indices ρ, for which there exists $u \in \mathcal{U}_\rho$, making $\dot{x} = [A + u]x$ asymptotically stable for some $x_0 \neq 0$. The radius r_L^* is the dual of the stability radius r_L in a precise sense, see Proposition 6, below.

On the other hand, the instability radius $\tilde{r}_L(A)$ makes sense for unstable matrices A (i.e. there is $\lambda \in \sigma(A)$ with $\operatorname{Re}\lambda > 0$) and describes the lower bound of these indices ρ for which one obtains asymptotic stability for all initial values $x_0 \neq 0$.

The concepts $r_L, r_L^*, \tilde{r}_L$ introduced so far, depend on the choice of the initial value in the following way: For r_L (and r_L^*) only existence of a point x_0 with stable (or unstable) solution is considered, while for $\tilde{r}_L$ the system has to be stable for all x_0, but the corresponding $u(\cdot)$ may still depend on x_0.

Remark 1. One may also introduce the following additional uniform concepts, cp. [9]:

$$\hat{\kappa}_\rho(A) = \sup_{u \in \mathcal{U}_\rho} \inf_{x_0 \neq 0} \lambda(x_0, u), \quad \hat{\kappa}_\rho^*(A) = \inf_{u \in \mathcal{U}_\rho} \sup_{x_0 \neq 0} \lambda(x_0, u) \tag{2.10}$$

and the corresponding radii

$$\hat{r}_L(A) = \inf\{\rho \geq 0;\ \hat{\kappa}_\rho(A) \geq 0\}, \quad \hat{r}_L^*(A) = \inf\{\rho \geq 0;\ \hat{\kappa}_\rho^*(A) \leq 0\} \tag{2.11}$$

For a stable matrix A, $\hat{r}_L(A)$ describes the lower bound of those indices ρ for which there is $u \in \mathcal{U}_\rho$ making $\dot{x} = [A + u]x$ exponentially unstable for all $x_0 \neq 0$; similarly for $\hat{r}_L^*(A)$. The following inequalities follow directly from the definitions:

$$r_L(A) \leq \hat{r}_L(A), \quad r_L^*(A) \leq \tilde{r}_L(A) \leq \hat{r}_L^*(A) \tag{2.12}$$

□

The theory of (in-)stability radii in the above set-up relates to specific problems of (open-loop) exponential destabilizability and stabilizability of bilinear control systems: One simply considers u as a control function, which is to be chosen such that $\lambda(x_0, u)$ is maximized or minimized. In the next two sections we recall some related facts from [9].

3 Bogolyubov's Projection and Lyapunov Exponents

In this section, we show how the Lyapunov exponents can be obtained from the projected system on the sphere. This is a standard argument for stochastic differential equations (cp. e.g. [6]). We denote the Euclidean norm on $\mathbb{K}^d = \mathbb{R}^d$ or $\mathbb{C}^d$ by $|\cdot|$.

Any linear differential equation

$$\dot{x} = M(t)x, \quad x(0) = x_0 \in \mathbb{K}^d \tag{3.1}$$

can be described in polar coordinates by its angular and radial part. A straightforward application of the chain rule shows that the angular part $s(t) = \frac{x(t)}{|x(t)|}$ is decoupled and looks like

$$\begin{aligned} \dot{s} &= h(M, s) = \{M(t) - \frac{1}{2}s^*(t)[M(t) + M^*(t)]s(t)\,\mathrm{Id}\}s(t) \\ s(0) &= s_0 = \frac{x_0}{|x_0|} \in S^{d-1} \end{aligned} \tag{3.2}$$

on the sphere $S^{d-1} = \{x \in \mathbb{R}^d; |x| = 1\}$. Note that a linear vector field is homogeneous, thus it suffices to consider the system on projective space $\mathbb{P} := \mathbb{P}^{d-1}$, obtained by identifying opposite points on the sphere $\mathbb{S} = S^{d-1}$. Recall that $\mathbb{S}$ and $\mathbb{P}$ are compact manifolds.

The radial part becomes

$$|x(t, x_0)| = |x_0| \exp\Big[\int_0^t q(M(\tau), s(\tau, s_0))\, d\tau\Big] \tag{3.3}$$

with $q(M, s) := \frac{1}{2}s^*(M + M^*)s = \mathrm{Re}\{s^*Ms\}$. Thus we obtain for the corresponding Lyapunov exponent

$$\lambda(x_0, M) = \limsup_{t\to\infty} \frac{1}{t} \int_0^t q(M(\tau), s(\tau))\, d\tau. \tag{3.4}$$

Hence in the time-varying case, the Lyapunov exponents are the long term averages of the function q, evaluated along the trajectories. For time-invariant situations, the Lyapunov exponents (i.e. the real parts of the eigenvalues) are the averages of q in the direction of the eigenvectors. This fact allows some simple estimates of stability radii, compare Proposition 1 below. To be more specific, let M be a constant $d \times d$-matrix and denote by E_j, $j = 1, \dots, k$ the eigenspaces and let $\lambda_j = \mu_j + i\nu_j \in \sigma(M)$ be the corresponding eigenvalues. Let $\mathbb{P}E_j$ be the projection of (the nonzero elements of) E_j onto $\mathbb{P}$. Let $s_j \in \mathbb{P}E_j$ be an eigenvector and denote the solution of the differential equation (3.2) in $\mathbb{P}$ with initial value s_j by $s(\cdot, s_j)$. For $\mathbb{K} = \mathbb{C}$, the corresponding trajectory is always a point in $\mathbb{P}$; for $\mathbb{K} = \mathbb{R}$, the corresponding trajectory may be a point or forms a "circle" in $\mathbb{P}$ depending on whether $\lambda_j \in \mathbb{R}$ or not. In the latter case, $s(\cdot, s_j)$ is periodic, say with period T_j, and defines an occupation measure σ_j on the circle whose density is $\frac{1}{|h(M,s)|}$. If $\lambda_j \in \mathbb{R}$ set $\sigma_j = \delta_{s_j}$, the Dirac measure at s_j. One now has

$$\begin{aligned} \mu_j &= \lambda(s_j, M) \\ &= \begin{cases} \dfrac{1}{T_j} \displaystyle\int q(M,s)\, d\sigma_j = \dfrac{1}{T_j} \int_0^{T_j} q(M, s(\tau, s_j))\, d\tau & \text{for } s(\cdot, s_j) \text{ a circle} \\ \displaystyle\int q(M,s)\, d\sigma_j = q(M, s_j) & \text{for } s(\cdot, s_j) \text{ a point.} \end{cases} \end{aligned}$$

Note that $q(M, s)$ can be positive somewhere on the circle even if μ_j is negative and vice versa. These formulas show that the exponential growth rates of solutions to time-invariant linear equations are obtainable by evaluating q in eigenspaces. Formula (3.4) implies that q is also closely connected with exponential growth rates of solutions to time-varying linear equations. More precisely, define for a stable matrix A and $\{U_\rho : \rho \geq 0\}$ as above

$$\rho_*(A) := \inf\{\rho; \text{there is } u \in U_\rho \text{ with } \max q(A + u, s) \geq 0\}$$

Then by (3.4)

$$\rho_*(A) \leq r_L(A). \tag{3.5}$$

The following proposition establishes a connection between the (unstructured) real and complex stability radii $r_{\mathbb{R}}$ and $r_{\mathbb{C}}$ of a (stable) matrix A — as defined in [14] — and the function q: Recall that for $\mathbb{K} = \mathbb{R}, \mathbb{C}$

$$r_{\mathbb{K}}(A) = \inf\left\{\rho > 0 : \begin{array}{l} \text{there is } D \in M(d,d,\mathbb{K}) \text{ with } \|D\|_2 \leq \rho \text{ such} \\ \text{that } A + D \text{ is not assymptotically stable} \end{array}\right\}.$$

where $\|\cdot\|_2$ denotes the operator norm induced by the Euclidean norm on $\mathbb{K}^d$. Recall also that $\mathbb{S}$ denotes the sphere in $\mathbb{R}^d$, and that $\mathbb{P}$ denotes the corresponding projective space.

Proposition 1. *Let A be a stable matrix in $M(d,d;\mathbb{R})$. Then*

$$0 \leq \min_{s \in \mathbb{S}} |q(A,s)| \leq r_{\mathbb{C}} \leq r_{\mathbb{R}} \leq \max_{s \in \mathbb{S}} |q(A,s)|$$

PROOF. One easily sees (cp.e.g. [15, Lemma 4.1]) that

$$0 < r_{\mathbb{C}} \leq r_{\mathbb{R}} \leq |\mu_n|,$$

where μ_n is the smallest real part of an eigenvalue of A. Clearly, we also have

$$|\mu_n| \leq \max_{s \in \mathbb{P}} |q(A,s)| = \max_{s \in \mathbb{S}} |q(A,s)|.$$

In order to prove the remaining inequality

$$\min |q(A,s)| \leq r_{\mathbb{C}},$$

we note that $\min |q(A,s)| = 0$, iff there is $s_0 \in \mathbb{P}$ with $q(A,s_0) = 0$; and for A stable, it suffices to consider the case $q(A,z) \leq \delta < 0$ for all z in the complex projective space $\mathbb{P}_{\mathbb{C}}$. We use the characterization of $r_{\mathbb{C}}$ from [15, Proposition 3.3]:

$$r_{\mathbb{C}}^2 = \min_{z \in \mathbb{P}_{\mathbb{C}}} \{\|Az\|^2 + \langle A_u z, z\rangle^2\},$$

where $A_u = \frac{1}{2}(A - A^*)$. But for $z \in \mathbb{P}_{\mathbb{C}}$

$$\|Az\|^2 + \langle A_u z, z\rangle^2 = \langle \frac{1}{2}(A + A^*)z, z\rangle^2.$$

Hence

$$r_{\mathbb{C}} \geq \min_{z \in \mathbb{P}_{\mathbb{C}}} |q(A,z)|.$$

But in $\mathbb{R}^2$ and $\mathbb{C}^2$ one easily checks

$$\min_{s \in \mathbb{P}} |q(A,s)| = \min_{z \in \mathbb{P}_{\mathbb{C}}} |q(A,z)|.$$

Q.E.D.

4 Extremal Lyapunov Exponents and the Corresponding Control Sets

In this section we relate the extremal Lyapunov exponents κ, κ^* etc. of bilinear control systems to controllability properties on the projective space $\mathbb{P}$. This discussion is based on [8, 9, 10].

Consider system $(2.3)_\rho$ as a bilinear control system with controls u in $\mathcal{U}_\rho$, where $U_\rho \subset M(d,d;\mathbb{R})$. Under Bogolyubov's projection, the system projected onto $\mathbb{P}$ can be described by

$$\begin{aligned} \dot{s}(t) &= h(u(t), s(t)) = [A + u(t) - \tfrac{1}{2}s^*(t)(A + u(t))s(t) \cdot \mathrm{Id}]s(t), \\ s(0) &= s = \tfrac{x_0}{|x_0|} \in \mathbb{P}. \end{aligned} \qquad (4.1)_\rho$$

Throughout this section we will use the assumption

$$\dim \mathcal{LA}\{h(u,\cdot) \mid u \in U_\rho\}(s) = d-1 \text{ for all } s \in \mathbb{P}, \text{ all } \rho > 0, \tag{H}$$

where $\mathcal{LA}$ denotes the Lie algebra of vector fields on $\mathbb{P}$ generated by $\{h(u,\cdot),$ $u \in U_\rho\}$. (H) implies (see e.g. [17]) that the system (4.1) is locally accessible on $\mathbb{P}$, i.e. for all $s \in \mathbb{P}$ and all $T > 0$, the set $\mathcal{O}^+_{\leq T}(s)$ has nonvoid interior in $\mathbb{P}$, where $\mathcal{O}^+_{\leq T}(s)$ is the set of points attainable in the time $0 \leq t \leq T$ from s,

$$\begin{aligned}\mathcal{O}^+_{\leq T} := \{y \in \mathbb{P};\ &\text{there exists a piecewise constant control}\\ &u \in \mathcal{U}_\rho \text{ s.t. the corresponding trajectory}\\ &\text{of (4.1) satisfies } \psi(t,s,u) = y \text{ for some } 0 \leq t \leq T\}.\end{aligned}$$

In fact, for any "reasonable" choice of U_ρ, $(4.1)_\rho$ is an analytic system and thus has maximal integral manifolds through each point $s \in \mathbb{P}$. (H) says that the whole space $\mathbb{P}$ is the maximal integral manifold.

Remark 2. (H) is always satisfied if $\operatorname{int} U_\rho \neq \emptyset$, which covers the unstructured case (the interior here is w.r.t. $M(d,d;\mathbb{R})$). For structured systems, the interplay between the dynamics A and the "uncertainty" U_ρ is crucial, see [3] and [4] for some details. Generically, (H) is satisfied, if U_ρ contains at least two matrices, see [3, Remark 2.3]. Note that, if (H) holds for some $\rho_1 > 0$, then it holds for all $\rho \geq \rho_1$. We remark also that — in the stochastic context — [21] presents a theory of Lyapunov exponents where (H) is violated. □

In general, (H) does not imply complete controllability of (4.1) on $\mathbb{P}$. However, there are topologically nice sets where controllability holds.

Let

$$\mathcal{O}^+(x) := \bigcup_{T>0} \mathcal{O}^+_{\leq T}(x).$$

Definition 1. A set $D \subset \mathbb{P}$ is called a control set for (4.1) if D contains more than one point, $D \subset \overline{\mathcal{O}^+(x)}$ for all $x \in D$ and D is maximal with respect to this property. A control set is called invariant if $\mathcal{O}^+(x) \subset D$ for all $x \in D$.

Theorem 2. *Assume that the projected systems* $(4.1)_\rho$ *in* $\mathbb{P}$ *satisfy* (H). *Then for each* $\rho > 0$:

(i) There is exactly one invariant control set C_ρ; *this set is closed,* $C_\rho = \bigcap_{s\in\mathbb{P}} \overline{\mathcal{O}^+_\rho(s)}$ *and* $\operatorname{int} C_\rho \neq \emptyset$, $C_\rho = \overline{\operatorname{int} C_\rho}$.

(ii) There is exactly one open control set C^-_ρ.

(iii) Either $C_\rho = C_\rho^-$ *(i.e.* $(4.1)_\rho$ *is controllable on* $\mathbb{P}$*) or* $C_\rho \cap \mathbb{C}_\rho^- = \emptyset$.

(iv) $C_{\rho_1} \subset C_{\rho_2}$ *and* $C_{\rho_1}^- \subset C_{\rho_2}^-$ *for* $\rho_1 \le \rho_2$.

We call C_ρ^- the minimal and C_ρ the maximal control set of $(4.1)_\rho$. A detailed theory of control sets for bilinear systems on $\mathbb{P}$ is developed in [10], where also the significance of the terms "maximal" and "minimal" becomes apparent. The proof of the theorem above can be found in [19] and [9].

Using the Lyapunov exponents of $(2.3)_\rho$, we define the spectrum of the projected system $(4.1)_\rho$ on the control sets C_ρ and C_ρ^- in the following way.

For $X \subset \mathbb{P}$, $s \in \mathbb{P}$ let

$$\begin{aligned} \lambda_\rho(s,X) &= \sup\{\lambda(s,u);\ u \in \mathcal{U}_\rho,\ \psi(t,s,u) \in X \text{ f.a. } t \ge 0\} \\ \lambda_\rho^*(s,X) &= \inf\{\lambda(s,u);\ u \in \mathcal{U}_\rho,\ \psi(t,s,u) \in X \text{ f.a. } t \ge 0\}, \end{aligned}$$

where $\psi(\cdot,s,u)$ denotes the corresponding trajectory of $(4.1)_\rho$, and let

$$\begin{aligned} \alpha_\rho &:= \inf\nolimits_{s\in C_\rho} \lambda^*(s,C_\rho), \qquad & \omega_\rho &:= \sup\nolimits_{s\in C_\rho} \lambda(s,C_\rho), \\ \alpha_\rho^- &:= \inf\nolimits_{s\in C_\rho^-} \lambda^*(s,C_\rho^-), \qquad & \omega_\rho^- &:= \sup\nolimits_{s\in C_\rho^-} \lambda(s,C_\rho^-). \end{aligned}$$

The minimal and maximal spectral interval of the system $(4.1)_\rho$ are given by

$$I_\rho^- := [\alpha_\rho^-, \omega_\rho^-], \quad \text{and} \quad I_\rho := [\alpha_\rho, \omega_\rho]$$

respectively. It turns out that three of the objects α_ρ, ω_ρ, α_ρ^-, ω_ρ^- can be "globally" defined, without a restriction on the whole trajectory. This is the decisive property which allows to associate a stability radius with them.

Proposition 3. *Suppose that the projection* $(4.1)_\rho$ *of system* $(2.3)_\rho$ *satisfies (H). Then*

$$\begin{aligned} (i)\ \omega_\rho &= \sup\nolimits_{s\in\mathbb{P}} \lambda_\rho(s,\mathbb{P}) = \kappa_\rho \\ (ii)\ \alpha_\rho &= \inf\nolimits_{s\in C_\rho} \lambda_\rho^*(s,\mathbb{P}) \\ (iii)\ \alpha_\rho^- &= \inf\nolimits_{s\in\mathbb{P}} \lambda_\rho^*(s,\mathbb{P}) = \kappa_\rho^* \end{aligned}$$

Furthermore, for $0 < \rho_1 \le \rho_2$, *one has*

$$\omega_{\rho_1} \le \omega_{\rho_2}, \quad \alpha_{\rho_1} \ge \alpha_{\rho_2}, \quad \alpha_{\rho_1}^- \ge \alpha_{\rho_2}^-.$$

PROOF. (i) is proved in [8], (ii) is obvious from invariance of C and (iii) is proved in [9]; then the last assertion is obvious. Q.E.D.

Remark 3. If $C_\rho \cap C_\rho^- = \emptyset$, then $\alpha_\rho^- \le \alpha_\rho$, $\omega_\rho^- \le \omega_\rho$, but $\omega_\rho^- > \alpha_\rho$ may occur; thus the intervals I_ρ and I_ρ^- may overlap. □

Remark 4. If, in the definition of ω_ρ^-, we remove similarly as in (i)–(iii) above the restriction on the trajectory, we obtain $\sup_{s\in C_\rho^-} \lambda(s,\mathbb{P})$ which, however, equals ω_ρ. Hence for ω_ρ^-, this restriction is crucial. □

The proposition above shows in particular, that ω_ρ and α_ρ^- coincide with the previously defined κ_ρ and κ_ρ^*, respectively. We obtain the following result, which establishes the connection to stability radii.

Corollary 4. *Suppose that the projection* $(4.1)_\rho$ *of system* $(2.3)_\rho$ *satisfies* (H). *Then*

$$\begin{aligned} (i)\ r_L(A) &= \inf\{\rho \geq 0; \omega_\rho \geq 0\} \\ (ii)\ r_L^*(A) &= \inf\{\rho \geq 0; \alpha_\rho^- \leq 0\} \\ (iii)\ \tilde{r}_L(A) &= \inf\{\rho \geq 0; \alpha_\rho \leq 0\}. \end{aligned}$$

PROOF. (i) and (ii) are immediate from Proposition 3. (iii) follows from [8, Theorem 4.2 and the proof of Theorem 5.1]. Q.E.D.

5 Some Properties of Lyapunov (In-)Stability Radii

In this section we analyze the Lyapunov stability radius and discuss its relation to other stability radii. In [13], a stability radius based on Bohl exponents for complex perturbations was discussed. Our first result in this section shows that for *real* perturbations the concepts of stability radius based on Bohl and on Lyapunov exponents coincide.

Recall that $\{U_\rho; \rho \geq 0\}$ is an increasing family of subsets of $\mathbb{R}^d$ and $\mathcal{U}_\rho$ is defined by (2.1). For equation (2.3) and fixed $u \in \mathcal{U}_\rho$ define the (upper) Bohl exponent as (cp. [13, Proposition 2.2])

$$k_B(u) = \limsup_{s, t-s \to \infty} \frac{1}{t-s} \log \|\Phi_u(t,s)\|,$$

where $\Phi_u(t,s)$ is the fundamental matrix of (2.3) with $\Phi_u(s,s) = \mathrm{id}$. Let the Bohl stability radius $r_B(A)$ of (2.3) w.r.t. U_ρ be defined by

$$r_B(A) = \inf\{\rho \geq 0;\ \sup_{u \in \mathcal{U}_\rho} k_B(u) \geq 0\}.$$

Theorem 5. *Suppose that for all* $\rho > 0$ *the projection* $(4.1)_\rho$ *of system* $(2.3)_\rho$ *satisfies* (H). *Then the Bohl and Lyapunov stability radii* $r_B(A)$ *and* $r_L(A)$ *of* A *with respect to* U_ρ *coincide, i.e.*

$$r_B(A) = r_L(A).$$

PROOF. The inequality $r_B(A) \leq r_L(A)$ follows from $k_B(u) \geq \lambda(x_0, u)$ for all $x_0 \neq 0$, $u \in \mathcal{U}$. We will show that the converse inequality follows from Lemma 5.4 in [8].

First observe that for all $t \geq s$ and all $u \in \mathcal{U}$

$$\Phi_u(t,s) = \Phi_{u(\cdot+s)}(t-s, 0)$$

where $u(\cdot + s)$ is the shifted control function

$$u(\cdot + s)(\tau) = u(\tau + s), \quad \tau \in \mathbb{R}.$$

Hence for all $u \in \mathcal{U}$

$$\begin{aligned} &\limsup_{s, t-s \to \infty} \frac{1}{t-s} \log \|\Phi_u(t,s)\| \\ = &\limsup_{s, t-s \to \infty} \frac{1}{t-s} \log \|\Phi_{u(\cdot+s)}(t-s, 0)\| \\ \leq &\limsup_{s, t-s \to \infty} \frac{1}{t-s} \log \sup_{v \in \mathcal{U}} \|\Phi_v(t-s, 0)\|. \end{aligned}$$

Hence

$$\sup_{u \in \mathcal{U}} k_B(u) \leq \limsup_{t \to \infty} \frac{1}{t} \log \sup \|\Phi_v(t, 0)\| =: \delta$$

where the supremum at the right hand side is taken over all piecewise constant $v \in \mathcal{U}$. Now Lemma 5.4 in [8] claims that $\delta = \kappa$, hence

$$\sup_{u \in \mathcal{U}} k_B(u) = \kappa = \sup_{\substack{u \in \mathcal{U} \\ x_0 \neq 0}} \lambda(x_0, u),$$

which immediately implies the assertion. Q.E.D.

Next we note the following duality result.

Proposition 6. *Suppose that for all $\rho > 0$ the projection $(4.1)_\rho$ of system $(2.3)_\rho$ satisfies (H). Then the stability radius $r_L(A)$ of A w.r.t. U_ρ satisfies*

$$r_L(A) = r_L^*(-A^T)$$

where $r_L(-A^T)$ is the stability radius of $-A^T$ w.r.t. $-U_\rho^T$, $\rho > 0$.

PROOF. For the extremal Lyapunov exponents we have [9] that ω_ρ of the system (2.3) equals $-\alpha_\rho^-$ for the following time reversed control system

$$\dot{x}(t) = -[A + u(t)]^T x(t), \quad t \geq 0, \quad u \in \mathcal{U}_\rho. \tag{5.1$_\rho$}$$

Hence the assertion follows from Corollary 4. Q.E.D.

The result above shows that it suffices to concentrate on r_L and $\tilde{r}_L$, what we will do in the sequel.

Remark 5. The duality concerning the Lyapunov exponents of $(2.3)_\rho$ and $(5.1)_\rho$ referred to above is not valid in general for individual control functions u. It is a property of bilinear control systems satisfying (H), and relies on the fact, established in [8, Theorem 4.2 and Theorem 5.1], that for fixed ρ the supremal exponential growth rate $\kappa_\rho = \omega_\rho$ can approximately be attained by using controls and trajectories having a common period. By the way, note that $r_L(A)$ w.r.t. U_ρ is the same as $r_L(A^T)$ w.r.t. U_ρ^T. □

Remark 6. From Theorem 2(iii) we obtain immediately $r_L^* = \tilde{r}_L$ if $C_\rho \cap C_\rho^- \neq \emptyset$ for $\rho = r_L^*$. We will see later on in Section 6 that $\tilde{r}_L$ indeed depends crucially on the control structure of the projected system $(4.1)_\rho$. □

Next we discuss the effect of unbounded parameter variations. We separate the bounded and the unbounded uncertainties:

$$\dot{x} = [A + u(t) + v(t)]x(t), \quad x(0) = x_0 \in \mathbb{R}^d \qquad (5.2)_\rho$$

where $\{U_\rho^b, \rho \geq 0\}$ is an increasing family of compact sets in $M(d,d;\mathbb{R})$, $v(t) = \sum_{i=1}^p v_i(t)A_i$, $v_i(t) \in \mathbb{R}$, $A_i \in M(d,d;\mathbb{R})$, $i = 1,\ldots,p \leq d^2$. Thus $U_\rho = U_\rho^b \times \{\sum_{i=1}^p v_iA_i;\ v_i \in \mathbb{R} \text{ for } i = 1,\ldots,p\}$ (note that we do not allow here "one-sided unbounded" uncertainties). Denote by

$$N_\rho = \{A + u + v;\ u \in U_\rho^b, v = \sum_{i=1}^p v_iA_i, v_i \in \mathbb{R} \text{ for } i = 1,\ldots,p\}$$

the possible constant right hand sides of $(5.2)_\rho$ and by

$$N^u = \{\sum_{i=1}^p v_iA_i, v_i \in \mathbb{R} \text{ for } i = 1,\ldots,p\}$$

the unbounded part. For a matrix $M \in M(d,d;\mathbb{R})$ define

$$M^0 := M - \frac{1}{d}\,\mathrm{trace}\, M \cdot \mathrm{Id}$$

and let $N_\rho^0 = \{M^0, M \in N_\rho\}$, $N^{u,0} = \{M^0, M \in N^u\}$. Define the systems group and semigroup generated by N_ρ as

$$\begin{aligned} \mathcal{G}_\rho &= \{e^{t_nB_n}\cdots e^{t_1B_1};\ B_i \in N_\rho,\ t_i \in \mathbb{R},\ i = 1,\ldots,n \in \mathbb{N}\} \\ \mathcal{S}_\rho &= \{e^{t_nB_n}\cdots e^{t_1B_1};\ B_i \in N_\rho,\ t_i \geq 0,\ i = 1,\ldots,n \in \mathbb{N}\}. \end{aligned}$$

Analogously denote the groups and semigroups generated by N^u, N_ρ^0, $N^{u,0}$ by $\mathcal{G}^u$, $\mathcal{G}_\rho^0$, $\mathcal{G}^{u,0}$. The following estimates for κ, κ^* are given in [2, Theorem 2.3].

Theorem 7. *Assume that the family of systems* $(5.2)_\rho$ *satisfies* (H) *for all* $\rho > 0$. *Then the following holds:*

1. *If* $\overline{\mathcal{G}}^u$ *is not compact (in* $M(d,d;\mathbb{R})$*), then* $\kappa_\rho^* = -\infty$, $\kappa_\rho = \infty$.

2. *If* $\overline{\mathcal{G}}^u$ *is compact, then* $-\infty < \kappa_\rho^* \leq \kappa_\rho < +\infty$.

   2.1.1 *If* $\mathcal{G}_\rho$ *is not compact, and* $\mathcal{G}_\rho^0$ *is not compact, then* $\kappa_\rho^* < \kappa_\rho$.

   2.1.2 *If* $\mathcal{G}_\rho$ *is not compact, but* $\mathcal{G}_\rho^0$ *is compact then*

   (i) if $\frac{1}{d}\,\mathrm{trace}(A+u) \equiv c$, *then* $\kappa_\rho^* = \kappa_\rho = c$

   (ii) if $\frac{1}{d}\,\mathrm{trace}(A+u)$ *is not constant, then*

$$\begin{aligned}\kappa_\rho^* &= \frac{1}{d}\min_{u\in U_\rho^b}\mathrm{trace}(A+u)\\ &< \frac{1}{d}\max_{u\in U_\rho^b}\mathrm{trace}(A+u) = \kappa_\rho.\end{aligned}$$

   2.2 *If* $\mathcal{G}_\rho$ *is compact, then* $\kappa_\rho^* = \kappa_\rho = 0$.

Note that the group $\mathcal{G}_\rho \subset \mathrm{Gl}(d,\mathbb{R})$ (the group of invertible $d \times d$-matrices) is compact, if there exists $T \in \mathrm{Gl}(d,\mathbb{R})$ such that $TN_\rho T^{-1} \subset \mathrm{so}(d,\mathbb{R})$, the skew symmetric matrices; similarly for the other groups.

Theorem 7(1) implies in particular that the system $(5.2)_\rho$ has $\kappa_\rho^* = -\infty$, $\kappa_\rho = +\infty$ unless there exists a basis in $\mathbb{R}^d$, in which the unbounded part gives rise only to rotations. If we are in case (2), then in this basis the unbounded part does not contribute to the Lyapunov exponents at all (since $s^*Bs = 0$ for skew symmetric B). Note also that if a system semigroup $\mathcal{S}_\rho$ satisfies $\overline{\mathcal{S}}_\rho$ is compact, then $\overline{\mathcal{G}}_\rho = \overline{\mathcal{S}}_\rho$ and — under (H) — the corresponding system is completely (exactly) controllable on $\mathbb{P}$, i.e. $C_\rho = C_\rho^- = \mathbb{P}$ (cp. [3, Corollary 3.2]). Hence the cases (2.1.2) and (2.2) above describe completely controllable situations on $\mathbb{P}$.

This theorem has some immediate implications for the analysis of r_L.

Corollary 8. *Assume that the family* $(5.2)_\rho$ *satisfies* (H) *for all* $\rho > 0$ *and that the stability radius of* A *with respect to* $\{U_\rho^b; \rho \geq 0\}$ *is positive. Then the following are equivalent:*

(i) $r_L(A) > 0$,

(ii) $\overline{\mathcal{G}}^u$ *is compact,*

(iii) there exists a basis in $\mathbb{R}^d$ *such that all elements of* N^u *are skew symmetric.*

PROOF. By Theorem 7(1), assertion (i) implies (ii), and by the remarks following Theorem 7, assertions (ii) and (iii) are equivalent. Finally (iii) implies (i), since here the unbounded part does not contribute to κ_ρ, in the basis chosen according to (iii). (Note by the way, that $r_L(A)$ w.r.t. $\{U_\rho\}$ is the same as $r_L(TAT^{-1})$ w.r.t. $\{TU_\rho T^{-1}\}$, $T \in \mathrm{Gl}(d,\mathbb{R})$. Q.E.D.

Another consequence of Theorem 7 is the characterization of infinite stability radii: Let $\{U_\rho^b; \rho \geq 0\}$ be defined by a norm, i.e. for given matrices $B \in M(d,m;\mathbb{R})$, $C \in M(k,d,\mathbb{R})$ let $U_\rho^b := \{BDC \in M(d,d;\mathbb{R}); \|D\| \leq \rho\}$, where $\|\cdot\|$ is any norm in $M(m,k;\mathbb{R})$ (compare the structured or unstructured case in [13, 16]). If the assumptions of Corollary 8 are met, then $r_L(A) = \infty$ iff Theorem 7.2.1.2(i) holds with $c < 0$ for all $\rho \in (0,\infty)$.

Next we consider the following "interval" case when the family $\{U_\rho; \rho \geq 0\}$ is given by

$$U_\rho = \{M \in M(d,d;\mathbb{R});\ -\rho \leq m_{ij} \leq \rho,\ (i,j) \in I \quad \text{and } m_{ij} = 0,\ (i,j) \in \underline{d} \times \underline{d} \setminus I\} \tag{5.3}$$

here I is a subset of $\underline{d} \times \underline{d} := \{1,\ldots,d\} \times \{1,\ldots,d\}$ such that m_{ii} is allowed to be nonzero for at least one $i \in \{1,\ldots,d\}$.

Corollary 9. *Assume that A is stable, and for all $\rho > 0$ and U_ρ given by (5.3), $\mathcal{G}_\rho$ is not compact, but $\mathcal{G}_\rho^0$ is compact. Then the corresponding stability radius of A satisfies*

$$r_L(A) = -\frac{1}{l}\,\mathrm{trace}\,A.$$

where l is the number of diagonal elements which are allowed non-zero in U_ρ.

PROOF. By Theorem 7 case (2.1.2), it holds for all $\rho > 0$ that

$$\kappa_\rho = \frac{1}{d}\max_{u \in U_\rho} \mathrm{trace}(A+u) = \frac{1}{d}[\mathrm{trace}\,A + \max_{u \in U_\rho} \mathrm{trace}\,u].$$

Hence $r_L(A) = \rho$ iff $\max_{u \in U_\rho} \mathrm{trace}\,u = -\,\mathrm{trace}\,A$. Now $\mathcal{G}_\rho^0$ compact and A stable (hence $\mathrm{trace}\,A < 0$) imply: For each $\rho > 0$ there is at least one $M \in U_\rho$ with $m_{ii} = \rho$ for l diagonal elements m_{ii}. Therefore $r_L(A) = -\frac{1}{l}\,\mathrm{trace}\,A$. Q.E.D.

Now let A be an unstable matrix. From Remark 6 and $r_L^* \leq \tilde{r}_L$, we have that $\tilde{r}_L(A) = 0$ if $\overline{\mathcal{G}}^u$ is not compact and $C_\rho \cap C_\rho^- \neq \emptyset$ for all $\rho > 0$.

Remark 7. Proposition 3 and Corollary 4 show that determining the time varying stability radius r_L means solving a parametrized optimal control problem:

One has to find

$$\inf\{\rho > 0; \sup_{u \in \mathcal{U}_\rho, s_0} \limsup_{t \to \infty} \frac{1}{t} \int_0^t q(A + u(\tau), s(\tau, s_0, u))\, d\tau \geq 0\}$$

where $s(\tau, s_0, u)$ means the solution of the projected system $(4.1)_\rho$ for initial state s_0 and control u. In [8] it is shown that in fact it suffices to take the supremum over all piecewise constant control functions u and corresponding trajectories having a common period $T > 0$, T arbitrary. Similarly for $\tilde{r}_L$.

□

Thus the problem of finding r_L can considerably be simplified; nevertheless it still requires optimization over function spaces. Thus contrary to what is known for the time invariant radii $r_{\mathbb{C}}$ and $r_{\mathbb{R}}$, we obtain explicit formulae for r_L only in very special cases, compare Theorem 7 and Corollary 9, and Example 8 in the next section. A general strategy is to solve (for fixed $\rho > 0$) the optimal control problem numerically and to extract the point(s) where $\kappa_\rho = \omega_\rho = 0$. This is demonstrated for the linear oscillator with uncertain restoring force in the next section. The computed Lyapunov exponents κ_ρ, $\rho \neq r_L$, may also be used for design purposes, see Section 8.

6 Examples

In this section we will discuss the relations between the Lyapunov stability radius defined above and the stability radius for matrices under constant complex- or real-valued perturbations. In particular, we will consider the linear oscillator with uncertainties in the restoring force.

We use the following set-up. Let $A \in M(d, d; \mathbb{R})$ be stable and let $B \in M(d, m; \mathbb{R})$, $C \in M(k, d; \mathbb{R})$ be given. Denote for $\rho \geq 0$ and $\mathbb{K} = \mathbb{R}$ or $\mathbb{C}$

$$U_\rho^{(2)} := \{BDC; D \in M(m, k, \mathbb{K}) \text{ with } \|D\|_2 \leq \rho\} \subset M(d, d, \mathbb{K})$$
$$U_\rho^{(m)} := \{BDC; D \in M(m, k, \mathbb{K}) \text{ with } \|D\|_m \leq \rho\} \subset M(d, d, \mathbb{K})$$

here $\|\cdot\|_2$ is the operator norm induced by the Euclidean norm on $\mathbb{K}^2$ and $\|\cdot\|_m$ is defined by

$$\|D\|_m = \max_{i,j} |d_{ij}|, \quad D = (d_{ij})_{i,j},$$

corresponding to the case of interval uncertainties. We denote the Lyapunov stability radii of A with respect to the families $\{U_\rho^{(2)}; \rho \geq 0\}$ and $\{U_\rho^{(m)}; \rho \geq$

$0\}$, where $U_\rho^{(2)}$, $U_\rho^{(m)} \subset M(d,d;\mathbb{R})$, by $r_L^{(2)}$ and $r_L^{(m)}$ respectively. We denote the stability radii of A with respect constant complex- and real-valued perturbations in $U_\rho^{(2)}$ and $U_\rho^{(m)}$ by

$$r_{\mathbb{C}}^{(2)}, \quad r_{\mathbb{R}}^{(2)}, \quad r_{\mathbb{C}}^{(m)}, \quad r_{\mathbb{R}}^{(m)}.$$

The objects studied in [13, 16] are $r_{\mathbb{C}}^{(2)}$ and $r_{\mathbb{R}}^{(2)}$; by [16, Proposition 2.2 and Proposition 5.2] $r_{\mathbb{C}}^{(2)}$ is also the Bohl stability radius with respect to complex-valued time-varying perturbations. Hence

$$r_{\mathbb{C}}^{(2)} \leq r_L^{(2)}$$

The inequality $\|D\|_m \leq \|D\|_2$ implies

$$r_{\mathbb{C}}^{(m)} \leq r_{\mathbb{C}}^{(2)}, \quad r_{\mathbb{R}}^{(m)} \leq r_{\mathbb{R}}^{(2)}, \quad r_L^{(m)} \leq r_L^{(2)}.$$

Furthermore the inequalities

$$r_{\mathbb{C}}^{(m)} \leq r_{\mathbb{R}}^{(m)}, \quad r_{\mathbb{C}}^{(2)} \leq r_{\mathbb{R}}^{(2)}, \quad r_L^{(m)} \leq r_{\mathbb{R}}^{(m)}, \quad r_L^{(2)} \leq r_{\mathbb{R}}^{(2)}$$

are obvious from the definitions. The following Examples 8 and 9 show that

$$r_L^{(m)} < r_{\mathbb{C}}^{(2)} = r_L^{(2)} = r_{\mathbb{R}}^{(2)}$$

and

$$r_{\mathbb{C}}^{(m)} = r_{\mathbb{C}}^{(2)} < r_L^{(m)} = r_L^{(2)} < r_{\mathbb{R}}^{(m)} = r_{\mathbb{R}}^{(2)}$$

are possible. We do not know whether $r_L^{(m)} < r_{\mathbb{C}}^{(m)}$ is possible.

We now start the analysis of two-dimensional examples.

Consider the matrix

$$D = \begin{pmatrix} d_1 & d_2 \\ d_3 & d_4 \end{pmatrix} \in M(2,2;\mathbb{R}).$$

To obtain explicit formulas for the projected system on $\mathbb{P}$, we write

$$s = \begin{pmatrix} \cos\varphi \\ \sin\varphi \end{pmatrix} \in \mathbb{P}, \quad \varphi \in [0,\pi).$$

Then one computes

$$\begin{aligned} h(D,\varphi) &= -d_2 \sin^2\varphi + d_3 \cos^2\varphi + \frac{1}{2}(d_1 - d_4)\sin 2\varphi \\ q(D,\varphi) &= \frac{1}{2}(d_1 + d_4) + \frac{1}{2}(d_1 - d_4)\cos 2\varphi + \frac{1}{2}(d_2 + d_3)\sin 2\varphi, \end{aligned}$$

here h and q are first defined as in (3.2) and (3.3), and then the induced maps with the second argument in $[0, \pi)$ are denoted by the same symbols.

Note that

$$\begin{aligned} &\|D\|_m = \max|d_i| \\ &\leq \|D\|_2 = \tfrac{1}{2}\textstyle\sum_{i=1}^4 d_i^2 + \sqrt{(d_1d_2 + d_3d_4)^2 + \tfrac{1}{4}(d_1^2 + d_3^2 - d_2^2 - d_4^2)^{1/2}} \\ &\leq 2\|D\|_m \end{aligned}$$

Hence it may not come as a surprise that the strict inequality $r_L^{(m)} < r_{\mathbb{C}}^{(2)}$ can occur, as the following example shows.

Example 8. Consider

$$A = \begin{pmatrix} -a & 0 \\ 0 & -a \end{pmatrix}$$

Clearly $q(A, \varphi) \equiv -a$, and

$$\max_{D \in U_\rho^{(m)}} \max_{\varphi \in \mathbb{P}} q(D, \varphi) = 2\rho.$$

The maximum is attained for $d_i = \rho$, $i = 1, \ldots, 4$ and $\varphi = \pi/4$. The eigenvalues and eigenvectors of

$$D = \begin{pmatrix} d & d \\ d & d \end{pmatrix}$$

are $\lambda_1 = 0$, $\lambda_2 = 2d$, $e_1 = \frac{3}{4}\pi$, $e_2 = \frac{1}{4}\pi$. Since $q(A + D, \varphi) = q(A, \varphi) + q(D, \varphi) = -a + q(D, \varphi)$, we have

$$\max_{D \in \mathcal{U}_\rho} \max_{\varphi \in \mathbb{P}} q(A + D, \varphi) = -a + 2\rho$$

and hence

$$r_L^{(m)} = \frac{a}{2}.$$

On the other hand

$$r_{\mathbb{C}}^{(2)} = r_{\mathbb{R}}^{(2)} = a \quad \text{(see [18, pp. 41])},$$

and hence

$$r_L^{(m)} < r_{\mathbb{C}}^{(2)} = r_L^{(2)} = r_{\mathbb{R}}^{(2)}.$$

Note that for the matrix D above $\|D\|_2 = 2d$. □

The next example shows the more surprising fact that

$$r_{\mathbb{C}}^{(m)} = r_{\mathbb{C}}^{(2)} < r_L^{(m)} = r_L^{(2)} < r_{\mathbb{R}}^{(m)} = r_{\mathbb{R}}^{(2)}$$

can occur.

Example 9 (The linear oscillator with uncertain restoring force).

Consider the linear oscillator described by

$$\ddot{y} + 2b\dot{y} + (1+u)y = 0$$

or with $x = (x_1, x_2)^T = (y, \dot{y})^T$

$$\dot{x} = \begin{pmatrix} 0 & 1 \\ -1 & -2b \end{pmatrix} x + u(t) \begin{pmatrix} 0 & 0 \\ -1 & 0 \end{pmatrix} x$$

where $u(t) \in [-\rho, \rho]$, $\rho \geq 0$, i.e. $U_\rho = \left\{ \begin{pmatrix} 0 & 0 \\ -\alpha & 0 \end{pmatrix}; \alpha \in [-\rho, \rho] \right\}$ and $b \in \mathbb{R}$ is a constant. This yields structured stability radii (cp. [16]) with

$$A = \begin{pmatrix} 0 & 1 \\ -1 & -2b \end{pmatrix}, B = \begin{pmatrix} 0 & 0 \\ 0 & 1 \end{pmatrix}, \text{ and } C = \begin{pmatrix} 0 & 0 \\ -1 & 0 \end{pmatrix}.$$

Since the perturbation is one-dimensional, the equalities $r_\cdot^{(m)} = r_\cdot^{(2)} = r_\cdot$ are obvious.

Projection onto the sphere gives in polar coordinates with $s = (\cos\varphi, \sin\varphi)^T$, $\varphi \in [0, 2\pi)$ (compare for this example [2, Section 6]):

$$\begin{aligned} \dot{\varphi}(t) &= -\sin^2\varphi(t) + [b^2 - 1 - u(t)]\cos^2\varphi(t) \\ q(A+u, \varphi) &= \frac{1}{2}(b^2 - u)\sin 2\varphi - b. \end{aligned}$$

We restrict our attention to the case where $0 \leq b \leq 2$. Then one easily finds

$$\min_\varphi |q(A, \varphi)| = -\frac{1}{2}b^2 + b.$$

Figure 1 presents the different stability radii for $b \in [0, 1]$ (for $b > 1$ we have $r_\mathbb{C} = r_\mathbb{R} = r_L = 1$), as well as two estimates (Note that in the following we suppress the dependence on the matrix A, instead we indicate by the argument b, that all entities depend on the actual value of the damping b).

(a) is the real stability radius $r_\mathbb{R}(b)(\equiv 1)$, which follows from elementary considerations;

(b) is the time varying radius $r_L(b)$, its computation is described in [2] as the level curve $\gamma = 0$ (cp.[9]);

(c) is the complex radius $r_\mathbb{C}(b)$, computed using the program `STABRAD`, see [18];

(d) is the estimate via $\min_\varphi |q(A, \varphi)| = b - \frac{1}{2}b^2$;

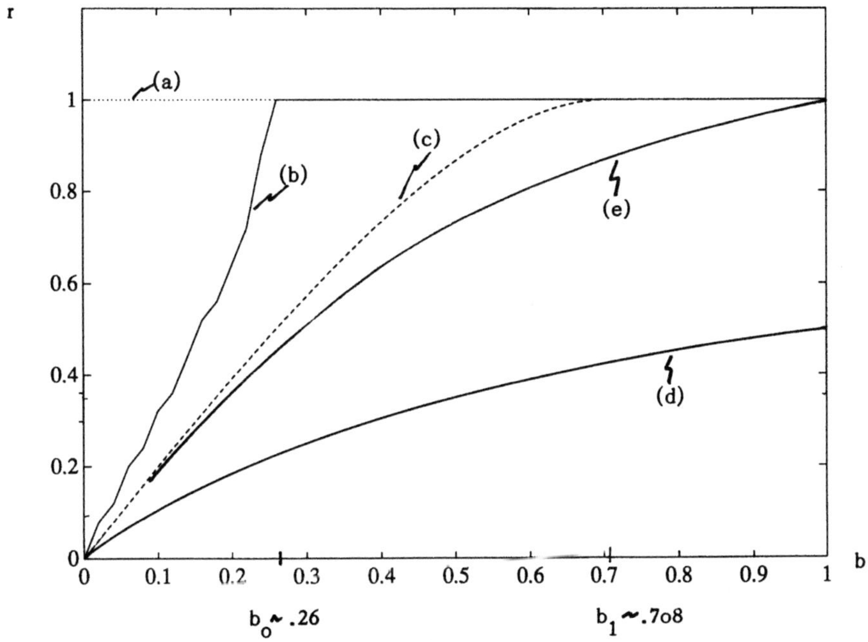

Figure 1. Stability radii for the linear oscillator

(e) is the estimate $\rho_*(b) = 2b - b^2$ according to (3.5).

□

For $\rho \downarrow 0$ and $0 \leq b \leq 1$ we have for the maximal Lyapunov exponent $\kappa_\rho(b)$

$$\kappa_\rho(b) = -b + \frac{1}{\pi\sqrt{1-b^2}}\rho + o(\rho)$$

see [2, Section 6]. Hence for b small one obtains

$$r_L(b) \sim \pi \cdot b$$

and also (by numerical evidence, compare Fig. 1)

$$r_{\mathbb{C}}(b) \sim 2b.$$

We have

$$0 < r_{\mathbb{C}}(b) < r_L(b) < r_{\mathbb{R}}(b) = 1 \text{ for } 0 < b < b_0$$
$$0 < r_{\mathbb{C}}(b) < r_L(b) = r_{\mathbb{R}}(b) = 1 \text{ for } b_0 \leq b < b_1$$
$$0 < r_{\mathbb{C}}(b) = r_L(b) = r_{\mathbb{R}}(b) = 1 \text{ for } b_1 < b.$$

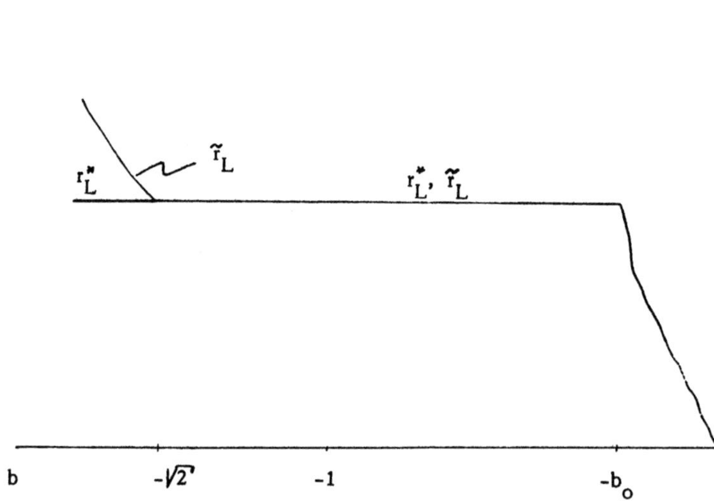

Figure 2. Instability radii for the linear oscillator

The Lyapunov stability radius $r_L(b)$ is monotone increasing in b for $b \geq 0$ (this follows directly from the explicit formulae for $\kappa(b)$ in [2, Section 6]). From the numerics presented in Figure 1, it also seems that $r_{\mathbb{C}}(b)$ is monotone. For the relative sizes we have from the numerical results:

$$\frac{r_{\mathbb{C}}(b)}{r_{\mathbb{R}}(b)} \in (0,1] \quad \text{monotone in } b,$$

$$\frac{r_{\mathbb{C}}(b)}{r_L(b)} \in [\delta,1] \quad \text{with } 0 < \delta \sim 0.5, \text{ not monotone in } b.$$

For the linear oscillator discussed above we can also compute r_L^* and $\tilde{r}_L$, now of course for $b < 0$. Figure 2 presents the curve $r_L^*(b)$, which is easily obtained from $r_L(b)$, compare Proposition 6. The curve for $\tilde{r}_L$ is more complicated, since it depends on the control structure of the projected system. It turns out that

$$C = C^- = \mathbb{P} \quad \text{iff} \quad \rho + 1 > b^2 \quad \text{cp. [9]}$$

Thus according to Remark 6, $r_L^* = \tilde{r}_L$ if $b \in [-\sqrt{2}, 0]$. For $b < -\sqrt{2}$, one has $r_L^* < \tilde{r}_L$. Figure 2 presents the curve $\tilde{r}_L(b)$.

Finally, we consider the uniform radii $\hat{r}_L$ and $\hat{r}_L^*$ introduced in Remark 1. If A is stable, i.e. $b > 0$, then

$$\hat{\lambda}_\rho(b) = \begin{cases} -b & \text{if } \dfrac{b^2-1}{\rho} < 1 \\ -b - \sqrt{-1-\rho+b^2} & \text{if } 1 \leq \dfrac{b^2-1}{\rho}, \end{cases}$$

hence $\hat{\lambda}_\rho < 0$ for all $\rho > 0$, i.e. $\hat{r}_L = \infty$. Thus uniform destabilization is not possible in this situation: For all $u \in \mathcal{U}_\rho$, $\rho \in (0, \infty)$ there always exists $x_u \neq 0$ such that $\lambda(x_u, u) < 0$.

By duality one gets for unstable A, i.e. $b < 0$, that $\hat{r}_L^* = \infty$.

7 Stochastic Stability Radii and Large Deviations

In this section we introduce two stochastic stability radii and discuss their relation to r_L. We also analyze how a stochastic system behaves, if its stochastic stability radius is larger than r_L: This indicates the "presence of large deviations". Since the case with unbounded uncertainty can be reduced to the bounded case, using Theorem 7, we restrict ourselves to bounded situations. In order to be able to compare stochastic stability radii with r_L, we use the following set-up:

Let M be a finite dimensional compact, connected (Riemannian) real analytic manifold and

$$D_\rho : M \to M(d, d; \mathbb{R})$$

a family of (real-) analytic maps for $\rho \geq 0$. Assume again that the family $U_\rho := D_\rho(M)$ is increasing. On M we consider two different classes of stochastic processes:

Stat: the stationary processes with values in M, i.e.
Denote by $D(M)$ the Skorohod space of piecewise continuous functions $\xi : [0, \infty) \to M$, right continuous with left hand limits, and by $\mathcal{B}(M)$ the Borel σ-algebra of $D(M)$. Let $\theta_t : D(M) \to D(M)$ be the shift on $D(M)$ for each $t \geq 0$, i.e. $\theta_t \xi(\cdot) = \xi(t + \cdot)$. Then every (strictly) stationary process with values in M and right continuous trajectories with left hand limits is given by a θ_t-invariant probability measure ν on $(D(M), \mathcal{B}(M))$, i.e. $\theta_t \nu = \nu$ for all $t \geq 0$. (This fact is known as Kolmogorov's construction.) In particular, if $\xi \in D(M)$ is a periodic function with period T, then $\eta(t) = \xi(t + \tau)$ is a stationary process with τ uniformly distributed in $[0, T]$.

Diff: the stationary, ergodic, nondegenerate diffusion processes on M, i.e.

$$d\xi_t = X_0(\xi_t)dt + \sum_{i=1}^{r} X_i(\xi_t) \circ dW_i,$$

where $X_0, \ldots, X_r$ are real analytic vector fields on M with

$$\dim \mathcal{LA}\{X_1, \ldots, X_r\}(m) = \dim M$$

for all $m \in M$. "$\circ$" denotes the symmetric (Stratonovič) stochastic integral.
Again the process ξ_t can be viewed as a flow Φ_t with flow-invariant ergodic measure ν, on the space $\Omega \times M$, where $(\Omega, \mathcal{B}(\Omega), P)$ is the Wiener space of continuous functions ω: $[0, \infty] \to \mathbb{R}^r$, with its Borel σ-algebra and the Wiener measure P. Here Φ_t is defined as $\Phi_t \colon \Omega \times M \to \Omega \times M$, $\Phi_t(\omega, p) = (\hat{\theta}_t\omega, \varphi(t, \omega, p))$, with $\hat{\theta}_t \colon \Omega \to \Omega$, $\hat{\theta}_t\omega(\cdot) = \omega(t + \cdot) - \omega(t)$, the Wiener shift, and $\varphi(t, \omega, p)$ the (pathwise) solution of the stochastic differential equation at time t with initial value $p \in M$.

For background information on differential equations driven by stationary processes see e.g. the survey paper [20].

For a matrix $A \in M(d, d; \mathbb{R})$ we now consider the systems

$$\dot{x} = [A + D_\rho(\xi_t)]x, \quad x(0) = x_0 \in \mathbb{R}^d, \tag{7.1$_\rho$}$$

again with assumption (H) as in Section 4 for the class Stat, and with the assumption

$$\begin{array}{c} \dim \mathcal{LA}\{X_0 + h, X_1, \ldots, X_r\}(m, s) = \dim M + d - 1 \\ \text{for all} (m, s) \in M \times \mathbb{P}^{d-1} \end{array} \tag{H'}$$

for the class Diff.

We define stochastic stability radii for the classes Stat and Diff in the following way:

The Lyapunov exponents of $(7.1)_\rho$ are

$$\lambda_\rho(x_0, \xi_t) = \lim_{t \to \infty} \frac{1}{t} \log |\varphi(t, x_0, \xi_t)|$$

where φ solves $(7.1)_\rho$.

By Oseledeč's Theorem these random numbers exist with probability 1. Define

$$\kappa_\rho(\xi_t) := \operatorname{ess\,sup} \sup_{x_0 \neq 0} \lambda_\rho(x_0, \xi_t)$$

where the essential supremum is taken w. r. t. the measure ν associated with ξ_t, and let the stability radius with respect to ξ_t be

$$r(\xi_t) := \inf\{\rho \geq 0;\ \kappa_\rho(\xi_t) \geq 0\}.$$

Define the stability radii w.r.t. the classes Stat and Diff

$$r_{\text{Stat}} := \inf\{\rho \geq 0;\ \sup_{\xi_t \in \text{Stat}} \kappa_\rho(\xi_t) \geq 0\}$$
$$r_{\text{Diff}} := \inf\{\rho \geq 0;\ \sup_{\xi_t \in \text{Diff}} \kappa_\rho(\xi_t) \geq 0\}.$$

Obviously $r_{\text{Stat}} \leq r_{\text{Diff}}$. Similarly, stochastic instability radii can be defined

$$\kappa_\rho^*(\xi_t) := \operatorname{ess\,inf} \inf_{x_0 \neq 0} \lambda_\rho(x_0, \xi_t)$$
$$\kappa_\rho^*(\xi_t, x_0) := \operatorname{ess\,inf} \lambda_\rho(x_0, \xi_t),$$

and

$$r^*(\xi_t) := \inf\{\rho \geq 0;\ \kappa_\rho^*(\xi_t) \leq 0\}$$
$$\tilde{r}(\xi_t) := \inf\{\rho \geq 0;\ \kappa_\rho^*(\xi_t, x_0) \leq 0 \text{ for all } x_0 \neq 0\},$$

and analogously for r^* and $\tilde{r}$ w.r.t. Stat and Diff using the inf's.

Remark 10. We have defined stochastic stability radii in complete analogy to deterministic ones, using the unifying concepts of Lyapunov exponents. It is important to notice that Lyapunov exponents of a stochastic, stationary process ξ_t do not only depend on the statistics of ξ_t. In fact, if e.g. the generator $L = X_0 + \frac{1}{2}\sum X_i^2$ of a stationary, ergodic, nondegenerate diffusion process ξ_t is given (and therefore the statistics are determined), there are in general many ways to take square roots of the second order part of L, and hence different choices for the vector fields in the associated stochastic differential equation. The Lyapunov exponents of ξ_t depend on the associated stochastic flow (the *multipoint motion* as described in the definition of the classes Stat and Diff), i.e. for diffusion processes on the vector fields $X_0, \ldots, X_r$. It is Oseledeč Multiplicative Ergodic Theorem, which shows that for stationary stochastic processes a theory of Lyapunov exponents can be developed in complete analogy to the deterministic Lyapunov regular case — with probability one, see [20] for more details and examples concerning this point. □

The stability radii r_{Stat} and r_{Diff} are defined via the almost sure Lyapunov exponents of the system $(7.1)_\rho$. We will also consider the Lyapunov exponents of the p-th moment $(0 < p < \infty)$

$$g_\rho(x_0, \xi_t, p) = \lim_{t \to \infty} \frac{1}{t} \log E(|\varphi(t, x_0, \xi_t)|^p), \quad \varphi \text{ solves } (7.1)_\rho,$$
$$g_\rho(\xi_t, p) = \sup_{x_0 \neq 0} g_\rho(x_0, \xi_t, p),$$

and the corresponding stability radii

$$r^{[p]}(\xi_t) = \sup\{\rho \geq 0;\ g_\rho(\xi_t, p) \leq 0\},\ 0 < p < \infty$$
$$r^{[\infty]}(\xi_t) = \sup\{\rho \geq 0;\ g_\rho(\xi_t, p) \leq 0 \text{ for all } 0 < \rho < \infty\}$$

and for the classes Stat and Diff, $0 < p < \infty$

$$r^{[p]}_{\text{Stat}} = \sup\{\rho \geq 0,\ \sup_{\xi_t \in \text{Stat}} g_\rho(\xi_t, p) \leq 0\}$$
$$r^{[p]}_{\text{Diff}} = \sup\{\rho \geq 0,\ \sup_{\xi_t \in \text{Diff}} g_\rho(\xi_t, p) \leq 0\},$$

analogously for $r^{[\infty]}_{\text{Stat}}, r^{[\infty]}_{\text{Diff}}$. In particular, in the engineering literature, p-th moment stability is often considered.

The situation for the class Stat is relatively simple:

Theorem 10. *Let $A \in M(d, d; \mathbb{R})$ be stable. Then under the assumptions above*

$$\frac{1}{p} r^{[p]}_{\text{Stat}} \geq r^{[\infty]}_{\text{Stat}} = r_{\text{Stat}} = r_L \text{ for all } p > 0,$$

where r_L is taken w.r.t. $U_\rho = D_\rho(M)$. If the initial values x_0 can be taken as bounded random variables, then furthermore $\frac{1}{p} r^{[p]}_{\text{Stat}} = r^{[\infty]}_{\text{Stat}}$ for all $p > 0$.

Proof. The key for the proof of this theorem is the fact that the extremal Lyapunov exponent ω_ρ (as well as $\alpha_\rho, \alpha_\rho^-$) can be approximated by Lyapunov exponents of periodic solutions of $(4.1)_\rho$, compare [8, Theorem 5.1], and [9, Theorem 4.2]. Now a periodic solution $s^0(t, u^0)$ gives rise to a stationary process ξ_t^0, as outlined in the description of the class Stat above, and $\lambda(s^0) = \kappa(\xi_t^0)$ by Corollary 4.4 in [2]. The equality $r_{\text{Stat}} = r_L$ follows from this.

Next, consider for given $x_0 \in \mathbb{R}^d$, $x_0 \neq 0$ and ξ_t in Stat the function $p \mapsto \frac{1}{p} g_\rho(x_0, \xi_t, p)$. This function is continuous and increasing on $(0, \infty]$. Denote its limit at $+\infty$ by $g_\rho(x_0, \xi_t, \infty)$, and let $g_\rho(\xi_t, \infty) = \sup_{x_0 \neq 0} g_\rho(x_0, \xi_t, \infty)$. Then we have by definition $g_\rho(x_0, \xi_t, p) \leq 0$ for all $0 < p < \infty$ iff $g_\rho(\xi_t, \infty) \leq 0$. Corollary 4.4 in [2] now says $r^{[\infty]}_{\text{Diff}} = r_{\text{Stat}}$, and the inequalities $r^{[\infty]}_{\text{Diff}} \geq r^{[\infty]}_{\text{Stat}}$ and $r^{[\infty]}_{\text{Stat}} \geq r_{\text{Stat}}$ are obvious, hence $r^{[\infty]}_{\text{Stat}} = r_{\text{Stat}}$.

The inequality $\frac{1}{p} r^{[p]}_{\text{Stat}} \geq r^{[\infty]}_{\text{Stat}}$ follows directly from the monotonicity of $\frac{1}{p} g_\rho(x_0, \xi_t, p)$ for $p > 0$.

Finally, if x_0 can be taken as a random variable, then we can associate with the periodic solution $s^0(t, u^0)$ a stationary ξ_t^0 and x_0 such that

$$E(|\varphi(t, x_0, \xi_t)|^p)^{\frac{1}{p}} = |\varphi(t, x_0, \xi_t)| \quad \text{w.p.1}$$

by defining x_0 accordingly to the uniform distribution on $[0,T]$, where T is the common period of u^0 and s^0. This proves the last statement of the theorem. Q.E.D.

Remark 11. In order to characterize the stochastic instability radii, we define the Lyapunov exponents of the p-th moment for $p<0$ as $g_\rho^*(\xi_t,p) = \inf_{x_0\neq 0} g_\rho(x_0,\xi_t,p)$, and

$$r^{[-\infty]}(\xi_t) = \inf\{\rho \geq 0;\ g_\rho^*(\xi_t,p) \leq 0 \text{ for all } -\infty < p < 0\},$$

similarly for the classes Stat (and Diff). Then the same techniques and the corresponding references as in the proof of Theorem 10 yield

$$\tilde{r}_{\text{Stat}} = \tilde{r}_L,$$
$$\frac{1}{p} r_{\text{Stat}}^{[p]} \geq r_{\text{Stat}}^{[-\infty]} = r_{\text{Stat}}^* = r_L^* \text{ for all } p<0,$$

where the equality holds again if x_0 can be chosen to be a (bounded) random variable. A p-th moment characterization of $\tilde{r}$ is obtained in the following way: Recall that by Proposition 3, the instability radius $\tilde{r}_L$ (and hence $\tilde{r}_{\text{Stat}}$) are obtained over the invariant control set $C \subset \mathbb{P}$. Define $g_\rho^*(\xi_t,p;C) = \inf_{x_0\in C} g_\rho(x_0,\xi_t,p)$, and similarly $r^{[-\infty]}(\xi_t;C)$, $r_{\text{Stat}}^{[-\infty]}(C)$. Then

$$\frac{1}{p} r_{\text{Stat}}^{[p]}(C) \geq r_{\text{Stat}}^{[-\infty]}(C) = r_{\text{Stat}}^*(C) = \tilde{r}_{\text{Stat}} \text{ for all } p<0,$$

where the equality holds under the same condition as above. □

The theory for the class Diff is considerably more complicated (and interesting), because we cannot use the periodic solutions of (4.1) as diffusion process that satisfy our assumptions for Diff. In fact, the Lie algebra assumption yields that every $\xi_t \in$ Diff has an invariant probability measure with C^∞ density, which is strictly positive on all of M. This implies (see e.g. [19, Proposition 3.2 and 3.4]) that all solutions of $(7.1)_\rho$, projected onto $\mathbb{P}$, enter the maximal control set C_ρ in finite time w.p.1 (even with finite expectation). Therefore, the Lyapunov exponents for $\xi_t \in$ Diff are independent of the initial value and they are attained in C_ρ. Furthermore, for each ξ_t and ρ the Lyapunov exponents are a.s. constant, and depend only on the (unique) invariant probability of the pair process (ξ_t, s_t), where s_t is the projection of $(7.1)_\rho$ onto $\mathbb{P}$ (see Hasminskii's formula (4.1) in [3]). We thus have for $\xi_t \in$ Diff

$$\lambda_\rho(x_0,\xi_t) = \lambda_\rho(\xi_t) \in \mathbb{R} \text{ for all } x_0 \neq 0, \text{w.p.1}$$

and

$$\lambda_\rho(\xi_t) \in I_\rho = [\alpha_\rho, \omega_\rho],$$

see [3] for proofs of these statements. We thus have

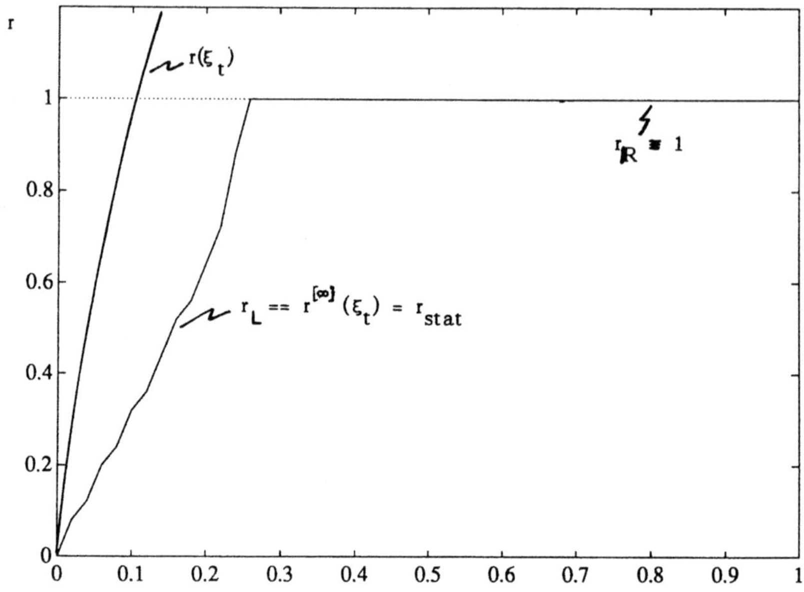

Figure 3. Stochastic stability radii for the linear oscillator

Theorem 11. *Let* $\xi_t \in$ Diff, *denote* $U_\rho = D_\rho(M)$. *Then under the assumptions above*

(i) $r(\xi_t) \geq r_{\mathrm{Diff}} \geq r_{\mathrm{Stat}} = r_L$ *for A stable.*

(ii) $\tilde{r}(\xi_t) = r^*(\xi_t) \geq \tilde{r}_{\mathrm{Diff}} = r^*_{\mathrm{Diff}} \geq \tilde{r}_{\mathrm{Stat}} = \tilde{r}_L$ *for A unstable.*

PROOF. (i) is clear from the definitions and Theorem 10. To show (ii), note first that for unstable matrix A and a given $\xi_t \in$ Diff we have $\tilde{r}(\xi_t) = r^*(\xi_t)$, because the Lyapunov exponents of $(7.1)_\rho$ are independent of the initial value x_0. This also means $\tilde{r}_{\mathrm{Diff}} = r^*_{\mathrm{Diff}}$. The rest follows from Remark 11 Q.E.D.

Remark 12. We conjecture that $r_{\mathrm{Diff}} = r_L$ and $r^*_{\mathrm{Diff}} = \tilde{r}_L$. For a characterization of $r(\xi_t) = r_L$, $\xi_t \in$ Diff, see Theorem 12 below. □

Theorem 11 and Example 9 show that $r_{\mathbb{C}}$ is in general not a good estimate for a stochastic stability radius. The linear oscillator shows that the real radius $r_{\mathbb{R}}$ is not suitable either:

Example 13. Consider again the linear oscillator from Example 9

$$\ddot{y}(t) + 2b\dot{y}(t) + [1 + D_\rho(\xi_t)]y(t) = 0$$

where $M = \mathbb{S}^1$, the one dimensional Euclidean sphere in $\mathbb{R}^2$, $D_\rho : \mathbb{S}^1 \to M(d, d; \mathbb{R})$ given by

$$D_\rho(m) = \rho \begin{pmatrix} 0 & 0 \\ -\cos m & 0 \end{pmatrix},$$

where ξ_t is the standard Brownian motion on $\mathbb{S}^1$. □

Figure 3 presents the stability radius $r(\xi_t)$ for this system, computed numerically via the law of large numbers (see "Hasminskii's formula", (4.1) in [3]). For b small, $r_L < r(\xi_t) < r_{\mathbb{R}}$, while for large b one has $r_{\mathbb{R}} < r(\xi_t)$. Note that for small b, $r(\xi_t, b)$ increases like const$\cdot\sqrt{b}$, compare [5].

We now turn to the analysis of the p-th moment stability radii. Throughout the rest of this section, which is based on the results in [2], we will assume that the following hypothesis (H''), which is slightly stronger than (H') holds:

$$\dim \mathcal{LA}\{X_0 + h + \tfrac{\partial}{\partial t}, X_1, \ldots, X_r\}(m, s, t) = \dim M + d \quad \text{for all } (m, s, t) \in M \times \mathbb{P}^{d-1} \times \mathbb{R}. \qquad (H'')$$

(H') and (H'') are equivalent e.g. if $M \times \mathbb{P}^{d-1}$ has a compact universal covering space (which is true e.g. if M is simply connected and $d > 2$).

First of all let us mention that

— $g_\rho(x_0, \xi_t, p) = g_\rho(\xi_t, p)$ for all $x_0 \neq 0$, $0 < p < \infty$,

— $g_\rho(\xi_t; \cdot) : \mathbb{R}_+ \to \mathbb{R}$ is convex, analytic with $g_\rho(\xi_t, 0) = 0$, $g_\rho'(\xi_t, 0) = \lambda_\rho(\xi_t)$.

Theorem 12. *Let $\xi_t \in$ Diff and suppose that A is stable. Then*

(i) $r(\xi_t) = \frac{1}{p} r^{[p]}(\xi_t) = r_L = r^{[\infty]}(\xi_t)$ *iff* $g_\rho(\xi_t, p) = p \cdot \lambda_\rho(\xi_t)$ *for all* $0 < p < \infty$. *This situation occurs exactly in the cases 2.1.2(i) and 2.2 of Theorem 7.*

(ii) $r^{[\infty]}(\xi_t) = r_L$ *for all* $\xi_t \in$ Diff.

The typical situation for deterministic and stochastic stability radii is characterized in Figure 4. (The reader should refer to the text for conditions, under which the strict inequalities hold. Note that $r_{\mathbb{C}}$, as defined in Section 6, is positive, if A is stable.) The corresponding moment Lyapunov function $g_\rho(\xi_t, \cdot)$ is indicated in Figure 5.

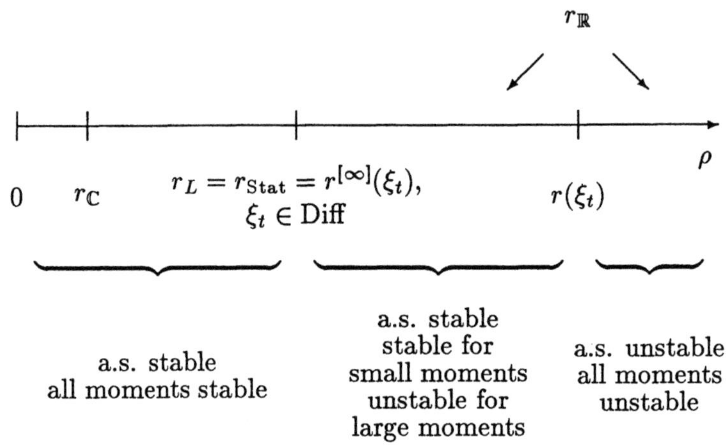

Figure 4. Typical order of deterministic and stochastic stability radii

If $r_L < \rho < r(\xi_t)$, then the system $(7.1)_\rho$ may "exhibit large deviations": Define the rate function $I_\rho(r, \xi_t) := \sup_{p\in\mathbb{R}}\{rp - g_\rho(\xi_t, p)\}$, denote

$$\gamma_\rho(\xi_t, p) = \begin{cases} \frac{1}{p} g_\rho(\xi_t, p) & \text{for } p \neq 0 \\ \lambda_\rho(\xi_t) & \text{for } p = 0 \end{cases}$$

and let

$$\gamma_\rho^*(\xi_t) = \lim_{p\to-\infty} \gamma_\rho(\xi_t, p), \quad \gamma_\rho(\xi_t) = \lim_{p\to+\infty} \gamma_\rho(\xi_t, p).$$

If $\gamma_\rho^*(\xi_t) < \gamma_\rho(\xi_t)$, then $I_\rho \geq 0$, finite exactly in the interval $(\gamma_\rho^*(\xi_t), \gamma_\rho(\xi_t))$ and strictly convex and analytic in this interval. The following is a consequence of a large deviations principle, suitable for our purposes:

Theorem 13. *Let $\xi_t \in$ Diff with (H'') holding. Then for all $\rho > 0$, the family of measures*

$$P_{t,\rho}(\cdot) = P\{\frac{1}{t} \log \frac{|\varphi(t, x_0, \xi_t)|}{|x_0|} \in \cdot\}$$

obeys a large deviations principle with rate function $I_\rho(r, \xi_t)$ (here $\varphi(t, x_0, \xi_t)$ is the solution of $(7.1)_\rho$). In particular for A stable:

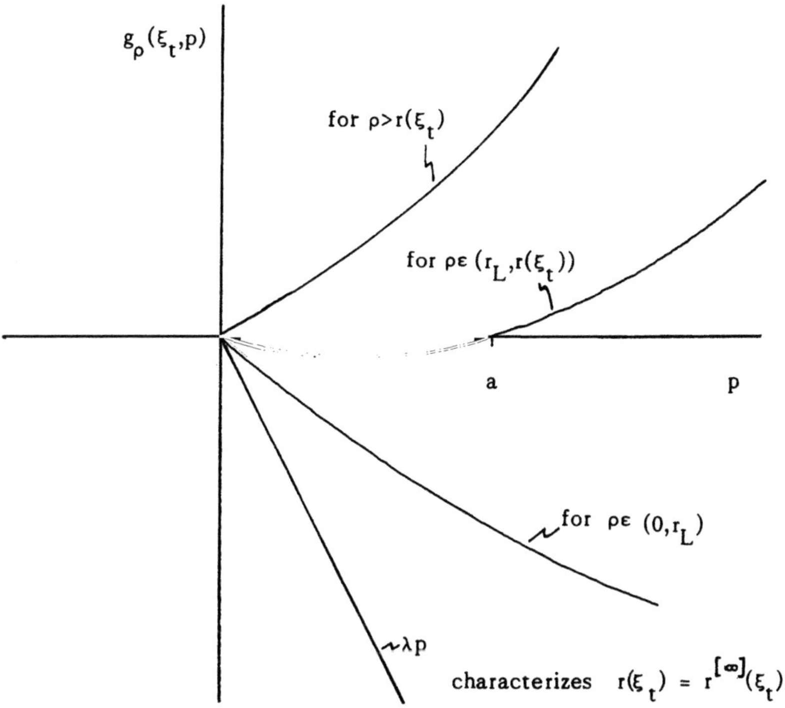

Figure 5. Typical moment Lyapunov functions for different ρ-values

(i) If $\rho \in (r_L, r(\xi_t))$, there exist constants $a > 0$, $k \geq 1$ such that for all $c > 0$ and all $0 < |x_0| < c$

$$\frac{1}{k}\left(\frac{|x_0|}{c}\right)^a \leq P\{\sup_{t\geq 0} |\varphi_\rho(t, x_0, \xi_t)| \geq c\} \leq k\left(\frac{|x_0|}{c}\right)^a.$$

(ii) If $\rho \in (0, r_L)$, then there is a constant $K \geq 1$ such that for all $x_0 \neq 0$

$$\sup_{t\geq 0} |\varphi(t, x_0, \xi_t)| \leq K|x_0|$$

with probability 1.

A precise formulation of a large deviations principle is given in [2].

This result can be interpreted as follows: If $\rho \in (r_L, r(\xi_t))$ then the system $(7.1)_\rho$ exhibits large deviations, i.e. for all $c > 0$ and all $t > 0$, we have $P\{\frac{1}{|x_0|}|\varphi(t, x_0, \xi_t)| \geq c\}$ is positive (and decays exponentially for $t \to \infty$), while for $0 < \rho < r_L$ this probability decays faster than any exponential function, in fact in this case $|\varphi(t, x_0, \xi_t)|$ is bounded for all

$t \geq 0$. Thus the theorem shows that for $\rho \in (r_L, r(\xi_t))$ the system is still stable with probability one, however arbitrarily large deviations in this range of ρ occur with positive probability.

Remark 14. It is worth noting that $r^{[\infty]}(\xi_t)$ does depend neither on the statistics nor on the dynamics of the diffusion process $\xi_t \in \text{Diff}$, since it always equals r_L and hence depends only on $U_\rho = D_\rho(M)$. □

Remark 15. For an unstable matrix A we know from Theorem 11 that $r^*_{\text{Diff}} = \tilde{r}_{\text{Diff}}$. A large deviations theory that ties the instability radius to the moment Lyapunov functions for $p < 0$ can be developed in complete analogy to Theorems 12 and 13, compare [2], in particular Section 3 therein. □

8 Remarks on Robust Design

In this section we draw some conclusions from the theory presented so far for a robust design technique, in which stability radii and Lyapunov exponents play a central role.

Stability radii can be viewed as a measure of robustness for the stability of a given system $\dot{x} = Ax$, depending on the nature of the uncertainty. Most of the results above deal with r_L, the Lyapunov radius for time varying deterministic disturbances, for a family of sets U_ρ of uncertainty matrices. This concept of stability radius appears to be important because

- information about uncertainties is usually given in terms of the size of possible variations of parameters in A,
- structured variations, i.e. dependent or independent disturbances in the parameters of A are covered by this model,
- complete ignorance about a parameter can be included as unbounded perturbation,
- no prior knowledge about the time dependence of the variation is assumed,
- stochastic, stationary perturbances lead to the same stability radii,
- stochastic, Markovian variations can be included in the theory in a consistent fashion, in particular, their stability radius for all moments leads to the same radius.

The disadvantage of r_L is that, in general, it can only be computed for a given system using numerical methods, i.e. by solving a parametrized optimal control problem for periodic trajectories.

The theory developed above, allows to incorporate information into a robust design which has two aspects:

- How large can the disturbances of A be, such that the system is still stable (this is covered e.g. by looking at stability radii),
- how fast does the system stabilize under a "typical variation" $D \in U_\rho$.

The second aspect is particularly important for the design, if different parameter constellations lead to the same stability radius. One then would choose the parameters in such a way that perturbations are damped in a maximal way. The notion of stability radii covers only the first aspect above. For the remainder of this section we discuss some strategies for this problem.

To outline the ideas let us separate the system parameters (i.e. the entrances of A) into three categories:

- structure parameters (not affected by uncertainties, not tunable by the designer, e.g. the first row of A in Example 9),
- tunable parameters (not affected by uncertainties, e.g. the damping b in Example 9),
- uncertain parameters (e.g. the restoring force in Example 9).

Note that the set of tunable parameters, say denoted by Θ, consists of constants, while the uncertain parameters are allowed to be time varying here. Let the range of the uncertainties as usual be given by U_ρ. Then the design problem may be formulated as follows: Given a family of systems

$$\dot{x} = A(\theta, u)x, \quad \theta \in \Theta, \quad u \in U_\rho, \quad \rho > 0,$$

find all parameters $\theta \in \Theta$ such that the stability radius (with respect to U_ρ) $r_L(\theta) \geq r_{\text{crit}}$, a critical radius, or such that $r_L(\theta)$ becomes maximal, if θ is chosen in Θ.

Additional design criteria can be given e.g. by deterministic or stochastic information on the uncertainty. Then one would not only maximize the (deterministic or stochastic) stability radius, determined by the corresponding Lyapunov exponents; one would also like to minimize Lyapunov exponents of "typical" perturbations.

As an illustration of these general remarks consider again the linear oscillator from Examples 9, 13:

$$\begin{pmatrix} \dot{x}_1(t) \\ \dot{x}_2(t) \end{pmatrix} = \begin{pmatrix} 0 & 1 \\ -1 - u(t) & -2b \end{pmatrix} \begin{pmatrix} x_1(t) \\ x_2(t) \end{pmatrix}$$

Consider $b \in [0, 2] =: \Theta \subset \mathbb{R}$ as a tunable parameter and u as an uncertain parameter, $u(t) \in [-\rho, \rho]$, $\rho \geq 0$, i.e. $U_\rho = \left\{ \begin{pmatrix} 0 & 0 \\ -\alpha & 0 \end{pmatrix}, \alpha \in [-\rho, \rho] \right\}$.

8.1 A strategy motivated by deterministic considerations

Figure 6 shows the level curves of the Lyapunov exponents $\kappa_\rho(b)$ as described in Example 9. Again, from the explicit formulae in [9], Section 6,

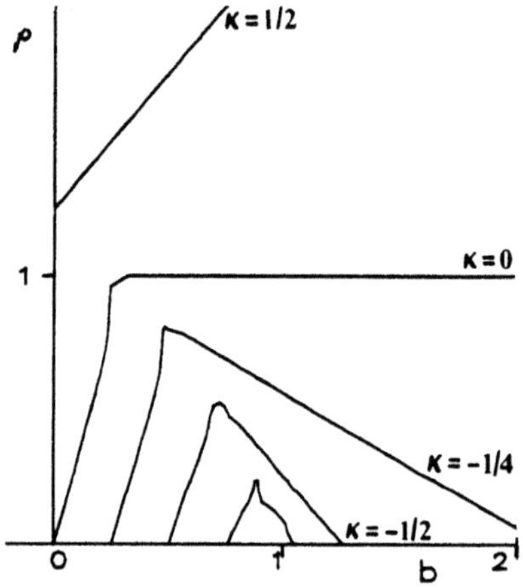

Figure 6. Level curves of κ for the linear oscillator

we obtain: For each $\rho \in (0,1)$, the curve $\kappa_\rho(b)$ has a unique minimum b_ρ, i.e. there is a unique damping value b_ρ such that (timevarying) uncertainties of size $U_\rho = [-\rho, \rho]$ are exponentially damped in an optimal way for b_ρ. Figure 7 shows the location of these values. Note that increased damping with b does not necessarily lead to faster damping of disturbances (overdamping) for fixed ρ. Note also that this strategy leads to optimal b_ρ values for $b_\rho \in (b_0, b_1)$, where the complex stability radius $r_\mathbb{C}$ is less than 1; i.e. maximization of $r_\mathbb{C}$ does not lead to optimality in this sense.

It follows from Section 7 that the same design is optimal, if one wants to dampen out, at a maximal rate, all moments of the solution $\varphi(t, x_0, \xi_t)$, caused by stationary ergodic nondegenerate diffusions with values in U_ρ or all stationary processes with values in U_ρ. We would like to mention that for $\rho > 1$, $\kappa_\rho(b)$ is strictly increasing in b.

Note that in this example the choice of the "tunable parameter" b may

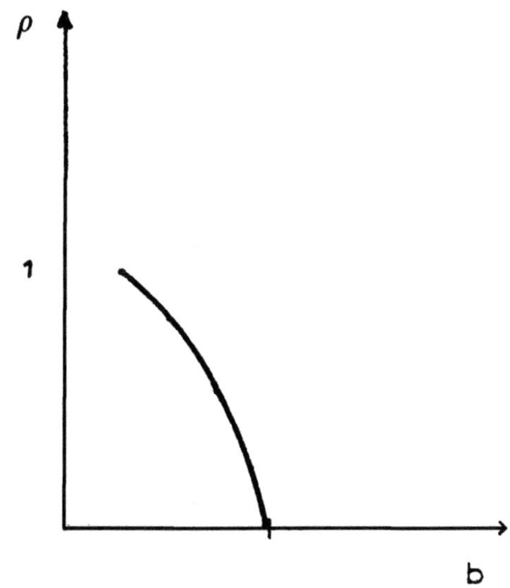

Figure 7. Optimal damping with timevarying uncertainties

be interpreted as the choice of a certain feedback: Define

$$A_0 = \begin{pmatrix} 0 & 1 \\ -1-u & 0 \end{pmatrix} \quad B = \begin{pmatrix} 0 \\ -2 \end{pmatrix} \quad C = (0 \ \ 1)$$

Then the choice of the tunable parameter b means choosing a feedback $F = b$ for the system

$$\dot{x} = A_0 x + Bu, \quad y = Cx$$

such that

$$\dot{x} = (A_0 + BFC)x = \begin{pmatrix} 0 & 1 \\ -1-u & -2b \end{pmatrix} x$$

has certain desired properties.

8.2 A strategy motivated by the stochastic a.s. Lyapunov exponent

Figure 8 presents typical level curves of a.s. Lyapunov exponents $\lambda_\rho(x_0, \xi_t; b)$ for a specific excitation ξ_t (see Example 13). Of course, these values were again obtained numerically via the law of large numbers. Recall that for $\xi_t \in \text{Diff}$ we have $\lambda_\rho(x_0, \xi_t; b) = \kappa_\rho(\xi_t, b)$, and thus we see from the numerical results, that the stochastic stability radius $r(\xi_t, b)$ increases to ∞ for $b \to \infty$ and for each $\rho \in (0, \infty)$ the curve $\kappa_\rho(\xi_t, b)$ has a unique minimum

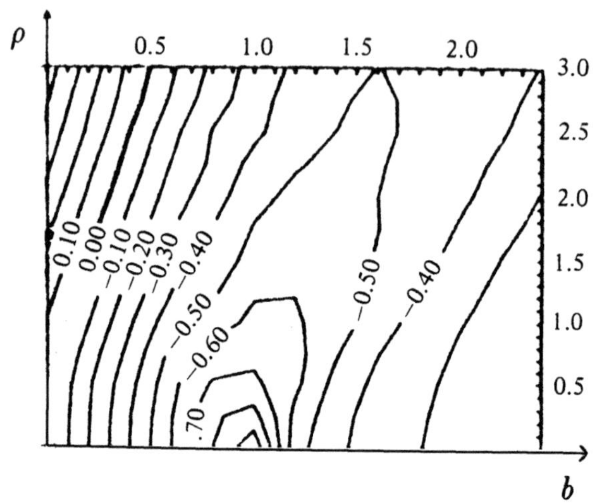

Figure 8. Level curves of a.s. Lyapunov exponents

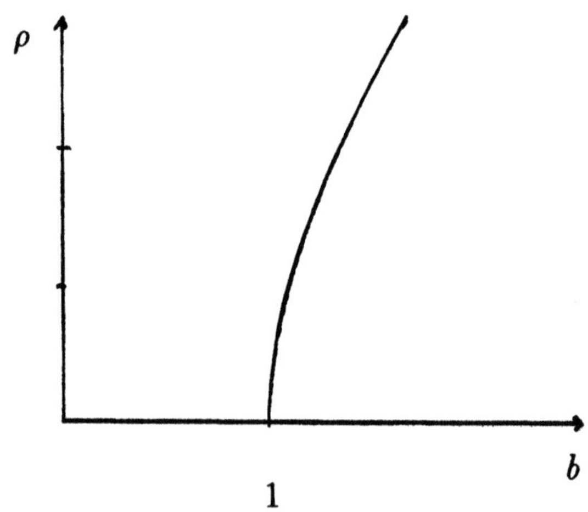

Figure 9. Optimal damping with stochastic uncertainties

$b_\rho(= b_\rho(\xi_t))$. These (numerical) findings are supported by the asymptotic expansions of stochastic Lyapunov exponents for small and large ρ's, see e.g. [5]. (For numerical methods for the computation of stochastic Lyapunov exponents we refer the reader to the work of Talay, e.g. [22].) The location of the optimal values b_ρ is shown in Figure 9. Obviously, the design strategies in 8.1 and 8.2 above do not agree. An optimal robust design depends not only on the requirements imposed on the system, but also on the prior information available on the uncertainties inherent in the system parameters.

Acknowledgement: The authors would like to thank Diederich Hinrichsen for introducing to them the theory of stability radii and for valuable discussions about this topic. We also thank two anonymous referees for the careful reading which lead to many improvements. The figures were produced by A. Carriquiry (2,6), B. Kelb (1,4) and K.U. Schaumlöffel (3,8).

REFERENCES

[1] L. ARNOLD, W. KLIEMANN, *Qualitative theory of stochastic systems,* in Probabilistic Analysis and Related Topics, Vol. 3, cd. Bharucha-Reid, Academic Press, New York 1983, 1–79.

[2] L. ARNOLD, W. KLIEMANN, *Large deviations of linear stochastic differential equations,* in Stochastic Differential Systems, ed. J.H. Engelbert, W. Schmidt, Lecture Notes in Control and Information Sciences Vol. 96, Springer-Verlag, 1987, 117–151.

[3] L. ARNOLD, W. KLIEMANN, E. OELJEKLAUS, *Lyapunov exponents of linear stochastic systems,* in [6], 85–125.

[4] L. ARNOLD, E. OELJEKLAUS, E. PARDOUX, *Almost sure and moment stability for linear Itō equations,* in [6], 129–159.

[5] L. ARNOLD, G. PAPANICOLAOU, V. WIHSTUTZ, *Asymptotic analysis of the Lyapunov exponent and rotation number of the random oscillator and applications,* SIAM J. Appl. Math., 46(1986), 427–450.

[6] L. ARNOLD, V. WIHSTUTZ (eds), *Lyapunov exponents,* Proceedings of a Workshop, Lecture Notes in Mathematics, Vol. 1186, Springer-Verlag, 1986.

[7] L. ARNOLD, W. WIHSTUTZ, *Lyapunov exponents: a survey,* in [6], 1–26.

[8] F. COLONIUS, W. KLIEMANN, *Infinite time optimal control and periodicity,* Appl. Math. Optim. 20 (1989), 113–130.

[9] F. Colonius, W. Kliemann, *Maximal and minimal Lyapunov exponents of bilinear control systems,* Preprint, 1990.

[10] F. Colonius, W. Kliemann, *Linear control semigroups acting on projective space,* Preprint, 1990.

[11] Ju. L. Daleckii, M. G. Krein, *Stability of Solutions of Differential Equations in Banach Spaces,* Amer. Math. Soc., Providence, R.I., 1974.

[12] W. Hahn, *Stability of Motion,* Springer Verlag, 1967.

[13] D. Hinrichsen, A. Ilchmann, A.J. Pritchard, *Robustness of time-varying linear systems,* J. Diff. Equations, 82 (1989), 219–250.

[14] D. Hinrichsen, B. Kelb, A. Linnemann, *An algorithm for the com- ıtation of the structured stability radius with applications,* Institut ıür Dynamische Systeme, Report 182, Universität Bremen 1987.

[15] D. Hinrichsen, A.J. Pritchard, *Stability radius of linear systems,* Systems and Control Letters, 7(1986), 1–10.

[16] D. Hinrichsen, A.J. Pritchard, *Stability radius for structured perturbations and the algebraic Riccati equation,* Systems and Control Letters, 8(1986), 105–113.

[17] A. Isidori, *Nonlinear Control Systems: An Introduction,* Lecture Notes in Control and Information Sciences, Vol. 72, Springer-Verlag, 1985.

[18] B. Kelb. Diplomarbeit, Universität Bremen, Bremen 1989.

[19] W. Kliemann, *Recurrence and invariant measures for degenerate diffusions,* Ann. Prob., 15(1987), 690–707.

[20] W. Kliemann, *Analysis of Nonlinear Stochastic Systems,* in Analysis and Estimation of Stochastic Mechanical Systems, ed. W. Schiehlen, W. Wedig, CISM Courses and Lectures No. 303, Springer-Verlag 1988, 43–102.

[21] H. Moormann, *Lyapunov-Exponenten einer linearen stochastischen Differentialgleichung (der degenerierte Fall),* Report Nr. 187, Institut für Dynamische Systeme, Universität Bremen, Bremen, 1988.

[22] D. Talay, *Approximation of upper Lyapunov exponents of bilinear stochastic differential systems,* INRIA-Sophia-Antipolis, Rapport de Recherche, no. 965, 1989.

[23] C. Van Loan, *How near is a stable matrix to an unstable matrix*, Contemporary Mathematics, 47(1985), 465–478.

Fritz Colonius
Institut für Mathematik
Universität Augsburg
Universitätsstraße 2
D-8900 Augsburg/West Germany

Wolfgang Kliemann
Department of Mathematics
400 Carver Hall
Iowa State University
Ames, Iowa 50011/USA

Topological Aspects of Robust Control

Filippo De Mari Raimund Ober

The paper discusses topologies that naturally arise in the study of robust control problems. A new 'robust topology' is defined. It is shown that this topology is Hausdorff.

1 Introduction

Robustness issues in control theory play an important role because each model of a real process has inherent uncertainties that have to be taken into account in any controller design. An underlying issue of robust control is the question of when two plants are close to one another from the point of view of robustness. The most important aspect of robustness is that of robust stability which is addressed in this paper. We are going to introduce a 'robust topology' in order to describe what we mean by two systems being close from the point of view of robust stability.

The purpose of the paper is to give a precise definition of this topology and to study some of its elementary properties. We are also going to compare this topology with the so called graph topology ([5]).

2 Definition of robust topology

In this section we are going to define a topology on sets of plants in such a way that we are able to study robustness issues with the help of this topology. By robustness we mean that if a plant P is stabilized by a controller K then all plants in a sufficiently small neighbourhood of P are also stabilized by the controller K.

Let now $\mathcal{P}$ be a set of plants and $\mathcal{K}$ a set of controllers, where we assume that for each plant $P_0 \in \mathcal{P}$ there is a controller $K_0 \in \mathcal{K}$ that stabilizes P_0.

The *robust topology* $\tau_{\mathcal{K}}(\mathcal{P})$ on the set of plants $\mathcal{P}$ which is induced by the set of controllers $\mathcal{K}$ is defined through the following prebase. For each controller $K \in \mathcal{K}$ let $\mathcal{P}(K)$ be the set of plants in $\mathcal{P}$ that are stabilized by the controller K. The family of sets $(\mathcal{P}(K))_{K \in \mathcal{K}}$ forms a prebase for the robust topology $\tau_{\mathcal{K}}(\mathcal{P})$ on the set $\mathcal{P}$.

Note that in a dual fashion we can define a topology on the set of controllers $\mathcal{K}$. Here we define a prebase on $\mathcal{K}$ by $(\mathcal{K}(P))_{P \in \mathcal{P}}$, where $\mathcal{K}(P)$ is the set of controllers in $\mathcal{K}$ that stabilize the plant $P \in \mathcal{P}$. We have to assume that for each controller $K \in \mathcal{K}$ there is at least one plant $P \in \mathcal{P}$ that is stabilized by K.

In order to be more concrete we are now going to recall one particular way of defining closed loop stability (see e.g. [5]). Let P be a multivariable transfer function of a continuous time finite dimensional system. By a right (left) coprime factorization of P we mean a factorization $P = ND^{-1}$ ($P = \tilde{D}^{-1}\tilde{N}$), where D, N ($\tilde{D}, \tilde{N}$) are asymptotically stable rational transfer functions, D ($\tilde{D}$) being square and invertible, such that the Bezout equation

$$XN + YD = I, \qquad (\tilde{N}\tilde{X} + \tilde{D}\tilde{Y} = I),$$

has a solution X, Y ($\tilde{X}, \tilde{Y}$), with X, Y ($\tilde{X}, \tilde{Y}$) being asymptotically stable proper transfer functions.

Consider the right coprime factorization $K = N_K D_K^{-1}$ (left coprime factorization $K = \tilde{D}_K^{-1}\tilde{N}_K$). Then K (internally) stabilizes P if and only if

$$\tilde{N}_K N + \tilde{D}_K D \qquad (\tilde{N} N_K + \tilde{D} D_K)$$

is unimodular, i.e. is invertible with asymptotically stable proper inverse (for precise definition of internal stability and result see [5]). Of particular importance in our context is that, given a continuous time plant with rational transfer function P of all (proper) rational controllers $\mathcal{K}(P)$ that stabilize P can be obtained via the so-called Youla parametrization

$$\begin{aligned}\mathcal{K}_p(P) &= \{(Y - R\tilde{N})^{-1}(X + R\tilde{D}) | R \in M(S), \|Y - R\tilde{N}\| \neq 0\} \\ &= \{(\tilde{X} + DS)(\tilde{Y} - NS)^{-1} | S \in M(S), \|\tilde{Y} - NS\| \neq 0\}\end{aligned}$$

where $P = D^{-1}N = \tilde{N}\tilde{D}^{-1}$ is a left respectively right coprime factorization with $X, Y, \tilde{X}, \tilde{Y} \in M(S)$ solving the corresponding Bezout equations

$$XN + YD = I \qquad \tilde{N}\tilde{X} + \tilde{D}\tilde{Y} = I.$$

The symbol $M(S)$ stands for the set of asymptotically stable proper rational transfer function matrices of appropriate dimensions. By asymptotic stability we here mean asymptotic stability as a continuous-time system. Essentially the same results hold for more general notions of stability (see

[5]). It has been shown in ([5], p. 115) that all strictly proper controllers $\mathcal{K}_{sp}(P)$ of a plant $\mathcal{P}$ are given by the same linear fractional map where the parameter space $M(S)$ is replaced by the ideal of all strictly proper functions in $M(S)$, and X and $\tilde{X}$ are chosen to be strictly proper. The set of strictly proper functions in $M(S)$ we denote by $M_{sp}(S)$.

Dually there is an analogous parametrization of the set of rational plants that are stabilized by a fixed rational controller.

In the next section we are going to use these results to show that the robust topology on the set of rational plants with rational controllers is in fact Hausdorff.

3 The robust topology is Hausdorff

In the previous section we have defined in an abstract way a topology on the set of plants with rational transfer functions. In order to be able to deal with such a topology in a satisfactory way we would hope that the topology is Hausdorff, i.e. that for each two points in the set we can find neighbourhoods which do not intersect. We will now show that this is indeed the case if the set of plants is $\mathcal{P} = M(\mathbb{R}(s))$ and the set of controllers is $\mathcal{K}_p = M(\mathbb{R}(s))$ or $\mathcal{K}_{sp} = M_{sp}(\mathbb{R}(s))$, where $M(\mathbb{R}(s))$ ($M_{sp}(\mathbb{R}(s))$ stands for the set of proper (strictly proper) rational transfer function matrices with fixed input and output dimensions.

We first need some preliminaries concerning the simultaneous stabilizability of two plants by a single controller. Let P_i, $i = 1, 2$ be two plants. By a *doubly coprime factorization* of P_i, $i = 1, 2$ we mean a choice of right and left coprime factorizations (N_i, D_i) and $(\tilde{N}_i, \tilde{D}_i)$, of P_i, $i = 1, 2$, together with matrices $X_i, Y_i, \tilde{X}_i, \tilde{Y}_i$, $i = 1, 2$, in $M(S)$ such that

$$L_i R_i := \begin{bmatrix} Y_i & X_i \\ -\tilde{N}_i & \tilde{D}_i \end{bmatrix} \begin{bmatrix} D_i & -\tilde{X}_i \\ N_i & \tilde{Y}_i \end{bmatrix} = I, \qquad i = 1, 2.$$

By abuse of language we will say 'let $L_i R_i = I$ be a doubly coprime factorization of P_i, $i = 1, 2$', as well as 'let $L_i R_i = I$ be a doubly coprime factorization of C_i', where $C_i = \tilde{X}_i \tilde{Y}_i^{-1} = Y_i^{-1} X_i$, $i = 1, 2$. We also set

$$L_1 R_2 = \begin{bmatrix} Y_1 D_2 + X_1 N_2 & -Y_1 \tilde{X}_2 + X_1 \tilde{Y}_2 \\ -\tilde{N}_1 D_2 + \tilde{D}_1 N_2 & \tilde{N}_1 \tilde{X}_2 + \tilde{D}_1 \tilde{Y}_2 \end{bmatrix} =: \begin{bmatrix} B_1 & \Delta_C \\ \Delta_P & B_2 \end{bmatrix}$$

We need to quote the following criterion on the simultaneous stabilizability of two plants.

Theorem 1. *([5]) Given two plants P_i, $i = 1, 2$, together with doubly coprime factorizations $L_i R_i = I$, $i = 1, 2$, then they are simultaneously stabilizable if and only if one of the following equivalent conditions holds:*

1. *There exists a* $M \in M(S)$ *such that* $B_1 + M\Delta_P$ *is unimodular.*

2. *At all real blocking zeros of* Δ_P *in the extended right half plane* $\mathbb{C}_{+e}$, $\det B_1$ *has the same sign. (A blocking zero of* Δ_P *is a value of the independent variable* s *at which all entries of the matrix are zero.)*

The following Proposition contains a result for the dual situation, i.e. given controllers C_i, $i = 1, 2$, does there exist a plant that is stabilized by both controllers.

Proposition 2. *Let* C_i, $i = 1, 2$, *be two controllers together with doubly coprime factorizations* $L_i R_i = I$, $i = 1, 2$. *Then* C_1 *and* C_2 *simultaneously stabilize a plant if and only if one of the following two equivalent conditions holds:*

1. *There exists a* $M \in M(S)$ *such that* $B_1 + \Delta_C M$ *is unimodular.*

2. *At all real blocking zeros of* Δ_C *in the extended right half plane* $\mathbb{C}_{+e}$, *the sign of* $\det B_1$ *remains constant.*

PROOF. We only prove *1.* The equivalence of *1.* and *2.* follows from the Remark on p. 125 and Corollary 6 on p. 118 in [5].

Assume first that C_1 and C_2 are both strictly proper. Then

$$\mathcal{P}(C_i) = \{(N_i + \tilde{Y}_i H)(D_i - \tilde{X}_i H)^{-1} | H \in M(S)\}, \qquad i = 1, 2.$$

Hence C_0 and C_1 simultaneously stabilize a plant if and only if there exist $H_1, H_2 \in M(S)$ such that

$$(N_1 + \tilde{Y}_1 H_1)(D_1 - \tilde{X}_1 H_1)^{-1} = (N_2 + \tilde{Y}_2 H_2)(D_2 - \tilde{X}_2 H_2)^{-1}.$$

Since both sides of the expression give a right coprime factorization of the same plant P, there exists a unimodular matrix U (see [5]) such that

$$(N_2 + \tilde{Y}_2 H_2) = (N_1 + \tilde{Y}_1 H_1)U, \qquad (D_2 - \tilde{X}_2 H_2) = (D_1 - \tilde{X}_1 H_1)U$$

We rewrite these expressions to obtain

$$\begin{bmatrix} D_2 & -\tilde{X}_2 \\ N_2 & \tilde{Y}_2 \end{bmatrix} \begin{bmatrix} I \\ H_2 \end{bmatrix} = \begin{bmatrix} D_1 & -\tilde{X}_1 \\ N_1 & \tilde{Y}_1 \end{bmatrix} \begin{bmatrix} I \\ H_1 \end{bmatrix} U$$

which is the case if and only if

$$\begin{bmatrix} Y_1 & X_1 \\ -\tilde{N}_1 & \tilde{D}_1 \end{bmatrix} \begin{bmatrix} D_2 & -\tilde{X}_2 \\ N_2 & \tilde{Y}_2 \end{bmatrix} \begin{bmatrix} I \\ H_2 \end{bmatrix} = \begin{bmatrix} I \\ H_1 \end{bmatrix} U$$

But this implies that

$$\begin{bmatrix} B_1 & \Delta_C \\ \Delta_P & B_2 \end{bmatrix} \begin{bmatrix} I \\ H_2 \end{bmatrix} = \begin{bmatrix} I \\ H_1 \end{bmatrix} U \qquad (*)$$

Thus C_1 and C_2 simultaneously stabilize a plant if and only if there exist $H_1, H_2 \in M(S)$ and a unimodular U such that the above equation holds.

Assume that (*) holds for suitable H_1, H_2, U. Then $B_1 + \Delta_C H_2 = U$. Conversely suppose that $B_1 + \Delta_C H_2 = U$ is unimodular for a suitable H_2. Then set $H_1 = (\Delta_P + B_2 H_2)U^{-1}$ and check that (*) holds.

To remove the assumption that C_1 and C_2 are strictly proper argue as in [5], p. 126–127. Q.E.D.

Before proving the main result of this section we need to establish the following technical Lemma (see Lemma 21, p. 96 in [5]for a similar result).

Lemma 3. *Let $\mathcal{I}$ be a nontrivial ideal in S, the set of asymptotically stable proper transfer functions. Let $N \in S^{p\times m}$, $Y \in S^{p\times p}$ and $H \in S^{m\times p}$ be such that*

1. *$[N, Y]$ has full row rank ($= p$).*
2. *$\det(Y + NH) = 0$ in S.*

Then there exists $G \in \mathcal{I}^{m\times p}$ such that $\det(Y + N(H + G)) \neq 0$.

PROOF. First observe that

$$[Y + NH, N] = [Y \;\; N] \begin{bmatrix} I & 0 \\ H & I \end{bmatrix}$$

so that $[Y + NH, N]$ has full rank p too. Select a nonzero minor m_p of $[Y + NH, N]$ obtained by considering a submatrix which minimizes the number of columns of N. Assume that this is achieved by selecting columns $j_1, \ldots, j_k$ of N and omitting columns $i_1, \ldots, i_k$ of $Y + NH$. Choose now $g \in \mathcal{I}$, $g \neq 0$, and put

$$g_{ji} = \begin{cases} g \text{ if } j = j_s \text{ and } i = i_s, \text{ for some } s = 1, \ldots, k \\ 0 \text{ otherwise.} \end{cases}$$

By the Binet-Cauchy formula we obtain

$$\det(Y+N(H+G)) = \det\left([Y + NH, N]\begin{bmatrix} I \\ G \end{bmatrix}\right) = \sum_K [Y+NH, N]_K \begin{bmatrix} I \\ G \end{bmatrix}_K$$

where K varies over all the increasing p-tuples $(\alpha_1, \ldots, \alpha_p)$ with $1 \leq \alpha_s \leq m + p$, $1 \leq s \leq p$. Observe that:

- Every K containing more than k rows of G is such that $\begin{bmatrix} I \\ G \end{bmatrix}_K = 0$.
- Exactly one p-minor of $\begin{bmatrix} I \\ G \end{bmatrix}$ containing k rows of G is nonzero. This is obtained by selecting rows $j_1, \ldots, j_k$ of G and omitting rows $i_1, \ldots, i_k$ of I. If K_0 is the corresponding p-tuple, then

$$\begin{bmatrix} I \\ G \end{bmatrix}_{K_0} = \pm g^k.$$

- Every K containing less than k column indexes of N is such that $[Y + NH, N]_K = 0$.

We therefore conclude that there is exactly one p-tuple K for which

$$[Y + NH, N]_K \begin{bmatrix} I \\ G \end{bmatrix}_K \neq 0,$$

namely,

$$\det(Y + N(H + G)) = [Y + NH, N]_{K_0} \begin{bmatrix} I \\ G \end{bmatrix}_{K_0} = \pm m_p g^k \neq 0.$$

Q.E.D.

We can now prove the main theorem of this paper.

Theorem 4. *Let*

$$\mathcal{K}(P_i) = \left\{ (Y_i - H\tilde{N}_i)^{-1}(X_i + H\tilde{D}_i) \| \|Y_i - H\tilde{N}_i\| \neq 0, H \in M(I) \right\},$$

$i = 0, 1$, *where* M *is either* $M(S)$ *or* $M_{sp}(S)$.

If $P_0 \neq P_1$ *are proper rational transfer functions of two plants, then there exist* $C_i \in \mathcal{K}(P_i)$, $i = 0, 1$, *such that* $\mathcal{P}(C_0) \cap \mathcal{P}(C_1) = \emptyset$.

PROOF. For convenience of notation we only prove the result for the case the set of controllers parametrized by $M(S)$. With the obvious modifications the result also holds if the set of parameters is $M_{sp}(S)$.

a.) We first assume that P_0 and P_1 are not simultaneously stabilizable.

Fix doubly coprime factorizations $L_i R_i = I$ of P_i, $i = 0, 1$. Since P_0 and P_1 are not simultaneously stabilizable, $\Delta_P = (-\tilde{N}_0 D_1 + \tilde{D}_0 N_1)$ has at least two real blocking zeros in $\mathbb{C}_{+e}$, say ξ_1, ξ_2, at which $\det(Y_0 D_1 + X_0 N_1) = \det B_1$ changes sign. In particular, since $\det B_1(\xi_j) \neq 0$, $j = 1, 2$, $B_1(\xi_j)$ is an invertible matrix, $j = 1, 2$. Thus we can find $\bar{H} \in M(S)$ such that

$$\bar{H}(\xi_j) = B_1(\xi_j)^{-1} \Delta_C(\xi_j). \qquad (**)$$

If $\det(\tilde{Y}_1 - N_1\bar{H}) = 0$, consider the ideal $\mathcal{I}$ of S consisting of those elements g for which $g(\xi_j) = 0$, $j = 1, 2$. Then by the lemma, there exists $G \in M(\mathcal{I})$, the set of matrices with entries in $\mathcal{I}$, such that $\det(\tilde{Y}_1 - N_1(\bar{H} + G)) \neq 0$. Observe that $(\bar{H} + G)(\xi_j) = \bar{H}(\xi_j)$, $j = 1, 2$. Hence a choice of $\bar{H}$ satisfying (**) can be made in such a way that $\det(\tilde{Y}_1 - N\bar{H}) \neq 0$. On the other hand

$$\mathcal{K}(P_1) = \{(\tilde{X}_1 + D_1 H)(\tilde{Y}_1 - N_1 H)^{-1} | H \in M(S), \det(\tilde{Y}_1 - NH) \neq 0\}.$$

Thus, with the obvious meaning of the symbols:

$$\begin{aligned}\Delta_C(\bar{H}) &= -Y_0\tilde{X}_1(\bar{H}) + X_0\tilde{Y}_1(\bar{H}) \\ &= -Y_0(\tilde{X}_1 + D_1\bar{H}) + X_0(\tilde{Y}_1 - N_1\bar{H}) \\ &= \Delta_C - B_1\bar{H}\end{aligned}$$

Now,

$$B_1(\xi_j)^{-1}\Delta_C(\bar{H})(\xi_j) = B_1(\xi_j)^{-1}\Delta_C(\xi_j) - \bar{H}(\xi_j) = 0, \qquad j = 1, 2,$$

implies that $\Delta_C(\bar{H})(\xi_j) = 0$, $j = 1, 2$. Thus we have found doubly coprime factorizations $L_0 R_0 = I$, $L_1(\bar{H})R_1(\bar{H}) = I$ relative to which Δ_C has two real blocking zeros in $\mathbb{C}_{+e}$ at which $\det B_1$ changes sign. Therefore C_0 and $C_1(\bar{H})$ stabilize no common plant.

b.) Now assume that P_0 and P_1 are simultaneously stabilizable. Choose right and left coprime factorizations $(\tilde{X}, \tilde{Y})$ and (X, Y) of a simultaneously stabilizing controller C of P_i, $i = 0, 1$. Then choose right coprime factorizations (N_i, D_i) and left coprime factorizations $(\tilde{D}_i, \tilde{N}_i)$ of P_i such that

$$\begin{bmatrix} Y & X \\ -\tilde{N}_i & \tilde{D}_i \end{bmatrix}\begin{bmatrix} D_i & -\tilde{X} \\ N_i & \tilde{Y} \end{bmatrix} = I \qquad i = 0, 1.$$

Then

$$\mathcal{K}(P_0) = \{(Y - T\tilde{N}_0)^{-1}(X + T\tilde{D}_0) | T \in M(S), \det(Y - T\tilde{N}_0) \neq 0\}$$

$$\mathcal{K}(P_1) = \{(\tilde{X} + D_1 R)(\tilde{Y} - N_1 R)^{-1} | R \in M(S), \det(\tilde{Y} - N_1 R) \neq 0\}$$

and so with the obvious meaning of the symbols:

$$\begin{aligned}B_1(T) &= Y_0(T)D_1 + X_0(T)N_1 \\ &= (Y - T\tilde{N}_0)D_1 + (X + T\tilde{D}_0)N_1 \\ &= I + T(-\tilde{N}_0 D_1 + \tilde{D}_0 N_1) \\ &= I + T\Delta_P\end{aligned}$$

and

$$\begin{aligned}\Delta_C(T, R) &= -Y_0(T)\tilde{X}_1(R) + X_0(T)\tilde{Y}_1(R) \\ &= -(Y - T\tilde{N}_0)(\tilde{X}_1 + D_1 R) + (X + T\tilde{D}_0)(\tilde{Y} - N_1 R) \\ &= -R + T - T\Delta_P R \\ &= T - (I + T\Delta_P)R \\ &= T - B_1(T)R.\end{aligned}$$

Now, since $P_0 \neq P_1$, $\Delta_P \neq 0$ in $M(S)$. Thus we can find two real ξ_j, $j = 1, 2$, in $\mathbb{C}_{+e}$ at which $\Delta_P(\xi_j) \neq 0$, $j = 1, 2$. Then we can find matrices T_{ξ_j}, $j = 1, 2$, with

$$\det B_1(T_{\xi_1})(\xi_1) = \det(I + T_{\xi_1}\Delta_P(\xi_1)) > 0$$

$$\det B_2(T_{\xi_2})(\xi_2) = \det(I + T_{\xi_2}\Delta_P(\xi_2)) < 0.$$

Finally, we can find $\bar{T} \in M(S)$ such that $\bar{T}(\xi_j) = T_{\xi_j}$, $j = 1, 2$. Applying the lemma we may assume that $\det(Y - \bar{T}\tilde{N}_0) \neq 0$. In particular $B_1(\bar{T})(\xi_j)$ is invertible and so we can find $\bar{R} \in M(S)$ such that

$$\bar{R}(\xi_j) = (B_1(\bar{T})(\xi_j))^{-1}(\bar{T}(\xi_j)), \qquad j = 1, 2,$$

and by the lemma, $\det(\tilde{Y} - N_1\bar{R}) \neq 0$. From

$$(B_1(\bar{T})(\xi_j))^{-1}\Delta(\bar{T}, \bar{R})(\xi_j) = (B_1(\bar{T})(\xi_j))^{-1}\bar{T}(\xi_j) - \bar{R}(\xi_j) = 0, \quad j = 1, 2.$$

it follows $\Delta(\bar{T}, \bar{R})(\xi_j) = 0$, $j = 1, 2$. Hence we have found doubly coprime factorizations $L_0(T)R_0(T)$ and $L_1(T)R_1(T)$ of P_0 and P_1 relative to which Δ_C has two real blocking zeros in $\mathbb{C}_{+e}$ at which $\det B_1$ changes sign. The corresponding controllers stabilize no common plant. Q.E.D.

We immediately have as a corollary that the robust topology as introduced in the previous section is Hausdorff.

Corollary 5. *If the set of plants and the set of controllers are given by $\mathcal{P} = M(\mathbb{R}(s))$ and $\mathcal{K}_p = M(\mathbb{R}(s))$ ($\mathcal{K}_{sp} = M_{sp}(\mathbb{R}(s))$), then $\tau_{\mathcal{K}_p}(\mathcal{P})$ ($\tau_{\mathcal{K}_{sp}}(\mathcal{P})$) is Hausdorff.*

In the same way we have that the dual topologies on the set of controllers are Hausdorff.

The theorem also has an interesting interpretation regarding model reduction. Since for every two plants however 'close' they are there is a controller that will stabilize one plant but not the other it follows that the general rule is confirmed whereby great care must be taken when a controller is designed on the basis of an approximate model. Note that a weaker form of the theorem was shown in [3].

4 Robust topology and graph topology

Vidyasagar defined the so called graph topology on the set of rational transfer functions. Similarly to our work the intention was to define a topology on the set of all plants that describes in topological terms what we normally mean by robustness. In contrast to our definition his approach does not only require robustness of the stability of the closed loop but also robustness of performance. In particular it is required that the closed loop is continuous in the H^∞ topology with respect to perturbations of the plant. (For more precise definitions and results see [5].) This implies that the robust topology defined here is weaker than the graph topology.

Example 1. We are going to consider some examples that show that on a set of plants $\mathcal{P}$ the topologies $\tau_{\mathcal{K}_p}(\mathcal{P})$ and $\tau_{\mathcal{K}_{sp}}(\mathcal{P})$ given by two different sets of controllers $\mathcal{K}_p$ and $\mathcal{K}_{sp}$ can be different. They also indicate the differences in definition between the robust topology and the graph topology.

Consider the sequence of plants given by

$$g_i(s) = \frac{i}{s+i}$$

Note that for all $i \in \mathbb{N}$ the transfer function $g_i(s)$ is asymptotically stable and that for all $s \in \mathbb{C}$,

$$\lim_{i\to\infty} g_i(s) = 1.$$

Since

$$\|g_i(s) - 1\|_\infty \geq |g_i(\infty) - 1| = 1,$$

the sequence $(g_i(s))_{i\geq 1}$ does not converge to 1 in the H^∞ topology on the right half plane. Since on the set of rational H^∞ functions the graph topology is equivalent to the H^∞ topology we have shown that the sequence does not converge in the graph topology.

Now define the topology $\tau_{\mathcal{K}_p}(\mathcal{P})$, by setting $\mathcal{P} = \mathbb{R}(s)$ and $\mathcal{K}_p = \mathbb{R}(s)$. We can see that the sequence of transfer functions also does not converge in the topology $\tau_{\mathcal{K}_p}(\mathcal{P})$ since the controller

$$k(s) = \frac{-2s + 0.5}{s - 1}$$

stabilizes the limiting system, i.e. the transfer function $g_0(s) = 1$, but it destabilizes $g_i(s)$ for all $i \in \mathbb{N}$.

If we however define the class of controllers $\mathcal{K}_{sp}$ to be strictly proper transfer functions we can show that the sequence $(g_i(s))_{i\geq 1}$ converges to $g_0(s) = 1$ in $\tau_{\mathcal{K}_{sp}}(\mathcal{P})$. Since $g_i(s)$ in H^∞ for all $i \in \mathbb{N}$, a coprime factorization $(n_i(s), d_i(s))$ of $g_i(s)$, $i \in \mathbb{N}$, is given by

$$n_i(s) = g_i(s), \qquad d_i(s) = 1,$$

$i \in \mathbb{N}$. Let $k(s) \in \mathcal{K}_{sp}$ be a strictly proper controller that stabilizes the limiting transfer function $g_0(s) = 1$. Write $n_k(s)d_k(s)^{-1}$ for a coprime factorization of $k(s)$, where $n_k(s)$ is strictly proper. We want to show that $k(s)$ stabilizes $g_i(s)$ for i sufficiently large. This will be the case if

$$g_i(s)n_k(s) + d_k(s)$$

is unimodular for large enough i. Note that

$$n_k(s) + d_k(s)$$

is unimodular since $k(s)$ stabilizes the transfer function $g_0(s) = 1$. Hence if we can show that

$$(g_i(s)n_k(s) + d_k(s)) - (n_k(s) + d_k(s)) = (g_i(s) - 1)n_k(s)$$

converges to zero in H^∞ we have proved the result since the unimodular functions are open in H^∞. But this is a simple consequence of the fact that $(g_i(s))_{i\geq 1}$ is a bounded sequence converging pointwise to $g_0(s) = 1$ and that $n(s)_k$ is strictly proper. □

It might be worthwhile pointing out at this point that it is relatively easy to characterize the graph topology of functions of fixed McMillan degree. It is essentially already shown in [5] that the graph topology on this set is equivalent to the topology on the set which is induced by the convergence of the coefficients of the transfer function. This topology which is highly nontrivial has been extensively studied by many authors (see e.g. [1], [4], [2]).

REFERENCES

[1] R. W. Brockett. Some geometric questions in the theory of linear systems. *IEEE Transactions on Automatic Control*, 21:449–454, 1976.

[2] U. Helmke. *Zur Topologie des Raumes linearer Kontrollsysteme.* PhD thesis, University of Bremen, 1982.

[3] A. Linnemann. Existence of controllers stabilizing the reduced-order model and not the plant. *Automatica*, 24:719, 1988.

[4] G. Segal. The topology of spaces of rational functions. *Acta Mathematica*, 143:39–72, 1979.

[5] M. Vidyasagar. *Control System Synthesis: A factorization approach.* MIT Press, 1985.

Filippo De Mari
Institut für Dynamische Systeme
Universität Bremen
Postfach 330 440
D-2800 Bremen 33
F.R.G.
filippo@mathematik.uni-Bremen.de
g09e@dhbrrz41.bitnet

Raimund Ober
Engineering Department
Cambridge University
Trumpington Street
Cambridge CB2 1PZ
GB
rjo@uk.ac.cam.eng.dsl

Robust Stabilization in the Gap Metric*

T.T. Georgiou M.C. Smith

This paper presents a brief overview of certain recent results on the application of the gap metric to robust stabilization of feedback systems. A detailed exposition of this work is reported in [8] – [11]. Here we present an explicit formula for the radius of gap ball uncertainty that a feedback system can tolerate, and we give a solution to the problem of robustness optimization in the gap metric. We also address the problem of robust stabilization under simultaneous plant and controller uncertainty. Finally we discuss an example of an infinite dimensional (delay) system and we give an explicit closed form expression for the optimally robust controller with respect to gap ball uncertainty.

1 Introduction

In this paper we consider the application of the gap metric to robust stabilization of feedback systems.

The gap metric was introduced into the control literature by Zames and El-Sakkary ([25] and [5]) as being appropriate for the study of uncertainty in feedback systems. In [21] Vidyasagar defined an alternative metric—the graph metric—in terms of normalized coprime factorizations. Both metrics are equivalent and thus induce the same topology. This topology is the weakest in which feedback stability is a robust property. The metrics define notions of distance in the space of (possibly) unstable systems which do not assume that the plants have the same number of poles in the right half plane (RHP).

*This work was supported in part by the NSF under Grant Nos. ECS-8705291 and MIP-8708811.

The problem of robustness optimization has been considered by a number of authors. Much work has been devoted to optimizing "unstructured" neighbourhoods in which the number of RHP poles remains fixed. This type of problem has been shown to be exactly solvable and, in fact, reduces to a standard Nehari problem. However, this approach cannot easily cope with poles on the imaginary axis. Moreover, the restriction on imaginary axis pole crossings is often unnatural.

Several authors have obtained sufficient conditions for robust stability in terms of the gap ([5], [26]) or the graph metric ([21], [23]). The problem of robustness optimization in the graph metric is currently unsolved and appears difficult. This can possibly be attributed to the fact that computation of the graph metric amounts to a constrained H_∞ optimization problem. This contrasts with the situation for the gap metric, which is computable in terms of two standard "2-block" H_∞ optimization problems [8].

Robust stabilization for perturbations in the gap metric was shown to be equivalent to robust stabilization for normalized coprime factor perturbations (see [9]). The problem of robust stabilization for general coprime factor uncertainty was introduced by Vidyasagar and Kimura [23]. In [12], Glover and McFarlane presented a theory for normalized coprime factor perturbations, and more recently, Curtain and her co-workers (see [2] and [4]) studied this type of uncertainty description for robust stabilization in the context of infinite dimensional systems.

In this paper we overview some recent results on the application of the gap metric to robust stabilization of feedback systems. In particular, in Theorem 1, we give an explicit expression for the radius of gap ball uncertainty which a given feedback system can tolerate. This result includes the case of simultaneous plant and controller uncertainty. In Theorem 1 (statements (b) and (c)) we also highlight a certain reciprocity between the case where the gap ball uncertainty is restricted to the plant and the case where the uncertainty is restricted to the controller. Theorem 2 shows that a ball of uncertainty in the gap metric, of a given radius, is equal to a ball of uncertainty, of the same radius, defined by perturbations of a normalized *right* coprime fraction, *providing* the radius is smaller than a certain quantity—this quantity is never less than the radius of the maximum gap ball about the plant which can be stabilized by some fixed controller. Theorem 3 is a restatement of Theorem 1 (a,b,d) in terms of coprime fractions. In Theorem 4 we quote a result from [12] on the structure of controllers that tolerate a given level of normalized coprime fraction uncertainty. The theory we outline allows explicit computation of the controller which is optimally robust with respect to gap ball uncertainty. For a fairly general class of infinite dimensional systems the required computations amount to solving a finite system of algebraic and transcendental equations (see [11]). This is demonstrated by an example (taken from [11]) of an infinite dimensional (delay) system for which we provide an explicit closed form expression for the optimal controller.

The research outlined in this paper is reported in [9] and [11]. Finally we mention that the question of well-posedness of stability for optimal robustness problems is investigated in [9] (see also [10]).

2 Preliminaries

In this section we review the definition of the gap metric and compile notation and some of the key results.

We consider a linear causal time-invariant system $\mathbf{P}$ as a (possibly unbounded) operator on H_2. The dimension of the input and output spaces will be suppressed for notational convenience. The transfer function $P(s)$ of the system admits right and left *weakly coprime* factorizations

$$P(s) = N(s)M(s)^{-1} = \tilde{M}(s)^{-1}\tilde{N}(s) \tag{1}$$

where $\binom{M}{N} \in H_\infty$ (resp. $(\tilde{M},\ \tilde{N}) \in H_\infty$) is co-outer (resp. outer) (cf. [18]). These factorizations can always be *normalized*, that is, chosen so that $\binom{M}{N}$ (resp. $(\tilde{M},\ \tilde{N})$) is inner (resp. co-inner):

$$M^*M + N^*N = I, \qquad \tilde{M}\tilde{M}^* + \tilde{N}\tilde{N}^* = I. \tag{2}$$

For convenience we will assume that $\binom{M}{N}$ and $(\tilde{M},\ \tilde{N})$ are analytic on a dense subset of the imaginary axis. This is a mild technical assumption. A weaker condition is also possible (cf. [9, Section 7]). This class of $p \times m$ linear time-invariant continuous-time systems will be denoted by $\hat{\mathcal{P}}_{p\times m}$.

The graph of the operator $\mathbf{P} \in \hat{\mathcal{P}}_{p\times m}$ is the subspace of $H_2 \times H_2$ consisting of all pairs (u, y) such that $y = \mathbf{P}u$. This is given by

$$\text{graph}(\mathbf{P}) = \begin{pmatrix} M \\ N \end{pmatrix} H_2 =: G_P \tag{3}$$

and is a closed subspace of H_2. Given the system $\mathbf{P}$ with coprime fraction $P = \tilde{M}^{-1}\tilde{N}$, we define the transposed system $\mathbf{P}^T$ to be the system with transfer function P^T and

$$\text{graph}(\mathbf{P}^T) = \begin{pmatrix} \tilde{M}^T \\ \tilde{N}^T \end{pmatrix} H_2 . \tag{4}$$

The *gap* between two (closed) subspaces K_1 and K_2 is defined as

$$\delta(K_1, K_2) := \|\mathbf{\Pi}_{K_1} - \mathbf{\Pi}_{K_2}\|. \tag{5}$$

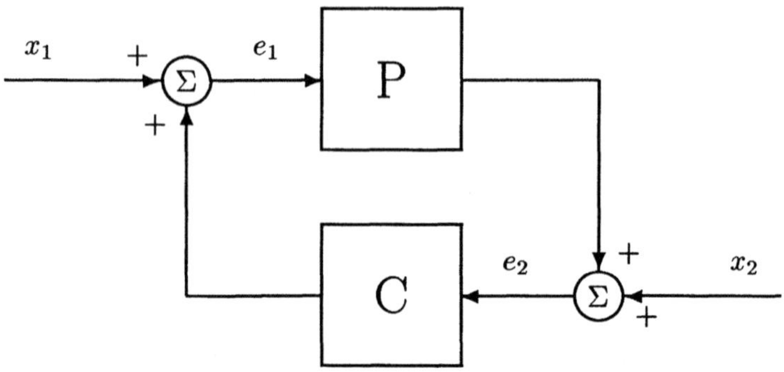

Figure 1. Standard feedback configuration

where $\mathbf{\Pi}_{K_i}$ denotes the orthogonal projection onto K_i. See [1], [13] and [14]. From (5) it follows that $\delta(\cdot,\cdot)$ is a metric. The *gap* between two systems $\mathbf{P}_1$ and $\mathbf{P}_2$ is defined to be the gap between their respective graphs G_{P_1} and G_{P_2} ([13], [25]). Given the normalized coprime factorizations $P_i = N_i M_i^{-1}$, $i = 1, 2$, for the transfer functions of $\mathbf{P}_1$ and $\mathbf{P}_2$, the following formula can be used to compute the gap:

$$\delta(\mathbf{P}_1, \mathbf{P}_2) = \max\left\{\vec{\delta}(\mathbf{P}_1, \mathbf{P}_2), \vec{\delta}(\mathbf{P}_2, \mathbf{P}_1)\right\} \tag{6}$$

where

$$\vec{\delta}(\mathbf{P}_1, \mathbf{P}_2) = \inf_{Q \in H_\infty} \left\| \begin{pmatrix} M_1 \\ N_1 \end{pmatrix} - \begin{pmatrix} M_2 \\ N_2 \end{pmatrix} Q \right\|_\infty . \tag{7}$$

See [8] for details. Finally we introduce the notation

$$\mathcal{B}(\mathbf{P}, b) := \left\{\mathbf{P}_1 \in \hat{\mathcal{P}}_{p \times m} : \delta(\mathbf{P}, \mathbf{P}_1) < b\right\} \tag{8}$$

to denote a gap ball of radius b about $\mathbf{P}$. Consider the feedback configuration of Fig. 1 where $\mathbf{P} \in \hat{\mathcal{P}}_{p \times m}$ and $\mathbf{C} \in \hat{\mathcal{P}}_{m \times p}$ are linear systems. The configuration, denoted by $[\mathbf{P}, \mathbf{C}]$, is defined to be stable if the operators $x_i \to e_j$ for $i, j = 1, 2$ are bounded. This is equivalent to

$$\begin{pmatrix} I & -C \\ -P & I \end{pmatrix}^{-1} = \begin{pmatrix} (I - CP)^{-1} & C(I - PC)^{-1} \\ P(I - CP)^{-1} & (I - PC)^{-1} \end{pmatrix} \tag{9}$$

being in H_∞.

A necessary and sufficient condition for $\mathbf{P}$ to be stabilizable is that there exist *strongly* coprime factorizations $P = NM^{-1} = \tilde{M}^{-1}\tilde{N}$ (see [18]);

i.e., that $\binom{M}{N}$ (resp. $(\tilde{M},\ \tilde{N})$) has a left (resp. right) inverse in H_∞. If $\mathbf{P}$ is stabilizable, there exist $U,\ V,\ \tilde{U},\ \tilde{V} \in H_\infty$ such that the following generalized Bezout equation holds:

$$\begin{pmatrix} \tilde{V} & -\tilde{U} \\ -\tilde{N} & \tilde{M} \end{pmatrix} \begin{pmatrix} M & U \\ N & V \end{pmatrix} = \begin{pmatrix} M & U \\ N & V \end{pmatrix} \begin{pmatrix} \tilde{V} & -\tilde{U} \\ -\tilde{N} & \tilde{M} \end{pmatrix} = \begin{pmatrix} I & 0 \\ 0 & I \end{pmatrix}. \tag{10}$$

Also the feedback configuration $[\mathbf{P}, \mathbf{C}]$ is stable if and only if C admits a factorization $C = XY^{-1}$ (or $C = \tilde{Y}^{-1}\tilde{X}$) where $\binom{X}{Y} = \binom{U}{V} + \binom{M}{N}Q$ (respectively $(\tilde{Y},\ \tilde{X}) = (\tilde{U},\ \tilde{V}) + Q(\tilde{M},\ \tilde{N})$) for some $Q \in H_\infty$.

3 Robust stabilization in the gap metric

In this section we overview several basic results on the application of the gap metric to robust stabilization of feedback systems.

Let $[\mathbf{P}, \mathbf{C}]$ be a stable feedback configuration and define

$$b_{P,C} := \left\| \begin{pmatrix} I \\ P \end{pmatrix} (I - CP)^{-1} (I\ ,\ C) \right\|_\infty^{-1}. \tag{11}$$

The quantity $b_{P,C}$ determines the precise maximal amount of uncertainty in the gap that a feedback system can tolerate. This is shown by the following theorem.

Theorem 1. **([9])** *Let $\mathbf{P} \in \hat{\mathcal{P}}_{p\times m}$ and $\mathbf{C} \in \hat{\mathcal{P}}_{m\times p}$. Then the following statements are equivalent:*

(a) $[\mathbf{P}, \mathbf{C}]$ is stable and $b \le b_{P,C}$,

(b) $[\mathbf{P}_1, \mathbf{C}]$ is stable for all $\mathbf{P}_1$ such that $\delta(\mathbf{P}, \mathbf{P}_1) < b$,

(c) $[\mathbf{P}, \mathbf{C}_1]$ is stable for all $\mathbf{C}_1$ such that $\delta(\mathbf{C}, \mathbf{C}_1) < b$,

(d) $[\mathbf{P}_1, \mathbf{C}]$ is stable for all $\mathbf{P}_1$ such that $\delta(\mathbf{P}^T, \mathbf{P}_1{}^T) < b$,

(e) $[\mathbf{P}, \mathbf{C}_1]$ is stable for all $\mathbf{C}_1$ such that $\delta(\mathbf{C}^T, \mathbf{C}_1{}^T) < b$,

(f) $[\mathbf{P}_1, \mathbf{C}_1]$ is stable for all $\mathbf{P}_1$ and $\mathbf{C}_1$ such that

$$\delta(\mathbf{P}, \mathbf{P}_1) + \delta(\mathbf{C}, \mathbf{C}_1) < b. \tag{12}$$

PROOF.
(a) $\Leftrightarrow$ (b) is shown in [9, Theorem 5].
(b) $\Leftrightarrow$ (c) is shown in [9, Theorem 6].
(b) $\Leftrightarrow$ (d) is shown in [9, Theorems 3′].
(c) $\Leftrightarrow$ (e) is analogous to (b) $\Leftrightarrow$ (d).
(a) $\Leftrightarrow$ (f) is shown in [9, Theorem 7].

An alternative proof of (a) $\Rightarrow$ (f) can be obtained from [26, Theorem 4.1]. Q.E.D.

Remark 1. The equivalence of (b) and (c) in the above theorem reveals a certain reciprocity between the plant and controller regarding the tolerance of uncertainty in a feedback system. The equivalence demonstrates the fact that $b_{P,C} = b_{C,P}$. □

Remark 2. The equivalence of (b) and (d) (similarly for (c) and (e)) is rather surprising because, as was shown in [9, Examples 5 and 6], the sets of plants in (b) and (d) (resp. (c) and (e)) are generally different in the matrix case. □

Remark 3. When both plant and controller deviate from their nominal values, the feedback system remains stable for all perturbations where the sum of the respective distances from the nominal plant and controller is less than or equal to $b_{P,C}$ (equivalence of (a) and (f)). Equation (12) is "tight" in case $\mathbf{P} = \mathbf{P}_1$ or $\mathbf{C} = \mathbf{C}_1$ but may not be tight in more general situations (cf. [9]). □

Let $P = NM^{-1}$ be a normalized right coprime fraction for $\mathbf{P}$ and define

$$\lambda(\mathbf{P}) := \inf_{\Re(s)>0} \sigma_{\min} \begin{pmatrix} M(s) \\ N(s) \end{pmatrix} \tag{13}$$

where $\sigma_{\min}$ denotes the minimum singular value. Consider the optimal robustness radius

$$b_{\text{opt}}(\mathbf{P}) := \max_{C \text{ stabilizing}} b_{P,C}.$$

It was shown in [9, Lemma 1] that

$$b_{\text{opt}}(\mathbf{P}) \leq \lambda(\mathbf{P}).$$

In [9] it was also shown that a ball of uncertainty in the gap metric, of a given radius, is equal to a ball of uncertainty, of the same radius, defined by normalized right coprime factor perturbations, *providing* the radius does not exceed $\lambda(\mathbf{P})$. This is the content of the following theorem.

Theorem 2. **([9, Lemmas 2 and 3])** *If $b \leq \lambda(\mathbf{P})$ then*

$$\mathcal{B}(\mathbf{P}, b) = \Big\{\mathbf{P}_1 \in \hat{\mathcal{P}}_{p\times m} : P_1 = (N + \Delta_N)(M + \Delta_M)^{-1}$$
$$\text{where } \Delta_N, \Delta_M \in RH_\infty \text{ and } \left\|\begin{pmatrix}\Delta_M \\ \Delta_N\end{pmatrix}\right\|_\infty < b\Big\}.$$

The following result is, in the light of Theorem 2, a restatement of Theorem 1 (a,b,d).

Theorem 3. *Let $\mathbf{P} \in \hat{\mathcal{P}}_{p\times m}$, $\mathbf{C} \in \hat{\mathcal{P}}_{m\times p}$, and let $P = NM^{-1}$ and $P = \tilde{M}^{-1}\tilde{N}$ be normalized factorizations. Then the following statements are equivalent:*

(a) $[\mathbf{P}, \mathbf{C}]$ *is stable and* $b \leq b_{P,C}$,

(g) $[\mathbf{P}_1,\mathbf{C}]$ *is stable for all* $\mathbf{P}_1$ *with transfer function* $P_1 = (N+\Delta_N)(M+\Delta_M)^{-1}$ *where* $\Delta_M, \Delta_N \in H_\infty$ *and* $\|\binom{\Delta_M}{\Delta_N}\|_\infty < b$,

(h) $C = \tilde{Y}^{-1}\tilde{X}$, $\tilde{Y}M - \tilde{X}N = I$ *and* $\|(\tilde{Y},\ \tilde{X})\|_\infty \leq \frac{1}{b}$,

(i) $[\mathbf{P}_1,\mathbf{C}]$ *is stable for all* $\mathbf{P}_1$ *with transfer function* $P_1 = (\tilde{M} + \Delta_{\tilde{M}})^{-1}(\tilde{N} + \Delta_{\tilde{N}})$ *where* $\Delta_{\tilde{M}}, \Delta_{\tilde{N}} \in H_\infty$ *and* $\|(\Delta_{\tilde{M}},\ \Delta_{\tilde{N}})\|_\infty < b$,

(j) $C = XY^{-1}$, $-\tilde{N}X + \tilde{M}Y = I$ *and* $\|\binom{X}{Y}\|_\infty \leq \frac{1}{b}$.

PROOF. Referring to Theorem 1 we have (a) ⇔ (b). That the latter condition is equivalent to (g) is shown in [9, Theorem 4].
(g) ⇔ (h) is shown in [23, Theorem 5].
(g) ⇔ (i) is shown in [9, Theorem 3].
(i) ⇔ (j) is analogous to (g) ⇔ (h). Q.E.D.

Remark 4. If $C = XY^{-1} = \tilde{Y}^{-1}\tilde{X}$ are taken to be normalized, analogous statements to (g) through (j) can be obtained where $\mathbf{P}$, $\mathbf{P}_1$, N, $\tilde{N}$, M and $\tilde{M}$ are interchanged with $\mathbf{C}$, $\mathbf{C}_1$, X, $\tilde{X}$, Y and $\tilde{Y}$ respectively. □

Remark 5. In view of Theorem 1, the sets of plants in (b) and (g) are the same (similarly for those in (d) and (i)). However those in (g) and (i) are in general different (cf. Remark 2 above). □

Remark 6. By replacing the fraction $C = XY^{-1}$ in Theorem 3, (j) with the expression $C = (U + MQ)(V + NQ)^{-1}$, $Q \in H_\infty$ we obtain

$$b_{\text{opt}}(\mathbf{P}) = \left(\inf_{Q\in H_\infty} \left\|\begin{pmatrix}U \\ V\end{pmatrix} + \begin{pmatrix}M \\ N\end{pmatrix}Q\right\|_\infty\right)^{-1}. \tag{14}$$

A dual expression for b_{opt} using a left factorization of the controller can be obtained from Theorem 3, (h). □

Define the operators

$$\mathbf{A} := \mathbf{\Pi}_{H_2 \ominus \binom{M}{N} H_2} \begin{pmatrix} U \\ V \end{pmatrix} |_{H_2}, \tag{15}$$

$$\mathbf{B} := \mathbf{\Pi}_{H_2} (-\tilde{N},\ \tilde{M}) |_{H_2 \ominus \binom{M}{N} H_2}, \tag{16}$$

$$\mathbf{B}_0 := \mathbf{\Pi}_{H_2} (-\tilde{N},\ \tilde{M}) |_{L_2}, \tag{17}$$

$$\mathbf{\Gamma} := \mathbf{\Pi}_{H_2^{\perp}} \begin{pmatrix} -\tilde{N}^* \\ \tilde{M}^* \end{pmatrix} |_{H_2}. \tag{18}$$

Using the Commutant Lifting Theorem [20] (cf. [9, proof of Theorem 2]), it follows that

$$b_{\text{opt}} = \|\mathbf{A}\|^{-1}. \tag{19}$$

Further, see [9], $\mathbf{B} = \mathbf{A}^{-1}$, $\mathbf{B}^* + \mathbf{\Gamma} = \mathbf{B}_0{}^*$, and $\mathbf{B}_0{}^*$ is an isometry. From these it follows that

$$b_{\text{opt}} = (1 - \|\mathbf{\Gamma}\|^2)^{1/2}. \tag{20}$$

This striking formula was originally obtained by Glover and McFarlane [12]. The norm $\|\mathbf{\Gamma}\|$ is the maximal singular value of the operator provided that the operator is compact. An important difference between the two operators $\mathbf{A}$ and $\mathbf{\Gamma}$ is that the former is not compact while the latter is compact in most cases of engineering interest. Another important result that was established in [12] draws the connection between the "sublevel" sets of the operators $\mathbf{A}$ and $\mathbf{\Gamma}$. This result is quoted below.

Theorem 4. **([12])** *Let U, V, M, N satisfy (10) and $Q \in H_\infty$. If*

$$\left\| \begin{pmatrix} U \\ V \end{pmatrix} + \begin{pmatrix} M \\ N \end{pmatrix} Q \right\|_\infty < \mu, \tag{21}$$

then

$$\left\| \begin{pmatrix} -\tilde{N}^* \\ \tilde{M}^* \end{pmatrix} - \mu^{-2} \left(\begin{pmatrix} U \\ V \end{pmatrix} + \begin{pmatrix} M \\ N \end{pmatrix} Q \right) \right\|_\infty < (1 - \mu^{-2})^{1/2}. \tag{22}$$

Conversely, if for some $\hat{U}$ and $\hat{V} \in H_\infty$

$$\left\| \begin{pmatrix} -\tilde{N}^* \\ \tilde{M}^* \end{pmatrix} + \begin{pmatrix} \hat{U} \\ \hat{V} \end{pmatrix} \right\|_\infty < (1 - \mu^{-2})^{1/2}, \tag{23}$$

then $\hat{Q} := \tilde{M}\hat{V} + \tilde{N}\hat{U}$ is invertible in H_∞ and $\|\binom{\hat{U}}{\hat{V}}\hat{Q}^{-1}\|_\infty < \mu$.

Proof. As in [12] (goes through in the infinite dimensional case).

Q.E.D.

Remark 7. The above result shows that compensators $C = (U + MQ)(V + NQ)^{-1}$, which achieve a certain robustness margin of μ^{-1}, relate directly to level sets of interpolating functions for the Nehari problem associated with the operator $\mathbf{\Gamma}$. This is especially convenient since the optimal interpolating functions for the Nehari problem can be constructed from the singular values and corresponding singular vectors of the Hankel operator $\mathbf{\Gamma}$ [17], [24]. For instance, consider the singular vector equations

$$\mathbf{\Gamma}x = \lambda y^* \tag{24}$$

$$\mathbf{\Gamma}y^* = \lambda x \tag{25}$$

where $x \in H_2$ and $y^* \in {H_2}^{\perp}$. If $\mathbf{P}$ is a scalar system then x is a scalar function and the Schmidt pair (x, y^*) corresponding to $\lambda = \|\mathbf{\Gamma}\|$ determines uniquely the solution of

$$\inf_{\binom{\hat{U}}{\hat{V}} \in H_\infty} \left\| \begin{pmatrix} -\tilde{N}^* \\ \tilde{M}^* \end{pmatrix} + \begin{pmatrix} \hat{U} \\ \hat{V} \end{pmatrix} \right\|_\infty . \tag{26}$$

Indeed (see [17, p. 104]), the infimum equals $\|\mathbf{\Gamma}\|$ and is achieved for

$$\begin{pmatrix} \hat{U} \\ \hat{V} \end{pmatrix} = \|\mathbf{\Gamma}\| y^* x^{-1} - \begin{pmatrix} -\tilde{N}^* \\ \tilde{M}^* \end{pmatrix}. \tag{27}$$

For the general vectorial case see [24]. □

4 Example

In [11] we studied the problem of computing the optimal robustness radius and the corresponding optimal controller for a class of infinite dimensional systems. Here we present an example quoted from [11] of a delay-system with transfer function $P(s) = e^{-hs}/(s - \sigma)$. For this system the theory allows for explicit computation of the optimal robustness margin, in terms of the parameters h and σ, as a solution of a certain transcendental equation. Moreover, we give a closed form expression for the optimal controller and the optimal robustness radius. The computations require only a finite system of algebraic computations and are close in spirit to the techniques in [3], [6], [15] and [19].

Consider a left normalized coprime fraction $P(s) = \tilde{M}^{-1}\tilde{N}$ and let $B := (-\tilde{N},\ \tilde{M})$. Then

$$B = \left(-\frac{e^{-hs}}{s + \gamma},\ \frac{s - \sigma}{s + \gamma} \right) \tag{28}$$

where $\gamma = \sqrt{1+\sigma^2}$. From Hartman's theorem, the corresponding Hankel operator $\mathbf{\Gamma} = \mathbf{\Pi}_{H_2{}^\perp} B^*$ is compact. From Remark 7, we need to compute the norm of $\mathbf{\Gamma}$ and an associated Schmidt pair (x, y^*). This is carried out in [11]. Below we summarize the conclusions.

Define

$$\rho := \frac{(1-\lambda^2)\gamma + \sigma}{\lambda^2}, \tag{29}$$

$$\mu := \sqrt{\frac{|\lambda^2\gamma^2 - 1|}{\lambda^2}}. \tag{30}$$

Then $\lambda > 0$ is a singular value of $\mathbf{\Gamma}$ if and only if

$$\lambda\gamma > 1 \quad \text{and} \quad e^{2h\mu} = \frac{\gamma-\mu}{\gamma+\mu}\frac{\rho+\mu}{\rho-\mu}, \tag{31}$$

$$\lambda\gamma < 1 \quad \text{and} \quad \tan(\mu h) = \frac{\mu\gamma - \mu\rho}{\rho\gamma + \mu^2}, \tag{32}$$

$$\lambda\gamma = 1 \quad \text{and} \quad h = \frac{\gamma - 2\sigma}{\gamma\sigma}. \tag{33}$$

Equations (31) and (32) are transcendental equations in the variable λ. If $\lambda_{\max}$ is the largest value among the solutions of these equations (together with $\lambda = \gamma^{-1}$ if (33) holds), then $\|\mathbf{\Gamma}\| = \lambda_{\max}$. After obtaining the norm of $\mathbf{\Gamma}$, we derive the associated Schmidt pair of singular vectors (cf. [11]). The transfer function of the optimally robust controller is $C = \hat{U}\hat{V}^{-1}$ where $\binom{\hat{U}}{\hat{V}}$ is determined from (27). In all three cases the form of the optimally robust controller turns out to be similar. The transfer function of the controller is

$$C_{\text{opt}} = -e^{-h\gamma}\frac{\gamma + \sigma + \phi(-\gamma)}{2\gamma}\frac{1}{\lambda^2(\gamma-\sigma) + (s+\sigma)\frac{\phi(s)-\phi(\gamma)}{s-\gamma}} \tag{34}$$

where $\lambda = \|\mathbf{\Gamma}\|$ and ϕ is a distributed delay given by

$$\phi_{\text{I}}(s) = \frac{1}{s^2-\mu^2}\left[(\gamma - s)\left\{\frac{\rho-\mu}{\gamma-\mu}e^{h\mu}\right\}e^{-hs} + (s-\rho)\right] \tag{35}$$

when $\lambda\gamma > 1$, by

$$\phi_{\text{II}}(s) = \frac{1}{s^2+\mu^2}\left[(\gamma - s)\Re\left\{\frac{\rho-j\mu}{\gamma-j\mu}e^{jh\mu}\right\}e^{-hs} + (s-\rho)\right] \tag{36}$$

$$= \frac{1}{s^2+\mu^2}\left[(\gamma - s)\frac{1-\lambda^2}{\lambda^4}\frac{(\sigma+\gamma)^2}{\rho\gamma+\mu^2}\cos(h\mu)e^{-hs} + (s-\rho)\right] \tag{37}$$

when $\lambda\gamma < 1$ (here $j = \sqrt{-1}$ and $\Re$ denotes "the real part of"), and by

$$\phi_{\text{III}}(s) = \frac{1}{s^2}\left[(\gamma - s)\frac{\rho}{\gamma}e^{-hs} + (s-\rho)\right] \tag{38}$$

when $\lambda\gamma = 1$. Clearly $\phi_{\mathrm{I}}(s)$ and $\phi_{\mathrm{II}}(s)$ coincide with $\phi_{\mathrm{III}}(s)$ when $\mu = 0$.

A detailed derivation of the above is given in [11]. Here in Fig. 2 we draw the value of the optimal robustness margin b_{opt} as a function of the delay h for $\sigma = \frac{1}{2}$. Figs. 3 and 4 show the magnitude and phase characteristics of the optimally robust controllers for $h \in \{0.1,\ 0.2,\ 0.3,\ 0.5\}$ and $\sigma = \frac{1}{2}$.

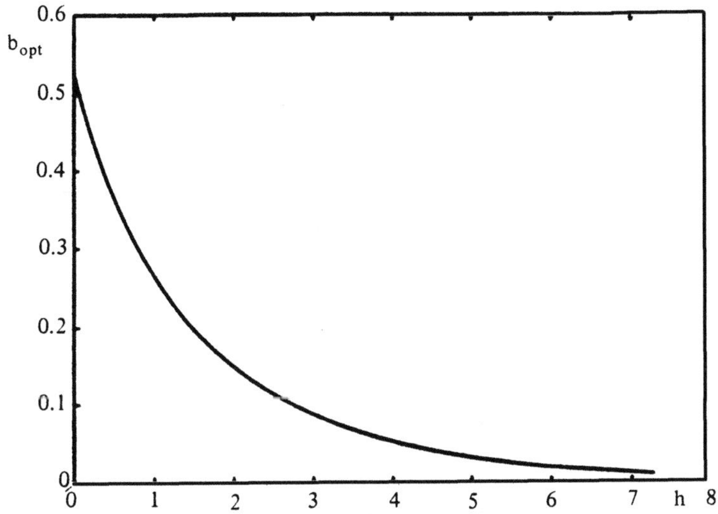

Figure 2. b_{opt} vs. h

REFERENCES

[1] N. I. Akhiezer and I. M. Glazman, *Theory of Linear Operators in Hilbert Space*, **1**, New York: Ungar, 1961.

[2] J. Bontsema and R. F. Curtain, Robust stabilization of a flexible beam model using a normalized coprime factorization approach, *preprint*, 1989.

[3] H. Berkovici, C. Foiaş, and A. R. Tannenbaum, On skew Toeplitz operators, *Operator Theory: Advances and Applications*, **32** (1988), 21–43.

[4] R. F. Curtain, Robust stabilization for infinite-dimensional linear systems using normalized coprime factorizations, *preprint*, presented at the 1989 MTNS, Amsterdam, June 1989.

[5] A. K. El-Sakkary, The gap metric: Robustness of stabilization of feedback systems, *IEEE Trans. on Automat. Contr.*, **30** (1985), 240–247.

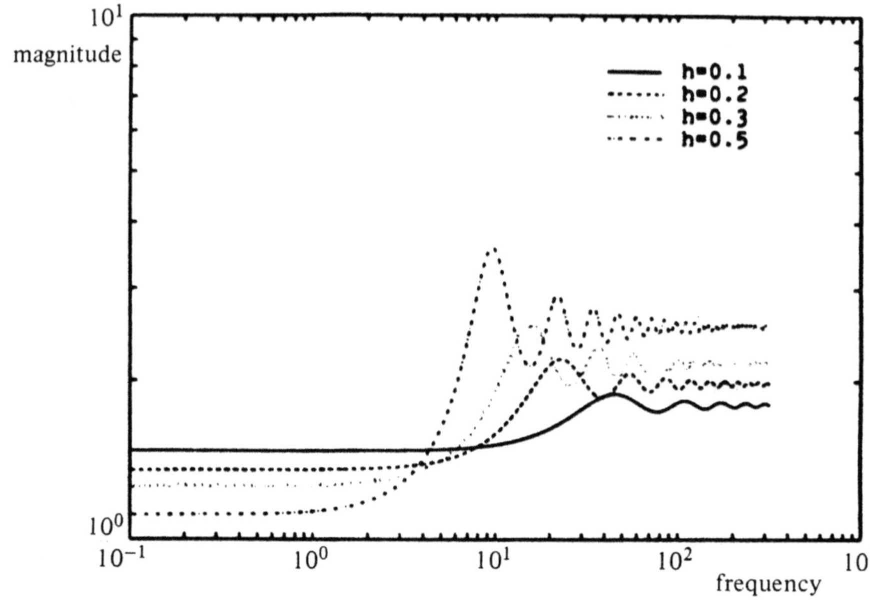

Figure 3. Magnitude plot of $C_{\mathrm{opt}}(j\omega)$

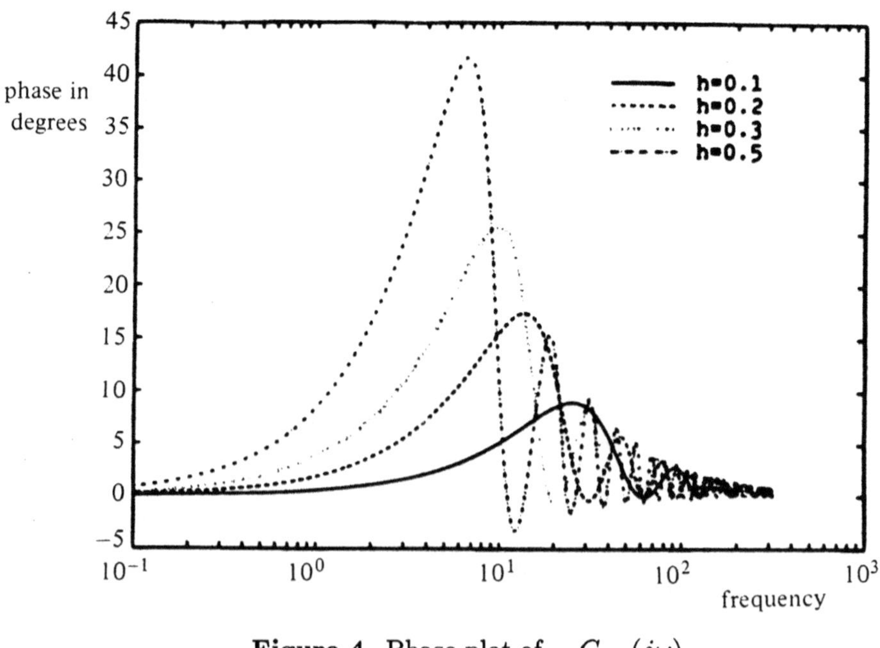

Figure 4. Phase plot of $-C_{\mathrm{opt}}(j\omega)$

[6] C. Foiaş, A. R. Tannenbaum, and G. Zames, Some explicit formulae for the singular values of certain Hankel operators with factorizable symbol, *SIAM J. Math. Anal.*, **19** (1988), 1081–1089.

[7] B. A. Francis, *A Course in H_∞ Control Theory*, Lecture Notes in Control and Information Sciences, **88**, Springer-Verlag: New York, 1987.

[8] T. T. Georgiou, On the computation of the gap metric, *Systems & Control Letters*, **11** (1988), 253–257.

[9] T. T. Georgiou and M. C. Smith, Optimal robustness in the gap metric, *IEEE Trans. on Automat. Contr.*, **35** (1990) to appear.

[10] T. T. Georgiou and M. C. Smith, W-stability of feedback systems, *Systems & Control Letters*, **13** (1989), 271–277.

[11] T. T. Georgiou and M. C. Smith, Robust stabilization in the gap metric: Controller design for distributed plants, *preprint*, submitted to the 1990 ACC, September 1989.

[12] K. Glover and D. McFarlane, Robust stabilization of normalized coprime factor plant descriptions with H_∞ bounded uncertainty, *IEEE Trans. on Automat. Contr.*, **34** (1989), 821–830.

[13] T. Kato, *Perturbation Theory for Linear Operators*, Springer-Verlag: New York, 1966.

[14] M.A. Krasnosel'skii, G.M. Vainikko, P.P. Zabreiko, Ya.B. Rutitskii, V.Ya. Stetsenko, *Approximate Solution of Operator Equations*, Groningen: Wolters-Noordhoff, 1972.

[15] H. Özbay, M. C. Smith, and A. R. Tannenbaum, Controller design for unstable distributed plants, *preprint*, submitted to the 1990 ACC, September 1989.

[16] L. B. Page, Bounded and compact vectorial Hankel operators, *Trans. Amer. Math. Soc.*, **150** (1970), 529–539.

[17] D. Sarason, *Function Theory on the Unit Circle*, Lecture Notes, Department of Mathematics, Virginia Polytechnic Institute and State Univ., June 1978.

[18] M. C. Smith, On stabilization and the existence of coprime factorizations, *IEEE Trans. on Automat. Contr.*, **34** (1989), 1005–1007.

[19] M. C. Smith, Singular values and vectors of a class of Hankel operators, *Systems & Control Letters*, **12** (1989), 301–308.

[20] B. Sz.-Nagy and C. Foiaş, *Harmonic Analysis of Operators on Hilbert Space*, North-Holland: Amsterdam, 1970.

[21] M. Vidyasagar, The graph metric for unstable plants and robustness estimates for feedback stability, *IEEE Trans. on Automat. Contr.*, **29** (1984), 403–418.

[22] M. Vidyasagar, *Control System Synthesis: A Factorization Approach*, Cambridge, Massachusetts: MIT Press, 1985.

[23] M. Vidyasagar and H. Kimura, Robust controllers for uncertain linear multivariable systems, *Automatica*, **22** (1986), 85–94.

[24] N. J. Young, The Nevanlinna-Pick problem for matrix-valued functions, *J. Operator Theory*, **15** (1986), 239-265.

[25] G. Zames and A.K. El-Sakkary, Unstable systems and feedback: The gap metric, *Proc. of the Allerton Conference*, pp. 380–385, Oct. 1980.

[26] S. Q. Zhu, M. L. J. Hautus and C. Praagman, Sufficient conditions for robust BIBO stabilization: Given by the gap metric, *Systems & Control Letters*, **11** (1988), 53–59.

Tryphon T. Georgiou
Department of Electrical Engineering
University of Minnesota
Minneapolis, MN 55455, USA
`f1georg@vax.ee.umn.edu`

Malcolm C. Smith
Department of Electrical Engineering
The Ohio State University
Columbus, OH 43210, USA
`malcolm@eagle.eng.ohio-state.edu`

Some Preliminary Results on Connecting Identification and Robust Control

G.C. Goodwin B.A. León de la Barra
R.J. Mazzaferri

An important consideration in control system design is that of model uncertainty. If one chooses to fix the nominal model, then in general, the associated description of the uncertainty will necessarily be very conservative. Alternatively, on line model estimation allows the nominal model to be adjusted as the system changes thereby offering the possibility for a less conservative estimate of the associated uncertainty. However this strategy depends on the model estimator having the capacity to provide information on both the nominal model and the associated uncertainty. Most of the existing paradigms for estimation give emphasis to the estimation of the nominal model but do not address the uncertainty issue.

This paper outlines preliminary results arising in the quantification of uncertainty in model estimation. This is achieved by using a Bayesian approach in which the posterior distribution covers both nominal model and the effects of undermodelling. It is also suggested how this information can be used to combine identification and robust control.

1 Introduction

Until recently, (non adaptive) robust control and adaptive control were viewed as two competing paradigms for dealing with the problem of uncertainty in control system design. However, there has recently been a recognition that, rather than being competitive, the methods of robust control and adaptive control have complementary features [11, 15, 16, 21].

In simple terms, robust control design [5] aims to optimize the system performance in the face of a specified level of uncertainty. On the other hand, traditional approaches to adaptive control [2, 8] emphasize the use of an on-line model estimator which can be used to update the model. To bring these two paradigms together, in principle, requires the model estimator to also yield information on uncertainty.

The importance of quantifying the effect of uncertainty in estimation has been recognized for some time and several preliminary ideas have been suggested for dealing with it. For example, Ljung [18] has developed qualitative measures of the effect of using restricted complexity models by expressing the usual least squares estimation criterion in the frequency domain using Parseval's theorem. Related work appears in [1, 4, 12].

More recently, attention has been focused on quantitative aspects of the distribution of errors in estimation using low order models. For example, La Maire and colleagues [17] have used Fourier analysis methods to obtain a bound on the unmodelled dynamics in estimation. Kosut [13, 14] has used simple inequalities to relate an a priori bound on the undermodelled impulse response to the associated parameter estimation errors. Related ideas for frequency response estimation have been described in [3].

In this paper we present a new conceptual framework for quantifying the effect of model uncertainty in estimation using a Bayesian formulation. This approach leads to insights into the design of an optimal estimator and also provides an alternative viewpoint of common assumptions used in (non adaptive) robust control.

2 System Description

To provide a basis for the subsequent development, we will describe the total system response via the following model which we term a Model Incorporating Uncertainty or MIU:

$$y(t) = T_1(\theta, \gamma)y(t) + T_2(\theta, \gamma)u(t) + \eta(t) \tag{1}$$

$$\eta(t) = G^y_\Delta(\gamma)y(t) + G^u_\Delta(\gamma)u(t) + v(t) \tag{2}$$

where

$y(t), u(t)$	denote the system output and input respectively.
$T_1(\theta,\gamma), T_2(\theta,\gamma)$	are nominal transfer functions depending on a parameter vector θ.
$G^y_\Delta(\gamma), G^u_\Delta(\gamma)$	are residual transfer functions.
$v(t)$	represents a noise process.
γ	is any linear operator, e.g. the Laplace Transform variable s in continuous time.

Some simple examples included in the above system description are:

Example 1. Say that a system has nominal transfer function $\frac{B(\theta,\gamma)}{A(\theta,\gamma)}$ where $B(\theta,\gamma)$, $A(\theta,\gamma)$ are polynomials in γ . Also, let us assume that the true system transfer function is $G_T(\gamma)$ then the description (1) becomes:

$$y(t) = \frac{B(\theta,\gamma)}{A(\theta,\gamma)}u(t) + \left[G_T(\gamma) - \frac{B(\theta,\gamma)}{A(\theta,\gamma)}\right]u(t) + v(t) \tag{3}$$

Thus, in the notation of the MIU :

$$T_1(\theta,\gamma) = 0 \ ; \quad T_2(\theta,\gamma) = \frac{B(\theta,\gamma)}{A(\theta,\gamma)} \tag{4}$$

$$G^y_\Delta(\gamma) = 0 \ ; \quad G^u_\Delta(\gamma) = \left[G_T(\gamma) - \frac{B(\theta,\gamma)}{A(\theta,\gamma)}\right] \tag{5}$$

Note that, in this example, $G^u_\Delta(\gamma)$ corresponds to the standard assumption [5] of additive unmodelled dynamics. □

Example 2. It is common in adaptive control [2, 8] to consider a system description which incorporates stochastic processes. However, model uncertainty is usually not included in these descriptions. A more complete description including both random noise and uncertainty is a model of the form:

$$A(\theta,\gamma)y(t) = B(\theta,\gamma)u(t) + C(\theta,\gamma)w(t) + \widetilde{M}(\gamma)u(t) \tag{6}$$

where A, B, C are polynomials in γ , $\widetilde{M}$ denotes a residual transfer function, and w denotes a 'white' innovations process.

However, in the model class (6), the estimation of the noise filter polynomial C represents a particular difficulty [18]. One way around this difficulty is to replace the unknown polynomial C by a known polynomial E which is 'close to' $C(\theta,\gamma)$. We can then expand $\frac{1}{C(\theta,\gamma)}$ in terms of $E(\gamma)$ as follows:

$$\frac{1}{C(\theta,\gamma)} = \frac{H(\gamma,\theta)}{E(\gamma)^r} + R(\gamma) \tag{7}$$

where

$$\text{degree } \{E(\gamma)^r\} - \text{degree } \{H(\gamma,\theta)\} = \text{degree } \{C(\gamma,\theta)\} \tag{8}$$

and $R(\gamma)$ denotes the residual in the expansion (7).

It is shown in [20] that the expansion given in (7) converges as $r \to \infty$ provided C and E are related by a positive real condition, that is :

$$\mathcal{R}e\left\{\frac{C}{E} - \frac{1}{2}\right\} > 0 \tag{9}$$

for all γ on the stability boundary.

Operating on (6) by $\frac{1}{C}$ and using (7) leads to a description of the form of the MIU where :

$$T_1(\theta,\gamma) = \tfrac{\bar{E}-\bar{A}}{\bar{E}} \; ; \; \bar{A} = AH \; ; \; \bar{E} = E^r \tag{10}$$

$$T_2(\theta,\gamma) = \tfrac{\bar{B}}{\bar{E}} \; ; \; \bar{B} = BH \tag{11}$$

$$G^u_\Delta(\gamma) = \left(BR + \frac{\widetilde{M}}{C}\right) \tag{12}$$

$$G^y_\Delta(\gamma) = -AR \tag{13}$$

and where $v = w$ is a white innovations process.

Thus, the description (6) is not only more realistic than the usual ideal ARMAX model, but we also see that it fits nicely into the MIU structure. A further feature of the form of (10) and (11) is that these transfer functions are linear in the unknown coefficients in $(\bar{E} - \bar{A})$ and $\bar{B}$. This property greatly simplifies the problem of estimating the parameters. We will exploit this property in our subsequent development. □

3 A Priori Information

Before proceeding to describe the estimator for the nominal model and associated uncertainty we need to specify more clearly the nature and extent of the prior knowledge available about the various terms in the MIU.

It can be seen that the MIU depends upon two basic quantities — the parameters θ and the residual transfer functions G^y_Δ and G^u_Δ. We will adopt a probabilistic viewpoint and assign a probability measure to the triplet v, θ and G_Δ. Typical prior information about these quantities is discussed below.

3.1 Noise Vector

A typical assumption in estimation [18] is that the noise sequence $v(t)$ is a filtered white noise sequence having known spectral properties and known amplitude distribution. We will adopt the same general description here. We will not emphasize this aspect since the novel feature of our analysis is the treatment of the residual transfer functions.

3.2 Parameter Vector

A very common assumption in (Non Adaptive) Robust Control is that the parameters are independent and lie in certain intervals [19]. This can be accommodated in a probabilistic model by assuming independent uniform prior distributions. A more general specification would allow for correlation in the parameter variations.

3.3 Residual Transfer Functions

The residual transfer function is more difficult to specify because it is a function in an infinite dimensional space. The description can be made in time or frequency domains. Obviously, a range of prior information is possible from simple specifications of amplitude to detailed specifications of the functional form.

A characterization of the prior knowledge about a function could include such features as magnitude, degree of smoothness, duration, area, etc. Some of these features are pictorially illustrated in Figure 1 for the impulse response.

Corresponding specifications can also be made in the frequency domain. Also, since the time and frequency domain functions constitute a transform pair, the specifications in the two domains are interrelated. For example, duration and smoothness are dual time and frequency domain properties. For example, the rougher the frequency response, the longer the duration of the impulse response and vice versa. Of course, depending on the specific problem, it may be more natural and indeed easier, to specify the prior uncertainty in either the time or frequency domains.

As an example, one possible frequency domain description would be to simply give an upper bound on the magnitude. This is a typical assumption in robust H^{∞} design [6]. More detailed specifications would include correlation between successive frequency points, giving a measure of smoothness. This would correspond to additional information about the time duration of the associated impulse response.

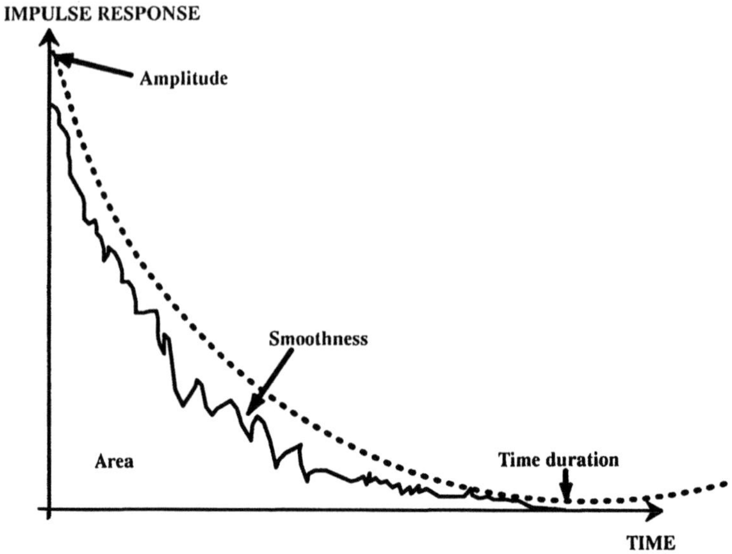

Figure 1. Time Domain Specification of Undermodelling.

3.4 Some Practical Issues

An important issue in the specification of prior information is the trade off between reality and computability in the subsequent estimation problem. For example, one may believe that the realistic specification of residual uncertainty is a hard bound in the frequency domain. This could be expressed by saying that there exists a parameter vector θ^* such that the residual transfer functions satisfy :

$$\operatorname*{Sup}_{\omega} |G_\Delta(\omega)| < \epsilon_2 \quad \text{for} \quad \omega \in [\omega_1, \omega_2] \tag{14}$$

$$\operatorname*{Sup}_{\omega} |G_\Delta(\omega)| < \epsilon_1 \quad \text{for} \quad \omega \notin [\omega_1, \omega_2] \tag{15}$$

where $[\omega_1, \omega_2]$ can be thought of as the desired design bandwidth. However, a constraint of the form given in (14), (15) (no matter how realistic) may be difficult to work with computationally. Indeed, these kinds of constraints will lead to non differentiable optimization problems in estimation. On the other hand, one could question the reality of (14), (15) since these allow arbitrary lack of smoothness which is perhaps unrealistic in any case. Thus alternative descriptions could be desirable if they lead to simpler computations without loosing the link with reality. This issue will be specifically explored in the example presented in the next section.

4 A Posteriori Information

Once the nature and extent of the prior knowledge about the parameters, noise and undermodelling is given, then, in principle, it is a simple matter to use Bayes' Rule to compute the joint a posteriori probability density for the parameters and undermodelling after data has been collected from the system. This gives a conceptually complete answer to the problem of estimating the nominal model and of quantifying the residual model uncertainty. The result can be expressed as:

$$p(\theta, G_\Delta|Y) = \frac{p(Y|\theta, G_\Delta)p(\theta, G_\Delta)}{p(Y)} \tag{16}$$

where

Y	denotes the data set (observations of the output y).
$p(Y\|\theta, G_\Delta)$	is the conditional probability density of the data given θ and G_Δ.
$p(\theta, G_\Delta)$	is the joint prior probability density of θ and G_Δ.
$p(Y)$	is the marginal probability density for the data.
$p(\theta, G_\Delta\|Y)$	is the joint a posteriori probability density of θ and G_Δ.

Example 3. As a trivial example, consider the following (non–dynamic) system.

$$y = \theta u + G_\Delta u^2 + v \tag{17}$$

We first consider independent uniform prior distributions for the triplet θ,G_Δ and v. In particular, we assume :

$$\theta \sim U\{1, 2\} \ , \ G_\Delta \sim U\{-0.2, 0.2\} \ , \ v \sim U\{-0.1, 0.1\} \tag{18}$$

where the notation $U\{a, b\}$ denotes a uniform probability density in the range $[a, b]$. For this case, and assuming $u = 1$ and one observation of y, (16) leads to the following expression for $p(\theta, G_\Delta|Y)$:

$$p(\theta, G_\Delta|Y=y) = \begin{cases} (0.5y^2 - 0.7y + 0.245)^{-1} & ; 0.7 < y \leq 0.9, \\ (0.2y - 0.16)^{-1} & ; 0.9 \leq y \leq 1.1, \\ (-0.5y^2 + 1.3y - 0.765)^{-1} & ; 1.1 \leq y \leq 1.3, \\ 12.5 & ; 1.3 \leq y \leq 1.7, \\ (-0.5y^2 + 1.7y - 0.365)^{-1} & ; 1.7 \leq y \leq 1.9, \\ (-0.2y - 0.44)^{-1} & ; 1.9 \leq y \leq 2.1, \\ (0.5y^2 - 2.3y + 2.645)^{-1} & ; 2.1 \leq y \leq 2.3, \\ 0 & ; \text{OTHERWISE} \end{cases} \tag{19}$$

The above description is valid for $(\theta, G_\Delta) \in [1,2] \times [-0.2, 0.2]$ and $y - 0.1 \leq \theta + G_\Delta \leq y + 0.1$.

The result in (19) is illustrated in Figures 2 and 3 for $y = 2.1$ and 1.5 respectively. The figures show that the hard bounds in the a priori distribution are mapped into corresponding hard bounds in the a posteriori distribution. Note, however, that the computation associated with (19) is large, even for this simple example. On the other hand, if we approximate (18) by Gaussian distributions, then the computation becomes very simple. For the Gaussian case, we replace (18) by :

$$\theta \sim N\{1.5, \frac{1}{12}\} \; , \; G_\Delta \sim N\{0, \frac{0.16}{12}\} \; , \; v \sim N\{0, \frac{0.04}{12}\} \tag{20}$$

where $N\{\mu, \sigma^2\}$ denotes a normal distribution with mean μ and covariance σ^2.

Based on these priors, then it is simple to show, using standard Gaussian manipulations, that the a posteriori density (for $u = 1$ and a single observation y) is :

$$\begin{aligned} p(\theta, G_\Delta|Y=y) = \frac{150}{\pi}\sqrt{\frac{3}{10}} \exp\{&-150(y - \theta - G_\Delta)^2 \\ &-6(\theta - 1.5)^2 - 37.5(G_\Delta)^2 + 5(y - 1.5)^2\} \end{aligned} \tag{21}$$

This function is illustrated in Figure 4 for the case $y = 1.5$. Comparing Figures 3 and 4, we see that the Gaussian description gives a good guide to the hard bound case. Moreover, the result in Figure 4 may be felt to be more realistic because of the smoother quantification of possible outcomes. Thus this example illustrates the trade off between realism and computational complexity mentioned earlier.

Finally, the relationship between prior and posterior information is illustrated in Figure 5. The lower density represents the prior distribution.

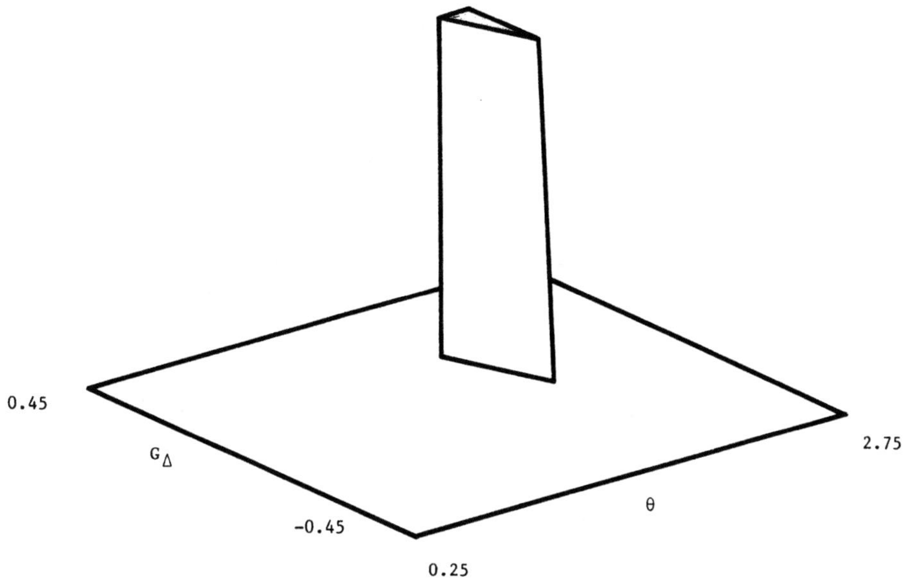

Figure 2. A Posteriori Probability Density for $y = 2.1$ and Uniform Prior Distribution.

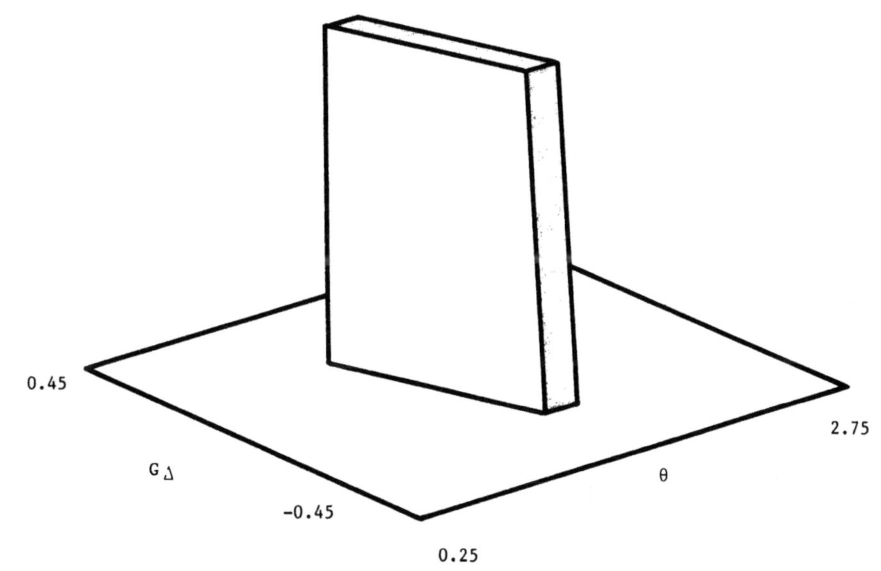

Figure 3. A Posteriori Probability Density for $y = 1.5$ and Uniform Prior Distribution.

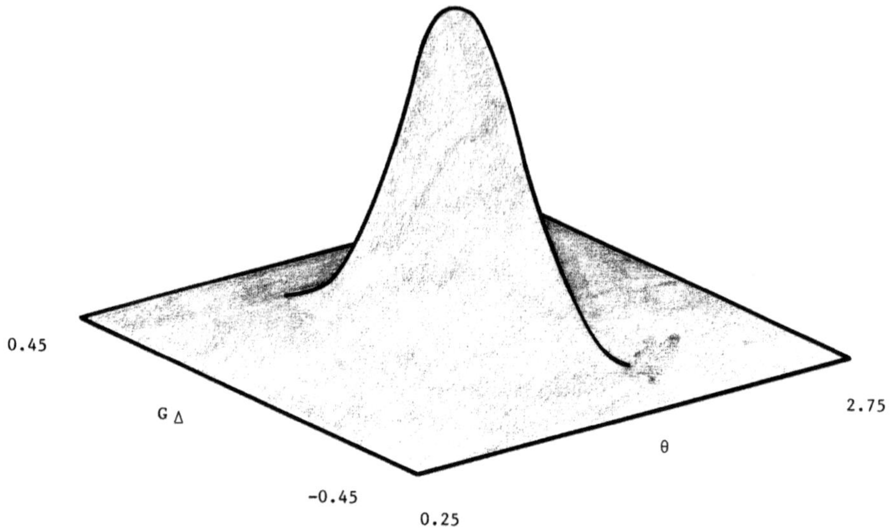

Figure 4. A Posteriori Probability Density for $y = 1.5$ and Gaussian Prior Distribution.

This is transformed into the higher, more concentrated density by application of Bayes' rule. This demonstrates the greater knowledge of model uncertainty afforded by the a posteriori description. □

In the next section, we simplify matters by assuming Gaussian prior distributions. These can be considered as approximations to uniform distributions or as capturing the true degree of belief in the uncertainty description.

5 Gaussian Priors

Based on the examples in sections 2 and 4, we assume that the true system description is given by:

$$y(t) = \phi(t)\theta + \psi(t)H_\Delta + n(t) \tag{22}$$

where

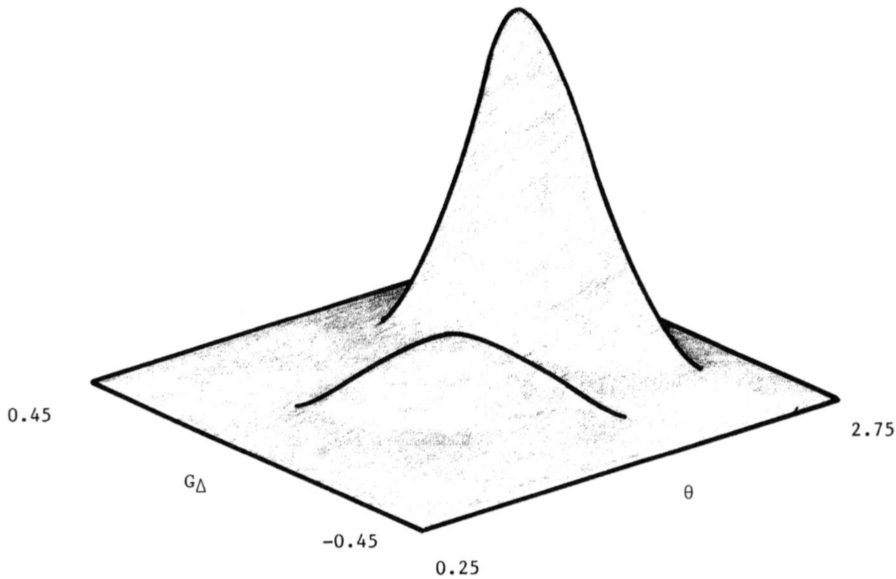

Figure 5. A Priori and A Posteriori Probability Densities for Gaussian Priors and $y = 2.3$.

$\phi(t)$ is a regression vector.
θ is a parameter vector.
$\psi(t)$ is a function of past inputs and outputs.
H_Δ represents the impulse responses of the unmodelled terms.
$n(t)$ denotes a white noise sequence.

For simplicity, we will take γ in the MIU description to be the unit delay operator q^{-1}. Also we will assume that the variables in (22) have the following specific form :

$$\phi(t) = \left[q^{-1}\frac{y(t)}{\bar{E}}, \ldots, q^{-nr}\frac{y(t)}{\bar{E}}; q^{-1}\frac{u(t)}{\bar{E}}, \ldots, q^{-nr}\frac{u(t)}{\bar{E}}\right] \tag{23}$$

$$\theta^T = \left[\bar{e}_1 - \bar{a}_1, \ldots, \bar{e}_{nr} - \bar{a}_{nr}; \bar{b}_1, \ldots, \bar{b}_{nr}\right] \tag{24}$$

$$\bar{E} = 1 + \bar{e}_1 q^{-1} + \cdots + \bar{e}_{nr} q^{-nr} \tag{25}$$

$$\bar{A} = 1 + \bar{a}_1 q^{-1} + \cdots + \bar{a}_{nr} q^{-nr} \tag{26}$$

$$\bar{B} = \bar{b}_1 q^{-1} + \cdots + \bar{b}_{nr} q^{-nr} \tag{27}$$

$$\psi(t) = [u(t-1), \ldots, u(t-m); y(t-1), \ldots, y(t-m)] \tag{28}$$

$$H_\Delta^T = [h_\Delta^u(1), \ldots, h_\Delta^u(m); h_\Delta^y(1), \ldots, h_\Delta^y(m)] \tag{29}$$

In (29) h^u_Δ,h^y_Δ are the vectors formed from the impulse responses of G^u_Δ and G^y_Δ respectively.

We assume independent Gaussian prior distributions for θ,H_Δ and $n(t)$. Thus we assume:

$$\begin{bmatrix} \theta \\ H_\Delta \\ N \end{bmatrix} \sim N\left\{\begin{bmatrix} \widehat{\theta}_0 \\ \widehat{H}_0 \\ \widehat{N}_0 \end{bmatrix}, \begin{bmatrix} C_\theta & 0 & 0 \\ 0 & C_h & 0 \\ 0 & 0 & C_N \end{bmatrix}\right\}$$
$$= N\left\{\begin{bmatrix} \widehat{\beta}_0 \\ \widehat{N}_0 \end{bmatrix}, \begin{bmatrix} C_\beta & 0 \\ 0 & C_N \end{bmatrix}\right\} \tag{30}$$

where

$$N^T = [n(1), n(2), \ldots, n(d)] \tag{31}$$

We assume that the prior mean of H_Δ and N are both zero. Note that the joint distribution of $Y = [y(1), \ldots, y(d)]^T$ and $\beta = (\theta^T, H_\Delta^T)^T$ is not Gaussian. However, due to the form of (22), the conditional distribution for β given Y is Gaussian. We define Φ, Ψ and C_N as follows:

$$\Phi^T = [\phi^T(1), \ldots, \phi^T(d)] \tag{32}$$
$$\Psi^T = [\psi^T(1), \ldots, \psi^T(d)] \tag{33}$$
$$C_N = \begin{bmatrix} \sigma^2 & \ldots & 0 \\ \vdots & \ddots & \vdots \\ 0 & \ldots & \sigma^2 \end{bmatrix} \tag{34}$$

We then have the following result:

Lemma 1. *The joint a posteriori probability density is given by:*

$$p\left(\begin{matrix} \theta \\ H_\Delta \end{matrix} \middle| Y\right) = N\left\{\widehat{\beta}, \widehat{C}_\beta\right\}$$
$$= N\left\{\begin{pmatrix} \widehat{\theta} \\ \widehat{H}_\Delta \end{pmatrix}, \begin{bmatrix} X & W \\ W^T & Z \end{bmatrix}\right\} \tag{35}$$

$$\widehat{\theta} = XC_\theta^{-1}\widehat{\theta}_0 + X\Phi^T C_N^{-1} Y + W\Psi^T C_N^{-1} Y \tag{36}$$
$$\widehat{H}_\Delta = W^T C_\theta^{-1}\widehat{\theta}_0 + W^T\Phi^T C_N^{-1} Y + Z\Psi^T C_N^{-1} Y \tag{37}$$

$$X = \alpha^{-1} + \alpha^{-1}\delta\left(\gamma - \delta^T\alpha^{-1}\delta\right)^{-1}\delta^T\alpha^{-1} \tag{38}$$
$$W = -\alpha^{-1}\delta\left(\gamma - \delta^T\alpha^{-1}\delta\right)^{-1} \tag{39}$$
$$Z = \left(\gamma - \delta^T\alpha^{-1}\delta\right)^{-1} \tag{40}$$

and

$$\alpha = C_\theta^{-1} + \Phi^T C_N^{-1}\Phi \tag{41}$$
$$\delta = \Phi^T C_N^{-1}\Psi \tag{42}$$
$$\gamma = C_h^{-1} + \Psi^T C_N^{-1}\Psi \tag{43}$$

PROOF. Follows from (22), (30) and (31) using Bayes' rule (16). Q.E.D.

6 Frequency Response Estimation

The formulae given in the previous section give the complete a posteriori distribution of the model. These formulae can be utilized in various ways. For example, in this section we use the formulae to evaluate the a posteriori distribution for the total system impulse response and total system frequency response. These are given in the following results. For simplicity we consider the case where the nominal impulse response can be expressed as a linear function ($\mathcal{B}\theta$) of the parameter vector θ and the total system impulse response is ($\mathcal{B}\theta + H_\Delta$) where H_Δ only includes the terms in h_Δ^u. This corresponds precisely to the situation in example 1 when A is known. More generally, the model covers any linear estimation problem in which a fixed set of basis functions ($\mathcal{B}$) is used. If this is not the case, a local linearized analysis can be performed.

Lemma 2. *The a posteriori distribution for the total system impulse response is Gaussian with mean $\widehat{H}_T$ and covariance $\widehat{C}_T$ where:*

$$\widehat{H}_T = \mathcal{B}\widehat{\theta} + \widehat{H}_\Delta \tag{44}$$
$$\widehat{C}_T = \mathcal{B}\alpha^{-1}\mathcal{B}^T + (\mathcal{B}\alpha^{-1}\delta - I)Z(\delta^T\alpha^{-1}\mathcal{B}^T - I) \tag{45}$$

PROOF. We note that the true impulse response is the sum of the nominal impulse response and unmodelled impulse response (as in (3) to (5)), i.e.

$$H_T = \mathcal{B}\theta + H_\Delta = \begin{bmatrix}\mathcal{B} & I\end{bmatrix}\begin{bmatrix}\theta \\ H_\Delta\end{bmatrix} \tag{46}$$

Hence

$$\widehat{H}_T = \begin{bmatrix}\mathcal{B} & I\end{bmatrix}\begin{bmatrix}\widehat{\theta} \\ \widehat{H}_\Delta\end{bmatrix}$$
$$= \mathcal{B}\widehat{\theta} + \widehat{H}_\Delta \tag{47}$$

and

$$\widehat{C}_T = [\,\mathcal{B}\; I\,]\,\widehat{C}_\beta \begin{bmatrix} \mathcal{B}^T \\ I \end{bmatrix} \tag{48}$$

Substituting (36),(37) into (47) gives (44). Similarly, substituting (35) into (48) gives (45). Q.E.D.

Note that the first term in the right side of (45) is the result that would be obtained if only noise were to be considered. The second term quantifies the additional variability due to undermodelling. The corresponding result for the frequency response is given in Lemma 3.

Lemma 3. *The a posteriori distribution for the total system frequency response is Gaussian with mean $\widehat{G}_T(j\omega)$ and variance $\widehat{S}_T(\omega)$ at frequency ω where:*

$$\widehat{G}_T(j\omega) = F(\omega)^T \widehat{H}_T \tag{49}$$

$$\widehat{S}_T(\omega) = F(\omega)^T \widehat{C}_T \left(F(\omega)^T\right)^{\#} \tag{50}$$

where # denotes conjugate transpose and

$$F(\omega)^T = \left[e^{-j\omega}, \ldots, e^{-j\omega d}\right] \tag{51}$$

PROOF. The frequency response at frequency ω, is simply the Fourier transform of H_T. Hence

$$G_T(j\omega) = F(\omega)^T H_T \tag{52}$$

Thus

$$\widehat{G}_T(j\omega) = F(\omega)^T \widehat{H}_T \tag{53}$$

and

$$\begin{aligned} \widehat{S}_T(\omega) &= E\left\{\left|G(j\omega) - \widehat{G}_T(j\omega)\right|^2\right\} \\ &= E\left\{F(\omega)^T \left[H_T - \widehat{H}_T\right] \left[H_T - \widehat{H}_T\right]^T \left(F(\omega)^T\right)^{\#}\right\} \end{aligned} \tag{54}$$

The result then follows immediately from Lemma 2. Q.E.D.

7 Implications in Robust Control

The results given in section 6 provide a mechanism for obtaining an estimated frequency response based on the a posteriori mean $\widehat{G}_T(j\omega)$ together with a measure of the uncertainty as provided by the a posteriori variance $\widehat{S}_T(\omega)$. This information has a simple graphical interpretation as shown on the polar plot in Figure 6.

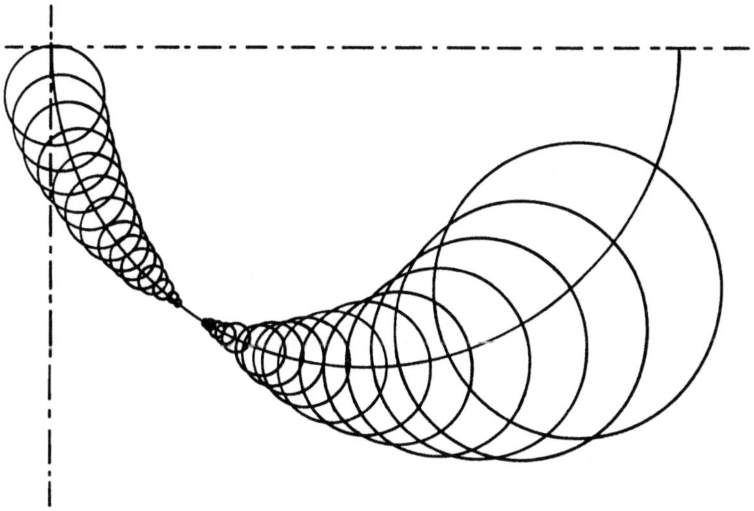

Figure 6. Estimated Frequency Response with Associated Uncertainty

In the figure,the solid line represents the estimated frequency response $\widehat{G}_T(j\omega)$, and the circles have radius one standard deviation, i.e. $\left(\widehat{S}_T(\omega)\right)^{1/2}$. Note that, as shown in section 6, the assumption of Gaussian priors makes the computation of $\widehat{G}_T(j\omega)$ and $\widehat{S}_T(\omega)$ particularly simple.

The kind of information displayed in Figure 6 suggests that there is the possibility of doing a robust control design based on the estimated model and associated uncertainty.

8 Conclusions

This paper has shown how the effects of undermodelling can be quantified in estimation. Particular emphasis has been given to a Bayesian approach and to the advantages of using Gaussian priors in the description of the undermodelling. The resultant quantification of uncertainty offers the potential for combining the paradigms of adaptive control and robust control.

REFERENCES

[1] B.D.O. Anderson, Moore, J.B., and Hawkes, R.M. Model approximations via prediction error identification. *Automatica*, Vol 14(No. 6):pp 615 – 622, 1978.

[2] K.J. Åström and Wittenmark, B. *Adaptive Control.* Addison Wesley, 1988.

[3] E.W. Bai, Fu, L., and Axfjord, A. On parameter and uncertainty estimation. Technical report, Department of Electrical & Computer Engineering, University of Iowa, 1989.

[4] P.E. Caines. Stationary linear and non–linear system identification and predictor set completeness. *IEEE Trans. on Automat. Contr.*, AC – 23:pp 583 – 594, 1978.

[5] P. Dorato, editor. *Robust Control.* IEEE Press, 1987.

[6] B.A. Francis. *A Course in H_∞ Control Theory*, volume 88. Springer Verlag, 1987.

[7] G.C. Goodwin, Mayne, D.Q., and Salgado, M.E. Uncertainty, information and estimation. In *IFAC Symposium on Adaptive Control and Signal Processing*, Glasgow, 1989.

[8] G.C. Goodwin and Sin, K.S. *Adaptive Filtering, Prediction and Control.* Prentice Hall, 1984.

[9] G.C. Goodwin and Salgado, M.E. Quantification of uncertainty in estimation using an embedding principle. In *Proc. 1989 ACC*, pages 1416 – 1421, Pittsburgh, 1989.

[10] G.C. Goodwin and Payne, R.L. *Dynamic System Identification : Experiment Design and Data Analysis.* Academic Press, 1977.

[11] A.J. Helmicki, Jacobson, C.A., and Nett, C.N. h_∞ identification of stable lsi systems : A scheme with direct application to controller design. In *Proc. 1989 ACC*, pages 1428 – 1434, Pittsburgh, 1989.

[12] P.V. Kabaila and Goodwin, G.C. On the estimation of the parameters of an optimal interpolator when the class of interpolators is restricted. *SIAM Journ. Contr. and Optim.*, Vol 18(No. 2):pp 121 – 144, 1980.

[13] R.L. Kosut. Adaptive control via parameter set estimation. *Int. Journal of Adapt. Contr. and Signal Proc.*, Vol 2(No. 4):pp 371 – 400, 1988.

[14] R.L. Kosut. Adaptive robust control via transfer function uncertainty estimation. In *Proc. 1988 ACC*, pages 349 – 354, Atlanta, 1988.

[15] J.M. Krause. Stability margins with real parameter uncertainty : Test data implications. In *Proc. 1989 ACC*, pages 1441 – 1445, Pittsburgh, 1989.

[16] J.M. Krause, Stein, G, and Khargonekar, P.P. Toward practical adaptive control. In *Proc. 1989 ACC*, pages 993 – 998, Pittsburgh, 1989.

[17] R.O. La Maire, Valavani, L., Athans, M., and Stein, G. A frequency domain estimator for use in adaptive control systems. In *Proc. 1987 ACC*, pages 238 – 244, Minneapolis, 1987.

[18] L. Ljung. *System Identification : Theory for the User.* Prentice Hall, 1987.

[19] M. Morari and Zafiriou, E. *Robust Process Control.* Prentice Hall, 1989.

[20] M.E. Salgado, Goodwin, G.C., and León de la Barra, B.A. Generalized expansion algorithm for identification of systems having coloured noise. Technical report, Department of Electrical Engineering & Computer Science, University of Newcastle, 1989.

[21] R.S. Smith and Doyle, J.C. Model invalidation : A connection between robust control and identification. In *Proc 1989 ACC*, pages 1435 – 1440, Pittsburgh, 1989.

G.C. Goodwin, B.A. León de la Barra and R.J. Mazzaferri
Department of Electrical Engineering and Computer Science
University of Newcastle
NSW 2308
AUSTRALIA

Adaptive Tracking for Scalar Minimum Phase Systems

U. Helmke D. Prätzel-Wolters S. Schmid

We present the concept of a universal adaptive tracking controller for classes of linear systems. For the class of scalar minimum phase systems of relative degree one, adaptive tracking is shown for reference signals, that are bounded solutions of linear differential equations. The controller requires no identification of the system parameters. Robustness properties are explored.

1 Introduction

In general terms, the basic problem of parameter adaptive control may be described as follows:

"Suppose a black box is given whose internal dynamics is only roughly understood; for example one might know only certain bounds on the (possibly time varying) system parameters but not the precise parameter values. Furthermore, some structural properties are supposed to be known, so that one can specify a *model class* the unknown system belongs to. One wants to control the system by an explicit controller which is capable of learning enough through the observed output $y(t)$ and input $u(t)$, resp., of the system, to achieve its control purpose." (cf. for example [11] for a precise formalization of this heuristic definition.)

Such a control objective might be for example that of *model reference adaptive control* (MRAC), where the output $y(\cdot)$ to be controlled is required to track the output $r(\cdot)$ of a prescribed linear model for a certain class of admissible control functions and initial conditions (see e.g. [6]).

Those MRAC-algorithms often work only under rather strong a priori assumptions on the systems to be controlled and due to the intrinsic computational complexity of the used identification scheme, these adaptive controllers are complicated both from a theoretical as well as computational point of view.

The purpose of this paper is to derive adaptive tracking controllers without explicit identification of the system parameters.

Particular algorithms for such simplified controllers have been proposed in [1] and [2]. In a more systematic framework adaptive controllers which require no explicit identification and which stabilize linear systems under very weak assumptions on the system parameters have been developed in [4], [7], [10], [12], [13] and [15]. These adaptive controllers are called *universal*, since they achieve their control objective for a whole prescribed class of linear systems and all possible initial conditions.

Previous work on universal adaptive controllers was mainly concerned with the adaptive stabilization problem. The purpose of this paper is to extend these results to the adaptive tracking problem:

"Derive for a given class of reference signals $r(\cdot) \in \mathcal{R}$ and a class Σ of linear systems an adaptive feedback controller such that for every system in Σ its output asymptotically tracks $r(\cdot)$."

For constant reference signals and ramp functions those controllers are analyzed in [8] and [9]. Furthermore [14] contains a design scheme for synchronizing the outputs of a finite number of scalar simple integral plants.

To formalize the above mentioned adaptive tracking problem for a larger class of reference signals let $\Sigma = \Sigma(m,p)$ be a *set of linear* time invariant *systems*

$$\dot{x}(t) = Ax(t) + Bu(t), \quad y(t) = Cx(t)$$
$$x(t) \in \mathbb{R}^n, \quad u(t) \in \mathbb{R}^m, \quad y(t) \in \mathbb{R}^p$$

with m, p given and n arbitrary.

Let

$$\mathcal{R} \subset \mathcal{C}_{\text{pc}}([0,\infty),\mathbb{R})$$

be a prescribed class of possible (piecewise continuous) *reference signals* $r(\cdot)$.

A *smooth universal adaptive tracking controller* (UATC) for $(\Sigma, \mathcal{R})$ consists of

— a parameter space $\mathbb{R}^q$ for the feedback gains $k(t)$.

— a smooth control law

$$u(t) = f(t, k(t), y(t), r(t))$$

where

$$f : \mathbb{R}^{1+q+2p} \longrightarrow \mathbb{R}^m \text{ is } \mathcal{C}^\infty.$$

— a parameter adaptation law

$$\dot{k}(t) = g(k(t), y(t), r(t))$$

where

$$g : \mathbb{R}^{q+2p} \longrightarrow \mathbb{R}^q$$

is smooth such that for any $(A, B, C) \in \Sigma(n, m, p)$, any initial data $x(0), k(0)$ and any $r(\cdot) \in \mathcal{R}$ a unique solution of the closed loop system

$$\begin{aligned} \dot{x}(t) &= Ax(t) + Bf(t, k(t), y(t), r(t)) \\ y(t) &= Cx(t) \\ \dot{k}(t) &= g(k(t), y(t), r(t)) \end{aligned} \tag{CL}$$

exists for all $t \geq 0$ and satisfies

- $k(\cdot) \in L_\infty([0, \infty), \mathbb{R}^q)$
- $\lim_{t\to\infty} e(t) = \lim_{t\to\infty}(y(t) - r(t)) = 0.$

In this paper we assume that the system class is the set $\Sigma(n, 1, 1)$ of all *scalar minimum phase systems* (A, b, c) *with relative degree one*:

- $m = p = 1$
- $cb \neq 0$
- $\det \begin{bmatrix} sI - A & b \\ c & 0 \end{bmatrix}$ a Hurwitz polynomial.

In our main Theorems 2.2 (sign cb known) and 3.2 (sign cb unknown) we derive UATC's which guarantee asymptotic tracking of reference signals, that are bounded solutions of known linear time invariant differential equations. In [5] the existence of those controllers is shown for arbitrary scalar (non minimum phase) systems.

2 Adaptive Tracking Without Switching

In the sequel let $\Sigma_+(n) \subset \Sigma(n, 1, 1)$ denote the subclass of scalar ($m = p = 1$), relative degree 1, minimum phase systems $(A, b, c) \in \mathbb{R}^{n\times n} \times \mathbb{R}^{n\times 1} \times \mathbb{R}^{1\times n}$ with

$$\det \begin{bmatrix} sI - A & b \\ c & 0 \end{bmatrix} \text{ a Hurwitz polynomial} \tag{2.1a}$$

$$cb > 0 \tag{2.1b}$$

The following robustness result is a straightforward extension of a well known adaptive stabilization theorem ($r \equiv 0$) for minimum phase systems, see e.g. [3], [4].

Proposition 2.1. *Let $(A,b,c) \in \Sigma_+(n)$ and $d(\cdot) \in L_p([0,\infty),\mathbb{R}^n)$, $p \in \mathbb{N}$. Then the solutions of the closed loop system*

$$\dot{x}(t) = (A - k(t)bc)x(t) + d(t) \tag{2.2a}$$
$$y(t) = cx(t) \tag{2.2b}$$
$$\dot{k}(t) = |y(t)|^p \tag{2.2c}$$

satisfy for all initial values $x(0) \in \mathbb{R}^n, k(0) \in \mathbb{R}$:

$$x(\cdot) \in L_p([0,\infty),\mathbb{R}^n) \text{ and } \lim_{t\to\infty} x(t) = 0 \tag{2.3a}$$
$$\lim_{t\to\infty} k(t) = k_\infty < \infty \text{ exists.} \tag{2.3b}$$

PROOF. By (2.2c), $k(t)$ is monotonically increasing. Let $0 < T_+ \leq \infty$ be the maximal time for which a solution $(x(t), k(t))$ of (2.2) exists. Suppose $k(t) \to +\infty$ as $t \to T_+$. Then, by the high gain theorem (or Tychonov's theorem) in [7], [13], the system $\dot{\xi}(t) = (A - k(t)bc)\xi(t)$ is exponentially stable. Thus, $x(\cdot) \in L_p([0,T),\mathbb{R})$ and therefore $k(\cdot) : [0,T_+) \to \mathbb{R}$ is bounded in contradiction to $k(t) \to \infty$. Thus, $\lim k(t)_{t\to T_+} < \infty$ exists. By a suitable state space coordinate transformation we can decompose (2.2a) (c.f. [15]):

$$\dot{x}_1(t) = A_1x_1(t) + A_2y(t) + d_1(t) \tag{2.4a}$$
$$\dot{y}(t) = (\alpha - \beta k(t))y(t) + A_3x_1(t) + d_2(t) \tag{2.4b}$$

where $A \in \mathbb{R}^{(n-1)\times(n-1)}$ has its eigenvalues in the open left half plane, $d_1(\cdot) \in L_p([0,\infty),\mathbb{R}^{n-1})$, $d_2(\cdot) \in L_p([0,\infty),\mathbb{R})$ and $\beta = cb > 0$. Since $y(\cdot) \in L_p([0,T_+),\mathbb{R})$ and $\dot{\xi} = A_1\xi$ is exponentially stable, $x_1(\cdot) \in L_p([0,T_+),\mathbb{R}^{n-1})$. (2.4) implies that $\dot{y} \in L_p([0,T_+),\mathbb{R})$ and $\dot{x}_1(\cdot) \in L_p([0,T_+),\mathbb{R}^{n-1})$. Thus, $\dot{x}(\cdot) \in L_p([0,T_+),\mathbb{R}^n)$ and $\lim_{t\to T_+} x_1(t) = 0$, $\lim_{t\to T_+} y(t) = 0$. In particular, $x(t)$ and $k(t)$ are bounded on $[0,T_+)$ and therefore $T_+ = \infty$. The result follows. Q.E.D.

Consider now the class $\Sigma_+(n)$ and let $\mathcal{R}_{p(s)}$,

$$p(s) = s^\ell + p_{\ell-1}s^{\ell-1} + \cdots + p_0 \in \mathbb{R}[s]$$

be the solution space of the differential equation

$$p\left(\frac{d}{dt}\right) r(\cdot) = 0. \tag{2.5}$$

The integer $\ell \in \mathbb{N}$ is called the *dynamic order* of the reference signals $r(\cdot) \in \mathcal{R}_{p(s)}$.

For the tracking problem it is now required that the tracking error $e(t) = y(t) - r(t)$ tends to zero if $t \to \infty$:

$$\lim_{t\to\infty} e(t) = 0. \tag{2.6}$$

It is possible to view the tracking problem (adaptive or non-adaptive) as a stabilization problem by inserting the dynamics of the reference signal into the dynamics of the given system. In most textbooks on control theory this idea (the internal model principle) of augmenting the system with a suitable *pre-filter of relative degree zero* is explained only for very simple reference signals such as constant reference signals or ramp functions. As we will see, this idea works also in the adaptive case, even for reference signals of *arbitrary* dynamic order ℓ.

Choose a monic Hurwitz polynomial $q(s)$ of the same degree ℓ as $p(s)$. Let (A, b, c, d), $d = 1$, be a state space realization of the transfer function $q(s)/p(s)$. u shall be the output of this system, the input is denoted by v. We obtain the following state space equation for the augmented system:

$$\dot{\hat{x}}(t) = \hat{A}\hat{x}(t) + \hat{b}v(t) \tag{2.7a}$$
$$y(t) = \hat{c}\hat{x}(t), \tag{2.7b}$$

where

$$\hat{A} = \begin{bmatrix} A & bc_r \\ 0 & A_r \end{bmatrix}, \quad \hat{b} = \begin{bmatrix} b \\ b_r \end{bmatrix}, \quad \hat{c} = \begin{bmatrix} c & 0 \end{bmatrix}, \quad \hat{x} = \begin{bmatrix} x \\ x_r \end{bmatrix}, \tag{2.7c}$$

and $(A, b, c) \in \Sigma_+$ denotes the system under consideration. Since $p(d/dt)r(\cdot) = 0$, also $\chi(d/dt)r(\cdot) = 0$, where χ is the characteristic polynomial of A. If we assume that $(\hat{A}, \hat{c})$ is observable, then $\hat{c}e^{\hat{A}t}$ is a fundamental system for the differential equation $\chi(d/dt)r(\cdot) = 0$, and $r(\cdot)$ can therefore be generated by $(\hat{A}, \hat{c})$ through an appropriate initial state $\bar{x}(0)$:

$$\dot{\bar{x}}(t) = \hat{A}\bar{x}(t) \tag{2.8a}$$
$$r(t) = \hat{c}\bar{x}(t) \tag{2.8b}$$

(This is in general not true for multivariable systems!) This yields

$$\frac{d}{dt}(\hat{x} - \bar{x})(t) = \hat{A}(\hat{x} - \bar{x})(t) + \hat{b}v(t) \tag{2.9a}$$
$$e(t) = \hat{c}(\hat{x} - \bar{x})(t) \tag{2.9b}$$

with $(\hat{A}, \hat{b}, \hat{c}) \in \Sigma_+(n + \ell)$, too.

Now choose for $p \in \mathbb{N}$ fixed:

$$v(t) = -k(t)e(t) \tag{2.10a}$$
$$\dot{k}(t) = |e(t)|^p \tag{2.10b}$$

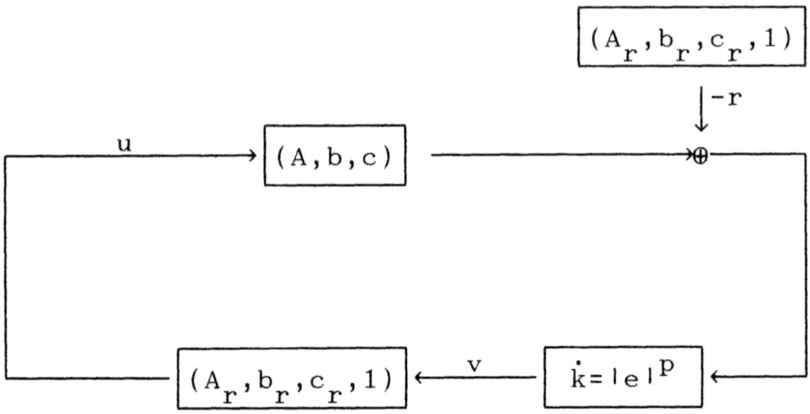

Figure 1. Universal adaptive tracking controller

This amounts to the following choice of f and g in the construction of an UATC for $\Sigma_+(n)$:

— feedback gains

$$z = \begin{bmatrix} x_r \\ k \end{bmatrix} \in \mathbb{R}^{\ell+1} \tag{2.11a}$$

— control law

$$\begin{aligned} u(t) &= f(t, z(t), y(t), r(t)) \\ &= c_r x_r(t) - k(t)(y(t) - r(t)) \\ &= (c_r, -e(t))z(t) \end{aligned} \tag{2.11b}$$

— parameter adaptation law

$$\dot{z}(t) = g(z(t), y(t), r(t)) = \begin{bmatrix} A_r x_r(t) - b_r e(t)k(t) \\ |e(t)|^p \end{bmatrix} \tag{2.11c}$$

Theorem 2.2. *Let $(A, b, c) \in \Sigma_+(n)$ and $p(s) \in \mathbb{R}[s]$ be a monic polynomial of degree ℓ. Let $q(s) \in \mathbb{R}[s]$ be any monic Hurwitz polynomial, $\deg q(s) = \deg p(s) = \ell$ and let $(A_r, b_r, c_r, 1)$ be a minimal realization of $q(s)/p(s)$.*

a) For all reference signals $r(\cdot) \in \mathcal{R}_{p(s)}$

$$p\left(\frac{d}{dt}\right) r(t) \equiv 0$$

and all disturbances $d(\cdot) \in L_p([0,\infty),\mathbb{R}^n)$, $p \in \mathbb{N}$, *the solutions of the closed loop system*

$$\begin{aligned} \dot{x}(t) &= Ax(t) + bc_r x_r(t) - bk(t)(y(t) - r(t)) + d(t) && (2.12a)\\ y(t) &= cx(t), \quad e(t) = y(t) - r(t) && (2.12b)\\ \dot{x}_r(t) &= A_r x_r(t) - b_r k(t)(y(t) - r(t)) && (2.12c)\\ \dot{k}(t) &= |y(t) - r(t)|^p && (2.12d) \end{aligned}$$

exist for $t \in [0,\infty)$ *and satisfy*

$$e(\cdot) \in L_p([0,\infty),\mathbb{R}) \text{ and } \lim_{t\to\infty} e(t) = 0 \qquad (2.13a)$$

$$\lim_{t\to\infty} k(t) = k_\infty \text{ exists.} \qquad (2.13b)$$

b) Let $p(s)$ *have all roots in the closed left half plane* $\overline{\mathbb{C}^-}$, *with simple roots (of multiplicity one) on the imaginary axis. Then (2.11) is a universal adaptive tracking controller for* $(\Sigma_+(n), \mathcal{R}_{p(s)})$, *i.e.*

$$\lim_{t\to\infty} e(t) = 0 \qquad (2.14a)$$

$$x_r(\cdot) \in L_\infty([0,\infty),\mathbb{R}^\ell) \qquad (2.14b)$$

$$\lim_{t\to\infty} k(t) = k_\infty < \infty \text{ exists.} \qquad (2.14c)$$

PROOF. Let

$$\hat{x} = \begin{bmatrix} x \\ x_r \end{bmatrix} \quad \hat{A} = \begin{bmatrix} A & bc_r \\ 0 & A_r \end{bmatrix} \quad \hat{b} = \begin{bmatrix} b \\ b_r \end{bmatrix} \quad \hat{c} = [\, c \ 0 \,].$$

Suppose $(\hat{A}, \hat{c})$ is observable and let $\bar{x}(t)$ defined by (2.8). Then (2.12) is equivalent to:

$$\begin{aligned} \frac{d}{dt}(\hat{x} - \bar{x})(t) &= (\hat{A} - k(t)\hat{b}\hat{c})(\hat{x} - \bar{x})(t) + \begin{bmatrix} d(t) \\ 0 \end{bmatrix} && (2.15a)\\ e(t) &= \hat{c}(\hat{x} - \bar{x})(t) && (2.15b)\\ \dot{k}(t) &= |e(t)|^p && (2.15c) \end{aligned}$$

By Proposition 2.1, $z(\cdot) = (\hat{x}(\cdot) - \bar{x}(\cdot)) \in L_p([0,\infty),\mathbb{R}^{n+\ell})$ and $\lim_{t\to\infty} z(t) = 0, \lim_{t\to\infty} k(t) < \infty$ exists.

This proves (a).

Suppose $(\hat{A}, \hat{c})$ is not observable, i.e. $p(s)$ and $n(s)$, the numerator of $g(s) = c(sI - A)^{-1}b = n(s)/d(s)$, have a common factor. Then, because $n(s)$ is Hurwitz, common factors of $p(s)$ and $n(s)$ are Hurwitz and the corresponding modes of $(A_r, b_r, c_r, d_r = 1)$ are asymptotically stable. Therefore the tracking result (a) remains also true in this case.

To prove (b), it remains to show (2.14b). By assumption on $p(s)$, the differential equation $\dot{\xi}_r = A_r\xi_r$ is stable (but not necessarily asymptotically stable!). Let $\bar{x}_r(\cdot)$ denote the second component of the vector $\bar{x}(\cdot)$ defined by (2.8). Thus, $\dot{\bar{x}}_r = A_r\bar{x}_r$, which implies (by stability) that $\bar{x}_r(\cdot)$ is bounded on $[0,\infty)$. By (a):

$$\lim_{t\to\infty} (x_r(t) - \bar{x}_r(t)) = 0$$

and therefore $x_r(\cdot) \in L_\infty([0,\infty),\mathbb{R}^\ell)$. Q.E.D.

The above result guarantees universal adaptive tracking of constant reference signals

$$r(t) \equiv r_0 \tag{2.16}$$

or of any finite superposition of periodic signals

$$r(t) = a_0 + \sum_{\nu=1}^{\ell} (a\cos\nu t + b\sin\nu t)\,. \tag{2.17}$$

In the first case (2.16), the order of the universal controller is 2, while for (2.17) the order of the controller is $2\ell+2$. Note that these bounds are independent of the order n of the plants $(A,b,c) \in \Sigma_+(n)$. A universal adaptive tracking controller for *constant* reference signals has been proposed by Mårtensson [8] and Miller and Davison [9]. We are not aware of any reference, where the problem of universal adaptive tracking is treated for higher order signals such as e.g. (2.17).

As an extension of the previous result we consider a *series coupling of systems* belonging to the class $\Sigma_+(n_i)$. Let $r_i(\cdot)$ be the reference signal for the i-th system with

$$\tilde{p}_i\left(\frac{d}{dt}\right) r_i(\cdot) = 0, \quad \tilde{p}_i(s) \in \mathbb{R}[s] \text{ monic of degree } \ell, \quad i = 1,\ldots,N \tag{2.18}$$

In order to construct an adaptive tracking controller for this series connection we define polynomials

$$p_1(s) := \tilde{p}_1(s) \tag{2.19}$$
$$p_{i+1}(s) := \operatorname{lcm}(\tilde{p}_i(s), \tilde{p}_{i+1}(s)) \tag{2.20}$$
$$i = 1,\ldots,N-1$$

and select local controllers (f_i, g_i), $i = 1,\ldots,N$ of the form (2.10) for the subsystems Σ_i. More precisely, for the first system (A_1,b_1,c_1) we choose (f_1,g_1) as in Theorem 2.2a). The tracking error $e_1(\cdot)$ then satisfies the prescribed tracking requirement $e_1(\cdot) \in L_p([0,\infty),\mathbb{R})$ and $\lim_{t\to\infty} e_1(t) = 0$. Now for the system $(i+1)$, $(A_{i+1},b_{i+1},c_{i+1})$, there is an external "input"

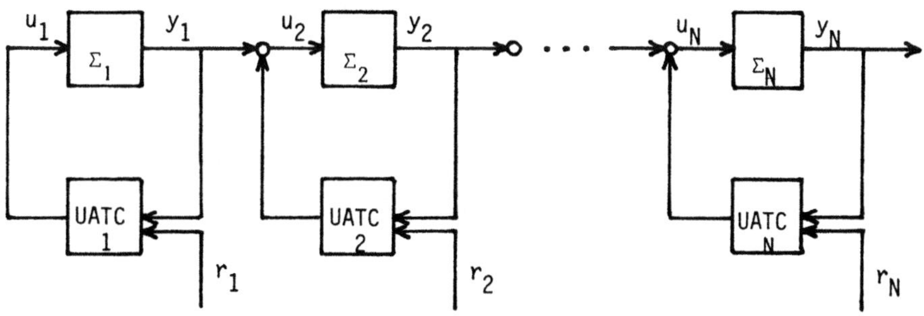

Figure 2. Series coupling of systems

$y_i(\cdot)$, the output of system i. Assume, tracking was performed successfully for system i, $e_i(\cdot) = y_i(\cdot) - r_i(\cdot) \in L_p([0,\infty),\mathbb{R})$, $\lim_{t\to\infty} e_i(t) = 0$. Let $(A^r_{i+1}, b^r_{i+1}, c^r_{i+1}, d^r_{i+1} = 1)$ be constructed as for Theorem 2.2 with $p_{i+1}(s)$ defined by (2.19)–(2.20). Then we obtain

$$\begin{aligned} \begin{bmatrix} \dot{x}_{i+1} \\ \dot{x}^r_{i+1} \end{bmatrix} &= \begin{bmatrix} A_{i+1} & b_{i+1}c^r_{i+1} \\ 0 & A^r_{i+1} \end{bmatrix} \begin{bmatrix} x_{i+1} \\ x^r_{i+1} \end{bmatrix} \\ &\quad + \begin{bmatrix} b_{i+1} \\ b^r_{i+1} \end{bmatrix} v_{i+1} + \begin{bmatrix} b_{i+1} \\ 0 \end{bmatrix} (e_i + r_i) \end{aligned} \tag{2.21a}$$

$$y_{i+1} = \begin{bmatrix} c_{i+1} & 0 \end{bmatrix} \begin{bmatrix} x_{i+1} \\ x^r_{i+1} \end{bmatrix} \tag{2.21b}$$

By observability of (A^r_{i+1}, c^r_{i+1}) and by (2.20), $r_i(\cdot)$ can be generated by

$$\begin{aligned} r_i &= c^r_{i+1}\bar{x}^r_{i+1} \\ \dot{\bar{x}}^r_{i+1} &= A^r_{i+1}\bar{x}^r_{i+1} \end{aligned}$$

hence

$$\begin{aligned} \dot{\hat{x}}_{i+1} &= \hat{A}_{i+1}\hat{x}_{i+1} + \hat{b}_{i+1}v_{i+1} + d_i \\ y &= \hat{c}_{i+1}\hat{x}_{i+1} \end{aligned}$$

with

$$\hat{x}_{i+1} = \begin{bmatrix} x_{i+1} \\ x^r_{i+1} + \bar{x}^r_{i+1} \end{bmatrix}, \quad \hat{A}_{i+1} = \begin{bmatrix} A_{i+1} & b_{i+1}c^r_{i+1} \\ 0 & A^r_{i+1} \end{bmatrix} \tag{2.22}$$

$$\hat{b}_{i+1} = \begin{bmatrix} b_{i+1} \\ b^r_{i+1} \end{bmatrix}, \quad d_i = \begin{bmatrix} b_{i+1} \\ 0 \end{bmatrix} e_i, \quad \hat{c}_{i+1} = \begin{bmatrix} c_{i+1} & 0 \end{bmatrix} \tag{2.23}$$

and $e_i(\cdot) \in L_p([0,\infty),\mathbb{R})$, $\lim_{t\to\infty} e_i(t) = 0$.

Assuming observability of $(\hat{A}_{i+1}, \hat{c}_{i+1})$, $r_{i+1}(\cdot)$ can be generated by

$$r_{i+1} = \hat{c}_{i+1}\bar{x}_{i+1}, \quad \dot{\bar{x}}_{i+1} = \hat{A}_{i+1}\bar{x}_{i+1}.$$

Thus, we obtain

$$\frac{d}{dt}\left(\hat{x}_{i+1} - \bar{x}_{i+1}\right) = \hat{A}_{i+1}\left(\hat{x}_{i+1} - \bar{x}_{i+1}\right) + \hat{b}_{i+1}v_{i+1} + d_i \qquad (2.24a)$$

$$e_{i+1} = \hat{c}_{i+1}\left(\hat{x}_{i+1} - \bar{x}_{i+1}\right) \qquad (2.24b)$$

with $d_i(\cdot) \in L_p([0,\infty), \mathbb{R}^{\nu_i})$, $\nu_i := \dim \hat{x}_{i+1}$.

Now consider the local control laws

$$v_{i+1}(t) = -k_{i+1}(t)e_{i+1}(t) \qquad (2.25a)$$

$$\dot{k}_{i+1}(t) = |e_{i+1}(t)|^p \qquad (2.25b)$$

$p \in \mathbb{N}$ fixed. This gives rise to local adaptive controllers (f_i, g_i) for (A_i, b_i, c_i) which are constructed in the same manner as for a single system, except for the modification that instead of the polynomials $\tilde{p}_i(s)$ the polynomials $p_i(s) = \operatorname{lcm}(\tilde{p}_{i-1}(s), \tilde{p}_i(s))$ are used.

Theorem 2.3. *Let $(A_i, b_i, c_i) \in \Sigma_+(n_i)$ and $\tilde{p}_i(s) \in \mathbb{R}[s]$ be monic polynomials of degree ℓ_i, $i = 1, \ldots, N$. Let $p_1(s) := \tilde{p}_1(s)$, $p_{i+1}(s) := \operatorname{lcm}(\tilde{p}_i(s), \tilde{p}_{i+1}(s))$, $i = 1, \ldots, N-1$ and choose monic Hurwitz polynomials $q_i(s) \in \mathbb{R}[s]$, $\deg q_i(s) = \deg_i p(s)$, and minimal realizations $(A_i^r, b_i^r, c_i^r, d_i^r = 1)$ of $q_i(s)/p_i(s)$, $i = 1, \ldots, N$.*

(a) For all reference signals $r_i(\cdot) \in \mathcal{R}_{p_i(s)}$

$$p\left(\frac{d}{dt}\right) r_i(t) \equiv 0$$

the solutions of the closed loop system for the series connection ($y_0 := 0$):

$$\dot{x}_i = A_i x_i + b_i c_i^r x_i^r + b_i y_{i-1} - b_i k_i e_i \qquad (2.26a)$$

$$y_i = c_i x_i, \quad e_i = y_i - r_i \qquad (2.26b)$$

$$\dot{x}_i^r = A_i^r x_i^r - b_i^r k_i e_i \qquad (2.26c)$$

$$\dot{k}_i = |e_i|^p, \qquad (2.26d)$$

$i = 1, \ldots, N$, exist for all $t \in [0,\infty)$ and satisfy ($i = 1, \ldots, N$):

$$e_i(\cdot) \in L_p([0,\infty), \mathbb{R}) \text{ and } \lim_{t\to\infty} e_i(t) = 0 \qquad (2.27a)$$

$$\lim_{t\to\infty} k_i(t) < \infty \text{ exists.} \qquad (2.27b)$$

(b) *Suppose that $p_i(s) \in \mathbb{R}[s]$, $i = 1,\ldots,N$, have all their roots in the closed left half plane, with simple roots of multiplicity one on the imaginary axis. Then (2.26) is the closed loop system of a (local) universal adaptive tracking controller for the series connection of $(\Sigma_+(n_i), \mathcal{R}_{\tilde{p}_i(s)})$, $i = 1,\ldots,N$, i.e.*

$$\lim_{t\to\infty} e_i(t) = 0 \tag{2.28a}$$

$$x_i^r(\cdot) \in L_\infty([0,\infty), \mathbb{R}^{\ell_i}) \tag{2.28b}$$

$$\lim_{t\to\infty} k_i(t) < \infty \text{ exists} \tag{2.28c}$$

for $i = 1,\ldots,N$. The total order of the universal adaptive tracking controller for the series connection is at most $\sum_{i=1}^{N}(\ell_{i-1} + \ell_i + 1)$, $\ell_0 := 0$, where ℓ_i are the dynamic orders of the reference signals $r_i(\cdot)$.

PROOF. The proof is an immediate consequence of Theorem 2.2, applied to the systems (2.26) for $i = 0,\ldots,N-1$. Q.E.D.

The following simulations show asymptotic tracking for a series coupling of 8 systems with sinusoidal or constant reference signals. In Fig. 3 the reference signals are: $r_i(t) = \sin\left(t + \frac{i-1}{4}\pi\right)$, $i = 1,\ldots,8$, $p_i(s) = s^2 + 1$ for $i = 1,\ldots,8$. The associated Hurwitz polynomial has been selected as $q_i(s) = (s+\pi)^2$. In Fig. 4 the values of the constant reference signals are $(r_1, r_2, r_3, r_4) = (5, -2, 2, -5)$ and the systems $\Sigma_1,\ldots,\Sigma_4$ are identical with the first 4 systems in Fig. 3. For both simulations the gain adaptation law is of the form $\dot{k}(t) = e(t)^2$. The transfer functions of the systems are:

$$\Sigma_1 : g_1(s) = \frac{s+1}{s^2 - 2s + 1}$$

$$\Sigma_2 : g_2(s) = \frac{s^3 + 4s^2 + 5s + 2}{s^4 - 5s^3 + 3s^2 + 4s - 1}$$

$$\Sigma_3 : g_3(s) = \frac{1}{s-1}$$

$$\Sigma_4 : g_4(s) = \frac{s^2 + 2s + 1}{s^3 + 2s^2 + 3s - 2}$$

$$\Sigma_5 : g_5(s) = \frac{s^4 + 4s^3 + 6s^2 + 4s + 1}{s^5 - s^4 - s^3 + s^2 - s}$$

$$\Sigma_6 : g_6(s) = \frac{1}{s+1}$$

$$\Sigma_7 : g_7(s) = \frac{s+1}{s^2 + 2s + 1}$$

$$\Sigma_8 : g_8(s) = \frac{s^2 + 4s + 4}{s^3 + 3s^2 + 2s - 1}$$

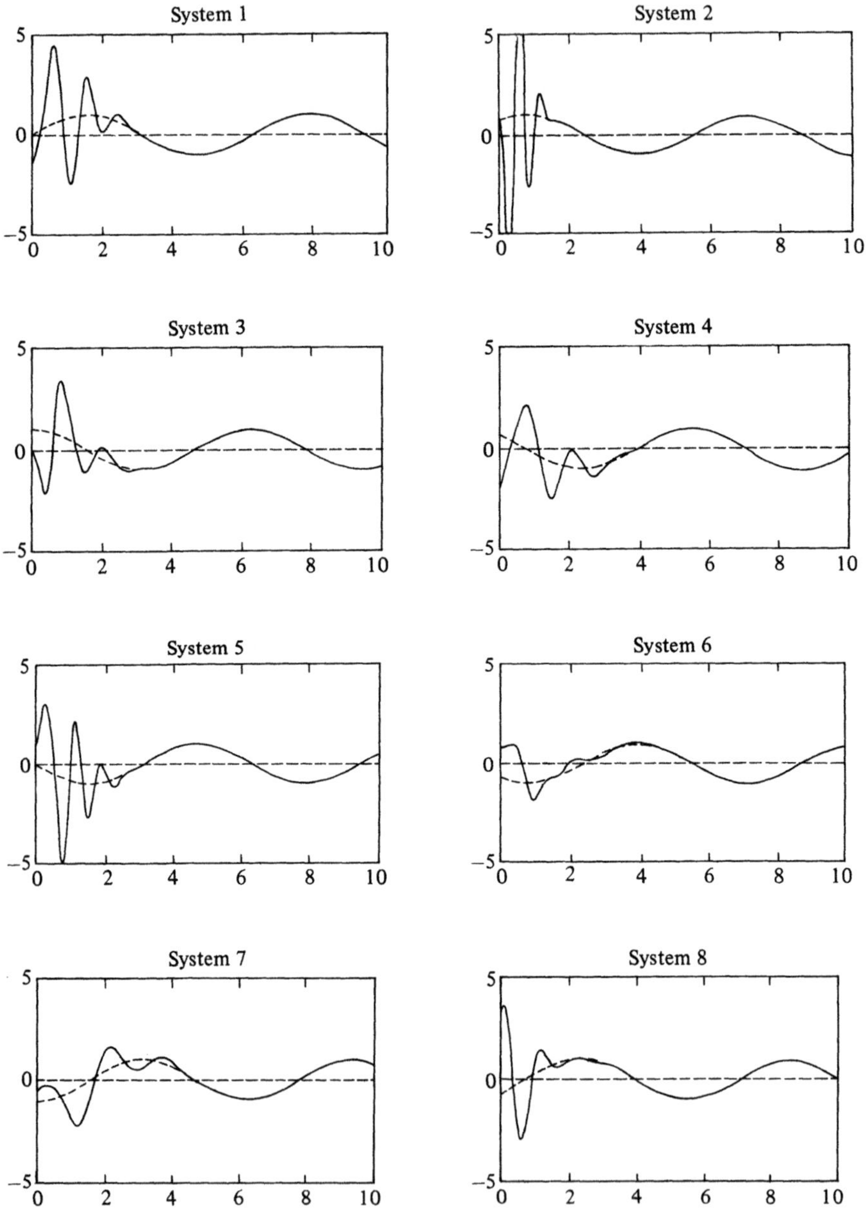

Figure 3. Simulation of series coupling without switching functions: periodic reference signals

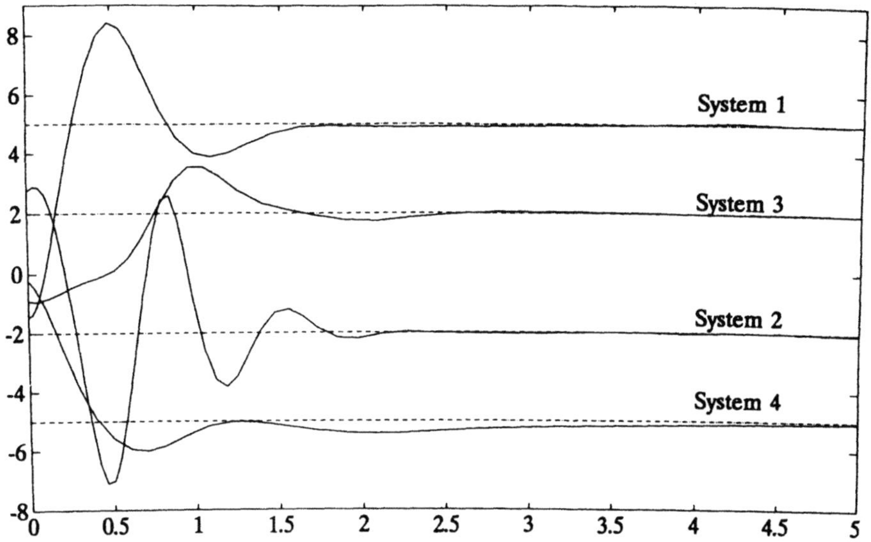

Figure 4. Series coupling: tracking of constant reference signals

3 Robust Adaptive Tracking With Switching Functions

If UATC's are to be constructed for the class $\Sigma(n,1,1)$, i.e. if $\operatorname{sgn} cb$ is not known, switching concepts become necessary. The following theorem shows that the Nussbaum type switching stabilizers are robust against L_2-disturbances. The proof is an obvious modification of the proof of Theorem 2 in [15].

Proposition 3.1. *Let $(A,b,c) \in \Sigma(n,1,1)$ be scalar minimum phase of relative degree 1 and $d(\cdot) \in L_2([0,\infty),\mathbb{R}^n)$. The solutions of the closed loop system*

$$\begin{aligned} \dot{x}(t) &= Ax(t) + bu(t) + d(t) \\ y(t) &= cx(t) \\ u(t) &= N(k(t))y(t) \\ \dot{k}(t) &= y^2(t) \end{aligned} \tag{3.1}$$

where $N(\cdot)$ is a Nussbaum type switching function, i.e.

$$\sup_{\eta>0} \frac{1}{\eta}\int_0^\eta N(\sigma)\,d\sigma = +\infty, \quad \inf_{\eta>0} \frac{1}{\eta}\int_0^\eta N(\sigma)\,d\sigma = -\infty \tag{3.2}$$

satisfy:

$$y(\cdot) \in L_2([0,\infty),\mathbb{R}) \text{ and } \lim_{t\to\infty} y(t) = 0 \tag{3.3a}$$
$$\exists k_\infty = \lim_{t\to\infty} k(t) < \infty \tag{3.3b}$$

The following Theorem is an immediate consequence of the above theorem and the by now familiar reasoning for the construction of an UATC in the non switching case (cf. Theorem 2.2).

Theorem 3.2. *Let $(A,b,c) \in \Sigma(n,1,1)$ be minimum phase with relative degree 1, $r(\cdot) \in \mathcal{R}_{p(s)}$ and $d(\cdot) \in L_2([0,\infty),\mathbb{R}^n)$. Let further $q(s)$ be Hurwitz, $\deg q(s) = \deg p(s) = \ell$ and $(A_r, b_r, c_r, 1)$ a minimal realization of $q(s)/p(s)$. Then the solutions $y(\cdot), k(\cdot)$ of the closed loop system:*

$$\dot{x}(t) = Ax(t) + bu(t) + d(t) \tag{3.4a}$$
$$y(t) = cx(t) \tag{3.4b}$$
$$\dot{x}_r(t) = A_r x_r(t) + b_r N(k)k(t)e(t) \tag{3.4c}$$
$$u(t) = [c_r, N(k)e(t)] \begin{bmatrix} x_r(t) \\ k(t) \end{bmatrix} \tag{3.4d}$$
$$\dot{k}(t) = e(t)^2 \tag{3.4e}$$

where $N(\cdot)$ is a Nussbaum type switching function, satisfy:

$$\lim_{t\to\infty} e(t) = \lim_{t\to\infty} (y(t) - r(t)) = 0$$
$$k_\infty = \lim_{t\to\infty} k(t) < \infty \text{ exists.}$$

If $p(s)$ has all roots in the (closed) left half plane with simple roots on the imaginary axis, then also

$$x_r(\cdot) \in L_\infty([0,\infty),\mathbb{R}^\ell)$$

and (3.4c)–(3.4e) is a UATC for $(\Sigma(n,1,1), \mathcal{R}_{p(s)})$.

Now consider the series connection in Fig. 2, where the subsystems (A_i, b_i, c_i) are allowed to belong to the larger class $\Sigma(n_i,1,1)$ instead of $\Sigma_+(n_i)$. Consider the local controllers

$$u_i(t) = [c_i^r, N(k_i(t))e_i(t)] \begin{bmatrix} x_i^r(t) \\ k_i(t) \end{bmatrix}$$
$$\dot{x}_i^r(t) = A_i^r x_i^r(t) + b_i^r N(k_i)k_i(t)e_i(t)$$
$$\dot{k}_i(t) = e_i(t)^2$$

for the i-th subsystems. Using Proposition 3.1, the same arguments as for Theorem 3.2 show that the local tracking errors $e_i(t)$ and gains $k_i(t)$ for the closed loop system satisfy (2.28a, c) and $x_i^r(\cdot) \in L_\infty([0,\infty),\mathbb{R}^{\ell_i})$ if the reference dynamics $\dot{\xi}_i = A_i^r \xi_i$ are stable (but not necessarily asymptotically stable!). The following simulations clearly demonstrate this behavior. The 8 systems are the same as in Fig. 3. The reference signals are $r_i(t) = \sin\left(4t + \frac{i-1}{4}\pi\right)$ and the switching function is $N(k) = k^2 \cos k$.

REFERENCES

[1] I. Bar-Kana, Adaptive Control — A Simplified Approach, in C. Leondes (ed.) Control and Dynamic Systems, Vol. 25, pp. 187–226, 1987

[2] I. Bar-Kana and H. Kaufmann, Global Stability and Performance of a Simplified Adaptive Algorithm, Int. J. Control, 42, pp. 1491–1505, 1985

[3] M. Corless, First order adaptive controllers for systems which are stabilizable via high gain output feedback, in Analysis and Control of Nonlinear Systems (ed. by Byrnes, Martin, Salks), pp. 13–16, North-Holland, 1988

[4] U. Helmke and D. Prätzel-Wolters, Stability and robustness properties of universal adaptive controllers for first order linear systems, Int. J. Control, Vol. 48 (3), pp. 1153-1182, 1988

[5] U. Helmke, D. Prätzel-Wolters and S. Schmid, Sufficient Conditions for Adaptive Stabilization and Tracking, Bericht der Arbeitsgruppe Technomathematik Nr. 38, Universität Kaiserslautern, 1989

[6] I.D. Landau, Adaptive Control — The Model Reference Approach, New York, Marcel Dekker, 1979

[7] B. Mårtensson, Adaptive stabilization, Doctoral Thesis, Lund Institute of Technology, 1986

[8] B. Mårtensson, Adaptive Stabilization without High-Gain. In C.I. Byrnes and A. Kurzansky, eds. Identification and Adaptive Control, pages 226–238, Berlin 1988, Lecture Notes in Control and Information Sciences, Springer Verlag

[9] D.E. Miller and E.J. Davison, A New Self-Tuning Controller to Solve the Servomechanism Problem, Proc. IEEE 26th Conf. on Decision and Control, 843-849, 1987

[10] A.S. Morse, An adaptive controller for globally stabilizing linear systems with unknown high-frequency gains, in Lecture Notes in Control and Information Science No. 62, Berlin, Springer Verlag, 1984

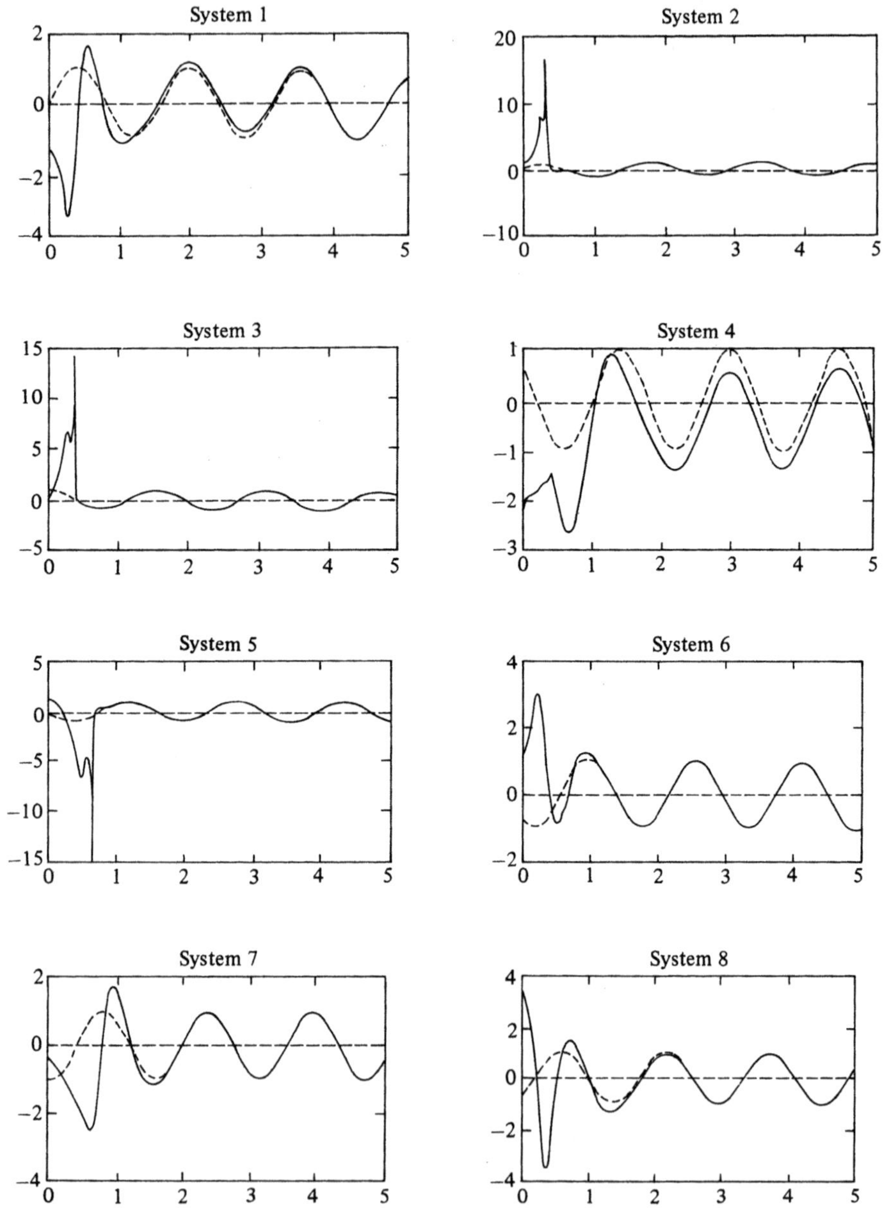

Figure 5. Series coupling: tracking of periodic signals by switching controllers

[11] A.S. Morse, Towards a unified theory of parameter adaptive control, Report No. 8807, EE, Yale University, 1988

[12] R.D. Nussbaum, Some remarks on a conjecture in parameter adaptive control, Systems and Control Letters 3 (1983), 243–246

[13] D.H. Owens, D. Prätzel-Wolters and A. Ilchmann, Positive real structure and high gain adaptive stabilization, IMA J. Math. Contr. Inf., 4, 1987

[14] H. Unbehauen and I. Vakilzadeh, Synchronization of n non-identical simple-integral plants, Int. J. Control, Vol. 50, No. 2, 543–574, 1989

[15] J.C. Willems and C.I. Byrnes, Global adaptive stabilization in the absence of information on the high frequency gain, in: Lecture Notes in Control and Information Science, No. 62 Berlin, Springer Verlag, 1984

U. Helmke
Naturwissenschaftliche Fakultät I — Mathematik —
Universität Regensburg
F. R. G.

D. Prätzel-Wolters and S. Schmid
Fachbereich Mathematik
Universität Kaiserslautern
F. R. G.

Real and Complex Stability Radii: A Survey

D. Hinrichsen A.J. Pritchard

In [14], [15] we introduced complex and real stability radii as robustness measures for stable matrices A under complex and real perturbations of the form $A \to A + D\Delta E$ where D, E are given and Δ is unknown. In this paper we give a survey of the results obtained so far. Although our main reference will be to continuous-time systems we will also present some results for the discrete-time case.

We first review the basic properties and characterizations of the stability radii emphasizing the remarkable differences between the complex and the real radius. In particular we discuss the role they play when the class of disturbances is enlarged to include time-varying, nonlinear and dynamic parameter perturbations. In a further generalization we consider arbitrary affine perturbations of the matrix A. Finally we study the problem of maximizing the complex stability radius by linear state feedback.

1 Introduction

In order to design a controller for a plant with complicated dynamic behaviour the engineer usually considers a simplified model. Then there arises the question of whether the controller designed on the basis of the model is robust enough to achieve the required performance when applied to the real system. Since a mathematical model never exactly represents the dynamics of a physical plant, the robustness issue is not only important in the context of model reduction but is a fundamental problem for the application of control theory in general. In fact, one of the basic goals of feedback control is to enhance the robustness of the system. Roughly speaking, a controller is said to be *robust* if it works well for a large class of perturbed models. If this class reflects the uncertain or neglected features of the plant then one can expect good performance of the controlled system.

In recent years robustness analysis and the problem of sensitivity reduction have received a good deal of attention. This is largely due to the success of the H^∞-approach (see [26], [6]) which proceeds from an input-output point of view and uses frequency domain techniques.

In [14], [15] we introduced a *state space approach* to robustness analysis. This was further developed in [8], [9] [10], [16], [17] and the purpose of this paper is to present an up to date survey of our published and unpublished work on the robustness of finite-dimensional time-invariant linear systems. Extension to time-varying and infinite-dimensional systems can be found in [7], [21].

We proceed as follows. In the next section we define the concept of stability radius for arbitrary stability regions and perturbation norms and discuss elementary properties. In Section 3 we illustrate the concept by relating the *unstructured* real and complex stability radii to the location of the spectrum of a matrix . Specializing to the open left half plane as the stability region and to *Euclidean* perturbation norms we then characterize the *complex* stability radius via parametrized algebraic Riccati equations and parametrized Hamiltonian matrices. We also present analogous characterizations in terms of symplectic matrix pencils for the discrete time case (Section 4). A reliable algorithm for the computation of the complex stability radius (in the continuous time case) is described.

General formulae for the *real* stability radius are analyzed in Section 5. These formulae provide insight but are difficult to evaluate. Considerable simplification occurs in the case of rank one perturbations for which computable expressions are obtained. These results can be applied to study various robustness measures for Hurwitz and Schur polynomials and establish a link between our state space analysis of robustness and the stability analysis of sets of polynomials which has received so much attention after the publication of Kharitonov's Theorem [18]. Since these applications have been summarized in [16] they will not been presented in this survey.

In Section 6 we analyze the stability of linear time-invariant state space systems when subjected to both real and complex *time-varying, nonlinear or dynamic* perturbations. It is shown that the complex stability radius does not change if the perturbation class is extended from static linear to these wider classes of perturbations, whereas the real stability radius depends on the specific perturbation class considered.

While all the above perturbations are of the *output feedback type*

$$A \to A + D\Delta E \tag{1}$$

(where D, E are given matrices), we investigate the effect of arbitrary affine perturbations of the system matrix in Section 7. More specifically we define and analyze stability radii with respect to "multi-perturbations" of the form

$$A \to A + \sum_{i=1}^{N} D_i \Delta_i E_i \tag{2}$$

where the matrices D_i, E_i are given. We indicate those aspects of the multi-perturbation theory which are natural counterparts of the single perturbation theory and also those aspects which we believe will require a new and different development.

In Section 8 we return to single affine complex perturbations and study the problem of maximizing the complex stability radius by linear state feedback. We show that the supreme achievable stability radius can be characterized via parametrized Riccati equations. The dependency on the parameters is investigated and the limiting behaviour described. Finally, the relationship between the supremum of the achievable stability radii and the distance of the system from non-stabilizability is discussed.

2 Definitions and basic properties

We consider linear time-invariant finite dimensional models of the form

$$\dot{x}(t) = Ax(t),\ t \in \mathbb{R}_+ \qquad \text{or} \qquad x(t+1) = Ax(t),\ t \in \mathbb{N} \tag{3}$$

which may be thought of as nominal feedback systems which are subjected to parameter perturbations. It is assumed that the required performance can be expressed in terms of pole locations. Let

$$\mathbb{C} = \mathbb{C}_g \,\dot{\cup}\, \mathbb{C}_b \tag{4}$$

be any partition of the complex plane into a "good" and a "bad" region where $\mathbb{C}_g$ is always supposed to be open and hence $\mathbb{C}_b$ is closed. Since we are interested in the effects of real and complex perturbations we have to consider both fields $\mathbb{K} = \mathbb{R}$ and $\mathbb{K} = \mathbb{C}$ of real and complex numbers. Throughout the paper we assume that the nominal system matrix is C_g-*stable*, i.e. it has a spectrum in the good part of the complex plane:

$$A \in \mathbb{K}^{n \times n}, \quad \sigma(A) \subset \mathbb{C}_g. \tag{5}$$

Of particular interest are the stability regions

$$\mathbb{C}_g = \mathbb{C}_- = \{s \in \mathbb{C}; \operatorname{Re} s < 0\} \quad \text{and} \quad \mathbb{C}_g = \mathbb{C}_1 = \{s \in \mathbb{C}; |s| < 1\}.$$

In the first part of the paper we suppose that the perturbed system matrix has the form $A + D\Delta E$, where $E \in \mathbb{K}^{q \times n}$ and $D \in \mathbb{K}^{n \times \ell}$ are given matrices defining the structure of the perturbations and $\Delta \in \mathbb{K}^{\ell \times q}$ is an unknown disturbance matrix. The perturbed systems

$$\dot{x}(t) = Ax(t) + D\Delta Ex(t),\ t \in \mathbb{R}_+ \ \text{or}\ x(t+1) = Ax(t) + D\Delta Ex(t),\ t \in \mathbb{N} \tag{6}$$

may be interpreted as closed loop systems with unknown static linear output feedback (Fig. 1).

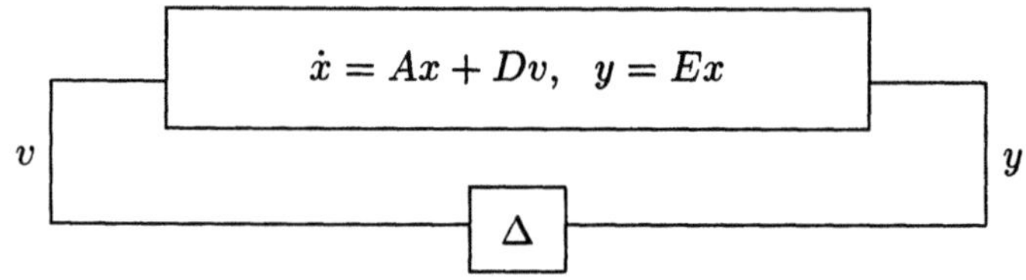

Figure 1. Feedback interpretation of the perturbed system

Note, however, that in applications D and E are chosen to reflect the structure of the model uncertainty, they do not represent input or output matrices in our context and hence controllability or observability assumptions for the pairs (A, D) or (A, E) are not appropriate. If $\|\cdot\|_{\mathbb{K}^\ell}$ and $\|\cdot\|_{\mathbb{K}^q}$ are given norms on $\mathbb{K}^\ell$ resp. $\mathbb{K}^q$ we measure the size of the perturbation matrix $\Delta \in \mathbb{K}^{\ell \times q}$ by the corresponding operator norm

$$\|\Delta\| = \max\{\|\Delta y\|_{\mathbb{K}^\ell}\,; y \in \mathbb{K}^q, \|y\|_{\mathbb{K}^q} \leq 1\}\,. \tag{7}$$

Remark 2.1. Note that not every affine perturbation of A can be represented in the form $A + D\Delta E$. In fact, if $(D\Delta E)_{ij} = 0$ for all $\Delta \in \mathbb{K}^{\ell \times q}$ then necessarily the i-th row vector of D or j-th column vector of E is zero. Hence if the (i, j)-th element of $A + D\Delta E$ does not depend on Δ then all the elements of the i-th row or j-th column remain unchanged under the perturbation. In particular it is not possible to represent (in this form) affine perturbations of A which affect exclusively the diagonal elements of A. More general multiple perturbation structures (2) which include any affine perturbation will be considered in Section 7. □

Definition 2.2. Given a partition (4) and a perturbation norm (7), the *stability radius* of $A \in \mathbb{K}^{n \times n}$ with respect to perturbations of the structure $(D, E) \in \mathbb{K}^{n \times \ell} \times \mathbb{K}^{q \times n}$ is defined by

$$r_{\mathbb{K}} = r_{\mathbb{K}}(A; D, E; \mathbb{C}_b) = \inf\left\{\|\Delta\|; \Delta \in \mathbb{K}^{\ell \times q}, \sigma(A + D\Delta E) \cap \mathbb{C}_b \neq \emptyset\right\} \tag{8}$$

If both structure matrices D, E are the identity matrix I_n (hence $q = \ell = n$) we obtain the *unstructured* stability radius

$$d_{\mathbb{K}}(A, \mathbb{C}_b) = r_{\mathbb{K}}(A; I, I; \mathbb{C}_b) = \min\left\{\|\Delta\|; \Delta \in \mathbb{K}^{n \times n}, \sigma(A + \Delta) \cap \mathbb{C}_b \neq \emptyset\right\} \tag{9}$$

$d_{\mathbb{K}}(A, \mathbb{C}_b)$ is the distance, within the normed space $(\mathbb{K}^{n\times n}, \|\cdot\|)$, between A and the set of $\mathbb{C}_g$-unstable matrices in $\mathbb{K}^{n\times n}$

$$\mathcal{U}_n(\mathbb{K}; \mathbb{C}_b) = \{X \in \mathbb{K}^{n\times n};\ \sigma(X) \cap \mathbb{C}_b \neq \emptyset\}, \tag{10}$$

see Fig. 2.

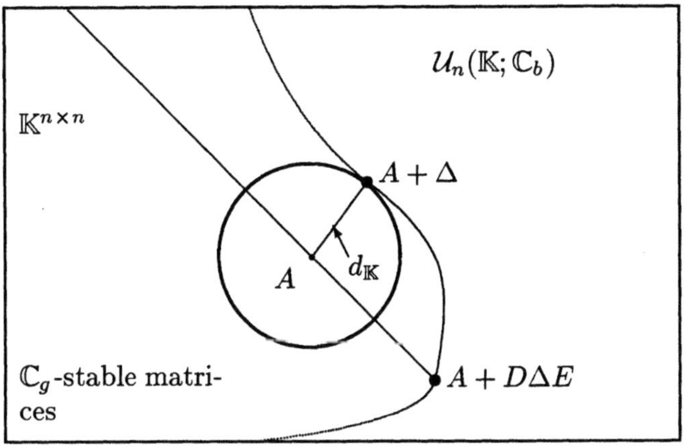

Figure 2. Unstructured and structured stability radius

If A, D, E are real, we obtain *two* stability radii, $r_{\mathbb{R}}$ or $r_{\mathbb{C}}$, according to whether real ($\mathbb{K} = \mathbb{R}$) or complex ($\mathbb{K} = \mathbb{C}$) perturbations are considered in (8). They are called, respectively, the *real* and the *complex* stability radii of A, and clearly

$$r_{\mathbb{R}}(A; D, E; \mathbb{C}_b) \geq r_{\mathbb{C}}(A; D, E; \mathbb{C}_b) \geq 0 \tag{11}$$

By definition, the infimum of the empty set is infinite so that $r_{\mathbb{K}} = r_{\mathbb{K}}(A; D, E; \mathbb{C}_b) = \infty$ iff there does not exist $\Delta \in \mathbb{K}^{\ell\times q}$ with $\sigma(A + D\Delta E) \cap \mathbb{C}_b \neq \emptyset$. On the other hand if $r_{\mathbb{K}} < \infty$ an easy compactness argument shows that there exists a *minimum norm destabilizing perturbation* Δ with $\|\Delta\| = r_{\mathbb{K}}$. So in this case we may replace the *infimum* in (8) by a *minimum*.

It is immediate from Definition 2.2 that the stability radius $r_{\mathbb{K}}$ is invariant under similarity transformations

$$r_{\mathbb{K}}(A; D, E; \mathbb{C}_b) = r_{\mathbb{K}}(TAT^{-1}; TD, ET^{-1}; \mathbb{C}_b) \tag{12}$$

Moreover if $\sigma(A+D\Delta E) \cap \operatorname{int} \mathbb{C}_b \neq \emptyset$ where $\operatorname{int} \mathbb{C}_b$ denotes the set of interior points of $\mathbb{C}_b$, then by continuity the same is true for all matrices in a small neighbourhood of $\Delta \in \mathbb{K}^{\ell \times q}$. Hence, if A is $\mathbb{C}_g$-stable

$$r_{\mathbb{K}}(A; D, E; \mathbb{C}_b) = r_{\mathbb{K}}(A; D, E; \partial\mathbb{C}_b) \tag{13}$$

where $\partial\mathbb{C}_b = \bar{\mathbb{C}}_g \cap \mathbb{C}_b$ is the boundary of $\mathbb{C}_b$.

An important role will be played by the transfer matrix associated with the triple (A, D, E)

$$G(s) = E(sI - A)^{-1}D \tag{14}$$

In fact we have

Proposition 2.3. *$\Delta \in \mathbb{K}^{\ell \times q}$ is destabilizing iff there exist $s_0 \in \mathbb{C}_b$ and a non-zero $y \in \mathbb{C}^q$ such that*

$$y = G(s_0)\Delta y. \tag{15}$$

PROOF: For any $\Delta \in \mathbb{K}^{\ell \times q}$, $\sigma(A + D\Delta E) \cap \mathbb{C}_b \neq \emptyset$ iff there exists $x \in \mathbb{C}^n$, $x \neq 0$ and $s_0 \in \mathbb{C}_b$ such that

$$(A + D\Delta E)x = s_0 x. \tag{16}$$

Since $\sigma(A) \subset \mathbb{C}_g$, $(s_0 I - A)$ is invertible so that (16) is equivalent to

$$x = (s_0 I - A)^{-1} D\Delta E x \tag{17}$$

where $Ex \neq 0$. Multiplying this equation with E from the left and setting $y = Ex$ the necessity of (15) follows. Conversely, if (15) holds we obtain (17) for $x = (s_0 I - A)^{-1} D\Delta y$, hence Δ is destabilizing. Q.E.D.

The above characterization of destabilizing perturbations works for both the real and the complex case. It immediately implies (see (13)) that

$$r_{\mathbb{K}}(A; D, E; \mathbb{C}_b) \geq \left[\max_{s \in \partial\mathbb{C}_b} \|G(s)\|\right]^{-1}. \tag{18}$$

In particular, we see that $r_{\mathbb{C}}(A; D, E; \mathbb{C}_b) = r_{\mathbb{R}}(A; D, E; \mathbb{C}_b) = \infty$ if $G(s) \equiv 0$.

We conclude this subsection with some remarks concerning the dependency of $r_{\mathbb{K}}(A; D, E; \mathbb{C}_b)$ on the data (A, D, E).

As we have seen above the unstructured stability radius $d_{\mathbb{K}}(A; \mathbb{C}_b)$ represents the distance, within the normed space $(\mathbb{K}^{n \times n}, \|\cdot\|)$, between A and the set $\mathcal{U}_n(\mathbb{K}; \mathbb{C}_b)$, see Fig. 2. This set of $\mathbb{C}_g$-unstable matrices is *independent* of (A, D, E). In contrast the structured stability radius $r_{\mathbb{K}}(A; D, E; \mathbb{C}_b)$ represents the distance, within the normed space $(\mathbb{K}^{\ell \times q}, \|\cdot\|)$, between the zero matrix $\Delta = 0$ and the *set of destabilizing perturbations* $\{\Delta \in \mathbb{K}^{\ell \times q}; \sigma(A + D\Delta E) \cap \mathbb{C}_b \neq \emptyset\}$. This set *depends* on the data (A, D, E). As a consequence one would expect differences in the behaviour of the structured and unstructured stability radii with respect to variations in the data. In fact, the *unstructured* stability radius — as a distance — satisfies the following estimate which is a consequence of the triangle inequality:

$$|d_{\mathbb{K}}(A, \mathbb{C}_b) - d_{\mathbb{K}}(\tilde{A}, \mathbb{C}_b)| \leq \|A - \tilde{A}\|. \tag{19}$$

In particular, $d_{\mathbb{K}} : A \mapsto d_{\mathbb{K}}(A, \mathbb{C}_b)$ is continuous on $\mathbb{K}^{n\times n}$. On the other hand there is no reason to expect the *structured* stability radius $r_{\mathbb{K}}(A; D, E; \mathbb{C}_b)$ to be continuous either as a function of A (with D, E fixed) or as a function of the structure matrices D, E with A fixed. Discontinuities should be expected at triples (A_0, D_0, E_0) for which the affine subspace of perturbed systems $\{A_0 + D_0\Delta E_0; \Delta \in \mathbb{K}^{\ell\times q}\}$ intersects the set $\mathcal{U}_n(\mathbb{K}; \mathbb{C}_b)$ of $\mathbb{C}_g$-unstable matrices *tangentially* at all points $A_0 + D_0\Delta_0 E_0$ where $\Delta_0 \in \mathbb{K}^{\ell\times q}$ is any minimum norm destabilizing perturbation, see Fig. 3.

Let $S_{n,\ell,q}(\mathbb{K}, \mathbb{C}_b)$ denote the set of $\mathbb{C}_g$-stable triples

$$S_{n,\ell,q}(\mathbb{K}, \mathbb{C}_b) = \{(A, D, E) \in \mathbb{K}^{n\times n} \times \mathbb{K}^{n\times \ell} \times \mathbb{K}^{q\times n}; \sigma(A) \subset \mathbb{C}_g\}$$

$S_{n,\ell,q}(\mathbb{K}, \mathbb{C}_b)$ is an open subset of $\mathbb{K}^{n\times n} \times \mathbb{K}^{n\times \ell} \times \mathbb{K}^{q\times n}$ and is provided with the corresponding induced topology. It is not difficult to construct counterexamples which show that there exist points in $S_{n,\ell,q}(\mathbb{R}, \mathbb{C}_b)$ at which $r_{\mathbb{R}}(\cdot)$ is *not* continuous [11]. However, we have

Proposition 2.4. **([11])** *The map*

$$\begin{aligned} r_{\mathbb{K}}(\cdot) : S_{n,\ell,q}(\mathbb{K}, \mathbb{C}_b) &\to \mathbb{R}_+ \\ (A, D, E) &\mapsto r_{\mathbb{K}}(A, D, E; \mathbb{C}_b) \end{aligned} \tag{20}$$

is continuous if $\mathbb{K} = \mathbb{C}$. *It is semicontinuous from below if* $\mathbb{K} = \mathbb{R}$.

The following definition is needed to give a sufficient condition for the the continuity of the real stability radius at a given triple (A, D, E).

Definition 2.5. Given $(A, D, E) \in \mathcal{S}_{n,\ell,q}(\mathbb{R}, \mathbb{C}_b)$, a perturbation matrix $\Delta \in \mathbb{R}^{\ell\times q}$ is said to be *strongly destabilizing* for (A, D, E), if for every $\varepsilon > 0$ there exists $\tilde{\Delta} \in \mathbb{R}^{\ell\times q}$, $\|\tilde{\Delta} - \Delta\| < \varepsilon$ such that $\sigma(A + D\tilde{\Delta}E) \cap \operatorname{int} \mathbb{C}_b \neq \emptyset$ where $\operatorname{int} \mathbb{C}_b$ denotes the set of interior points of $\mathbb{C}_b$.

There will not exist any strongly destabilizing perturbation if $\operatorname{int} \mathbb{C}_b = \emptyset$. But for all stability regions of practical interest we have

$$\mathbb{C}_b = \overline{\operatorname{int} \mathbb{C}_b}. \tag{21}$$

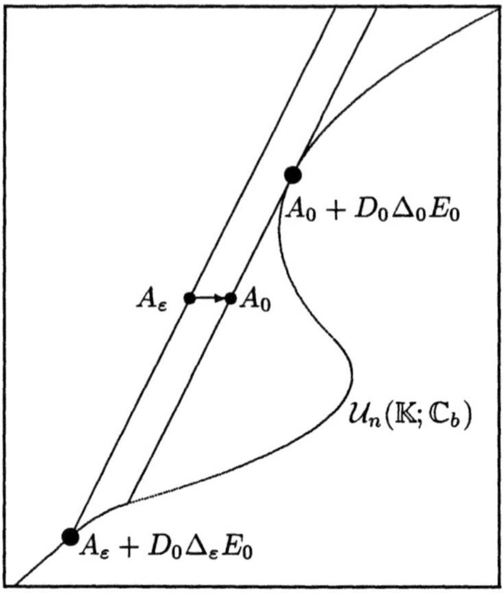

(i) Perturbed system matrix

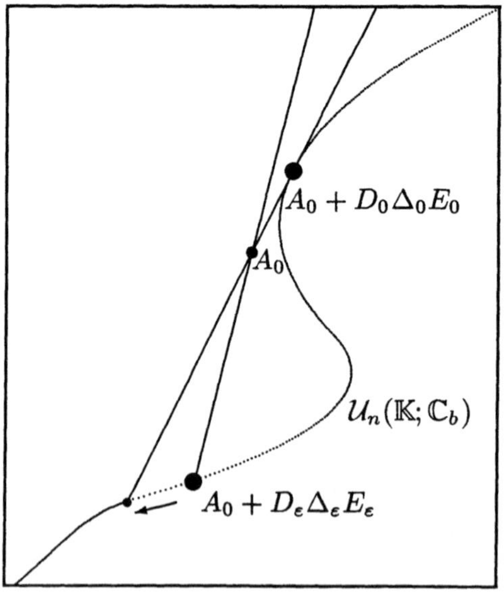

(ii) Perturbed structure matrices

Figure 3. Discontinuity of structured stability radii

Remark 2.6. An analogous definition can be given for complex perturbations. But since the roots $\lambda_i(A + D\Delta E)$ are algebraic functions of the entries of $\Delta \in \mathbb{C}^{\ell \times q}$ they are either constant or vary over an open subset of $\mathbb{C}$ if Δ varies over an open subset of $\mathbb{C}^{\ell \times q}$. *Hence, under condition (21), every destabilizing complex perturbation matrix automatically is strongly destabilizing.* □

This remark explains the continuity of $r_{\mathbb{C}}(\cdot)$ (Prop. 2.4) in view of the following

Proposition 2.7. **([11])**
$r_{\mathbb{R}}(\cdot)$ *is continuous at every triple* $(A_0, D_0, E_0) \in \mathcal{S}_{n,\ell,q}(\mathbb{R}, \mathbb{C}_b)$ *for which there exists a strongly destabilizing perturbation* $\Delta \in \mathbb{R}^{\ell \times q}$ *of minimum norm* $\|\Delta\| = r_{\mathbb{R}}(A_0; D_0, E_0; \mathbb{C}_b)$.

It may seem paradoxical that the structured real stability radius, introduced to measure robustness, is highly sensitive to perturbations of the data. However discontinuities can only occur when either the structure matrices D, E themselves are perturbed or the matrix A is perturbed in a way which is incompatible with the given structure. If the perturbation structure is *not* known *exactly* then the uncertainty is really *unstructured* and the *unstructured stability radius should be used.*

3 Unstructured stability radii

In this section we study the unstructured stability radius $d_{\mathbb{K}}(A, \mathbb{C}_b) = r_{\mathbb{K}}(A; I, I; \mathbb{C}_b)$ when $\mathbb{K}^{n \times n}$ is provided with the spectral norm (i.e. the operator norm with respect to the Hilbert norm on $\mathbb{K}^n$). Of particular interest are

$$d_{\mathbb{K}}^{-}(A) = d_{\mathbb{K}}(A, \mathbb{C} \setminus \mathbb{C}_{-}) \quad \text{and} \quad d_{\mathbb{K}}^{1}(A) = d_{\mathbb{K}}(A, \mathbb{C} \setminus \mathbb{C}_{1}) \tag{22}$$

which measure the distances of the systems (3) from instability. Since the smallest singular value of A, denoted by $s_{\min}(A)$, is equal to the distance of A from the set of singular matrices we have for any $A \in \mathbb{R}^{n \times n}$

$$d_{\mathbb{C}}^{-}(A) \leq d_{\mathbb{R}}^{-}(A) \leq s_{\min}(A). \tag{23}$$

By a similar argument, for any real A,

$$d_{\mathbb{C}}^{1}(A) \leq d_{\mathbb{R}}^{1}(A) \leq \min\left\{s_{\min}(I - A), s_{\min}(I + A)\right\}. \tag{24}$$

The smallest singular value of a matrix is easier to compute than its complex stability radius and since the complex stability radius in turn is much more easily determined than the real stability radius, it is natural to ask when equalities hold in (23) and (24). To answer this question it is useful to investigate the effect of similarity transformations on the unstructured stability radii.

It follows directly from the definition that *unitary* (resp. *orthogonal*) similarity transformations do not change the radii:

$$d_{\mathbb{K}}(A, \mathbb{C}_b) = d_{\mathbb{K}}(UAU^*, \mathbb{C}_b) \qquad \text{if} \quad UU^* = U^*U = I_n. \tag{25}$$

This[1] does *not* hold for *arbitrary* similarity transformations (see Prop. 3.8).

As a consequence of (25) one can show that the real and complex unstructured stability radii of a *normal* matrix coincide. They are both equal to the distance of the spectrum of A from the imaginary axis [14].

Proposition 3.1. *If A is a $\mathbb{C}_-$-stable real matrix and $AA^T = A^TA$ then*

$$d_{\mathbb{R}}^-(A) = d_{\mathbb{C}}^-(A) = \operatorname{dist}(\sigma(A), i\mathbb{R}). \tag{26}$$

The discrete-time counterpart is more difficult to prove, see [17].

Proposition 3.2. *If $A \in \mathbb{R}^{n\times n}$ is normal and $\mathbb{C}_1$-stable with spectral radius*

$$\rho(A) = \max_{\lambda\in\sigma(A)} |\lambda|$$

then

$$d_{\mathbb{R}}^1(A) = d_{\mathbb{C}}^1(A) = \operatorname{dist}(\sigma(A), \partial\mathbb{C}_1) = 1 - \rho(A) \tag{27}$$

As an immediate consequence of Props. 3.1, 3.2 we note:

Corollary 3.3. *If A is real, symmetric and $\mathbb{C}_-$-stable (resp. $\mathbb{C}_1$-stable), then equalities hold in (23) (resp. (24)).*

Since the real stability radius seems to be of prime importance in the robustness analysis of real systems (see, however, Section 6), one would like to know how conservative are — in general — the lower and upper bounds for $d_{\mathbb{R}}^-$, $d_{\mathbb{R}}^1$ given by the LHS and RHS of (23) and (24). Unfortunately they both can be very bad as we now illustrate by considering $d_{\mathbb{R}}^-(A)$, $d_{\mathbb{C}}^-(A)$ for $A \in \mathbb{R}^{2\times 2}$, see [9].

Proposition 3.4. *Let $A \in \mathbb{R}^{2\times 2}$ be $\mathbb{C}_-$-stable, then*

$$d_{\mathbb{R}}^-(A) = \min\{s_{\min}(A), \operatorname{dist}(\sigma(A), i\mathbb{R})\} = \min\{s_{\min}(A), |\operatorname{trace} A|/2\}$$

where $s_{\min}(A)$ is the smallest singular value of A.

---

[1]Note that in equation (12) the structure matrices D and E are transformed as well, whereas they are fixed to I_n in (25).

Example 3.5. Consider the normal matrices $A_\omega = \begin{bmatrix} -1 & \omega \\ -\omega & -1 \end{bmatrix}$. Then $s_{\min}(A_\omega) = \sqrt{1+\omega^2}$ and $d_{\mathbb{R}}^-(A_\omega) = d_{\mathbb{C}}^-(A_\omega) = 1$ for all $\omega \in \mathbb{R}$. In particular we see that the quotient $s_{\min}(A_\omega)/d_{\mathbb{R}}^-(A_\omega)$ is unbounded as $\omega \to \infty$. □

An analogous result holds true for the lower bound in (23).

Proposition 3.6. *For every $\gamma \in (0,1]$ there exists a $\mathbb{C}_-$-stable matrix $A \in \mathbb{R}^{2\times 2}$ such that $d_{\mathbb{C}}^-(A)/d_{\mathbb{R}}^-(A) = \gamma$.*

For a proof, see [9].

In the remaining part of this section we investigate the relationship between the spectrum of a system matrix and its robustness. Although Props. 3.1, 3.2 are satisfying results, they are restricted to normal matrices and normality is a non-generic property of matrices in $\mathbb{R}^{n\times n}$. For non-normal matrices $\operatorname{dist}(\sigma(A), i\mathbb{R})$ (or $\operatorname{dist}(\sigma(A), \partial\mathbb{C}_1)$) may be very poor indicators of robustness as the following example shows.

Example 3.7. Consider the $\mathbb{C}_-$-stable matrices

$$A_k = \begin{bmatrix} -k & k^3 \\ 0 & -k \end{bmatrix}, \quad k \in \mathbb{N}$$

If $D_k = \begin{bmatrix} 0 & 0 \\ k^{-1} & 0 \end{bmatrix}$, then $A_k + D_k$ is singular and hence $d_{\mathbb{R}}^-(A_k) \le \|D_k\| = k^{-1} \to 0$ as $k \to \infty$. On the other hand $\operatorname{dist}(\sigma(A_k), i\mathbb{R}) \to \infty$ as $k \to \infty$. □

This example illustrates a general fact. Each similarity orbit which is not reduced to a single matrix comes arbitrarily close to the set of unstable matrices [14].

Proposition 3.8. *Suppose that $A \in \mathbb{R}^{n\times n}$ is $\mathbb{C}_-$-stable and is not a multiple of the identity matrix. Then, for any $\varepsilon > 0$, there exist matrices $A_i = T_i A T_i^{-1}$, $i = 1,2$ similar to A such that $d_{\mathbb{R}}^-(A_1) < \varepsilon$ and $d_{\mathbb{R}}^-(A_2) > \operatorname{dist}(\sigma(A), i\mathbb{R}) - \varepsilon$.*

A discrete-time counterpart of this result can be found in [17].

Remark 3.9. It is easily seen that for a $\mathbb{C}_-$-stable matrix A

$$d_{\mathbb{C}}^-(A) \le d_{\mathbb{R}}^-(A) \le \operatorname{dist}(\sigma(A), i\mathbb{R}). \tag{28}$$

An analogous result also holds in the discrete time case but is more difficult to prove. A proof that

$$d_{\mathbb{C}}^1(A) \le \operatorname{dist}(\sigma(A), \partial\mathbb{C}_1) \tag{29}$$

is given [17]. □

In terms of realization theory, Prop. 3.8 together with Prop. 3.1 and (28) can be interpreted as follows. Let $H(s) \in \mathbb{R}(s)^{p\times m}$ be any given strictly proper transfer matrix with McMillan degree n and poles $\lambda_j \in \mathbb{C}_-$, $j = 1,\ldots,n$. Then there exist realizations (A, B, C) of $H(s)$ whose system matrices come arbitrarily close to instability. On the other hand there is no realization of $H(s)$ of larger distance from instability than $\min_{j=1,\ldots,n} |\operatorname{Re}\lambda_j|$ and this distance is achieved by any realization (A, B, C) with *normal* system matrix A (if it exists).

4 Complex stability radius

In this section we assume $\mathbb{K} = \mathbb{C}$ and $\sigma(A) \subset \mathbb{C}_g$. The following formula for the complex stability radius holds for arbitrary (open) stability regions $\mathbb{C}_g$ and arbitrary operator norms on $\mathbb{K}^{\ell\times q}$ [10], [15].

Proposition 4.1. *If $G(s) = E(sI - A)^{-1}D$ is the transfer matrix associated with (A, D, E) and A is $\mathbb{C}_g$-stable, then*

$$r_{\mathbb{C}}(A; D, E; \mathbb{C}_b) = \left[\max_{s\in\partial\mathbb{C}_b} \|G(s)\|\right]^{-1} \tag{30}$$

where $\|G(s)\|$ denotes the operator norm of $G(s) \in \mathbb{C}^{q\times\ell}$ and, by definition, $0^{-1} = \infty$.

The proposition shows that for the *complex* stability radius the inequality (18) is in effect an equality.

Remark 4.2. The stability radius $r_{\mathbb{C}}$ determines the magnitude of the smallest perturbation which destroys the $\mathbb{C}_g$-stability of the system. In applications it may be of interest to know not only the *critical magnitude* but also the *critical directions* of system perturbations. If $\|G(s_0)\| = \max_{s\in\partial\mathbb{C}_b} \|G(s)\| \not\equiv 0$, choose $v \in \mathbb{C}^\ell, \|v\|_{\mathbb{C}^\ell} = 1$ so that $\|G(s_0)v\| = \|G(s_0)\|$. By Hahn-Banach, there exists a linear form $y^* \in (\mathbb{C}^q)^*$ (the dual space of $\mathbb{C}^q$) of dual norm $\|y^*\|_{(\mathbb{C}^q)^*} = 1$ which is aligned with $G(s_0)v$ so that

$$y^*G(s_0)v = \|G(s_0)v\|_{\mathbb{C}^q} \tag{31}$$

It is easy to check that Δ defined by

$$\Delta = r_{\mathbb{C}}\, vy^* \tag{32}$$

has the norm $\|\Delta\| = r_{\mathbb{C}}$, satisfies (15) with $y = r_{\mathbb{C}}G(s_0)v$ and hence is a minimum norm destabilizing perturbation. If $\mathbb{C}^\ell$, $\mathbb{C}^q$ are provided with their usual Hilbert norms, one can choose $y^* = r_{\mathbb{C}}(G(s_0)v)^*$ so that

$$\Delta = r_{\mathbb{C}}^2\, vv^*G(s_0)^*. \tag{33}$$

Note that, even if A, D, E are all real, Δ will in general be complex and it will not be possible to find a destabilizing real perturbation of the same norm. If, however, $G(s_0)$ is real, then Δ may be chosen to be real so that $r_{\mathbb{C}} = r_{\mathbb{R}}$. □

For the rest of this section we restrict our considerations to the case where $\mathbb{C}_g = \mathbb{C}_-$ or $\mathbb{C}_g = \mathbb{C}_1$ and $\mathbb{C}^\ell$, $\mathbb{C}^q$ are provided with their usual Hilbert norms. Then formula (30) specializes to

$$r_{\mathbb{C}}^- = \left[\max_{\omega\in\mathbb{R}} \|G(\imath\omega)\|\right]^{-1} \quad , \quad r_{\mathbb{C}}^1 = \left[\max_{\theta\in[0,2\pi]} \|G(e^{\imath\theta})\|\right]^{-1} \tag{34}$$

In particular, we obtain the following formulae for the unstructured complex stability radii

$$d_{\mathbb{C}}^-(A) = \min_{\omega\in\mathbb{R}} s_{\min}(\imath\omega I - A) \quad , \quad d_{\mathbb{C}}^1(A) = \min_{\theta\in[0,2\pi]} s_{\min}(e^{\imath\theta} I - A) \tag{35}$$

As another consequence of Prop. 4.1 we note the following remarkable property of the complex stability radius $r_{\mathbb{C}}$ which is valid for arbitrary stability domains $\mathbb{C}_g$:

Corollary 4.3. (Decomposition property) *Let $A_1 \in \mathbb{C}^{n_1\times n_1}$ and $A_2 \in \mathbb{C}^{n_2\times n_2}$ be $\mathbb{C}_g$-stable matrices and let*

$$A = \begin{bmatrix} A_1 & 0 \\ 0 & A_2 \end{bmatrix}, \quad D = \begin{bmatrix} D_1 & 0 \\ 0 & D_2 \end{bmatrix}, \quad E = \begin{bmatrix} E_1 & 0 \\ 0 & E_2 \end{bmatrix}.$$

where D_i and E_i are of compatible formats. Then

$$r_{\mathbb{C}}(A; D, E; \mathbb{C}_b) = \min\{r_{\mathbb{C}}(A_1; D_1, E_1; \mathbb{C}_b), r_{\mathbb{C}}(A_2; D_2, E_2; \mathbb{C}_b)\} \tag{36}$$

In particular this corollary says that two $\mathbb{C}_g$-stable state space systems (A_1, D_1, E_1) and (A_2, D_2, E_2) cannot be destabilized by coupling them together with couplings of norm strictly smaller than both their *complex* stability radii. An analogous result does *not* hold for the *real* stability radius. Here we have only the trivial inequality:

$$r_{\mathbb{R}}(A; D, E; \mathbb{C}_b) \le \min\{r_{\mathbb{R}}(A_1; D_1, E_1; \mathbb{C}_b), r_{\mathbb{R}}(A_2; D_2, E_2; \mathbb{C}_b)\} \tag{37}$$

In fact, counterexamples against equality in (37) are easily constructed by making use of the following relationship between the complex and the real stability radius.

Proposition 4.4. ([11]) *For any real triple $(A, D, E) \in \mathcal{S}_{n,\ell,q}(\mathbb{R}, \mathbb{C}_b)$*

$$r_{\mathbb{R}}(A \oplus A; D \oplus D, E \oplus E; \mathbb{C}_b) = r_{\mathbb{C}}(A, D, E; \mathbb{C}_b) \tag{38}$$

Applying (38) and Proposition 3.6 it follows that, for any $\alpha > 0$, however large, and every $\varepsilon > 0$, however small, there exist stable systems (A_i, D_i, E_i) with $r_{\mathbb{R}}(A_i, D_i, E_i; \mathbb{C}_+) > \alpha$, $i = 1, 2$ which can be destabilized by interconnecting them with coupling gains of norms smaller than ε (see [11]). This fact indicates that $r_{\mathbb{C}}$ is more appropriate than $r_{\mathbb{R}}$ as a robustness measure in the context of decentralized control.

If $\mathbb{C}_g = \mathbb{C}_-$ or $\mathbb{C}_g = \mathbb{C}_1$, a time-domain characterization of the complex stability radius is obtained from Prop. 4.1 by considering the *input-output operator* associated with the continuous-time or discrete-time system represented by the triple (A, D, E) [15], [17]:

$$L : L^2(0,\infty;\mathbb{C}^\ell) \longrightarrow L^2(0,\infty;\mathbb{C}^q), \quad (Lv(\cdot))(t) = \int_0^t Ee^{A(t-s)}Dv(s)\,ds \tag{39}$$

$$L : l^2(\mathrm{N},\mathbb{C}^\ell) \longrightarrow l^2(\mathrm{N},\mathbb{C}^q), \quad (Lv(\cdot))(t) = \sum_{k=0}^{t-1} EA^{t-1-k}Dv(k) \tag{40}$$

Corollary 4.5. *If A is $\mathbb{C}_-$-stable (resp. $\mathbb{C}_1$-stable) then*

$$r_{\mathbb{C}}^-(A; D, E) = 1/\|L\| \qquad (\textit{resp. } r_{\mathbb{C}}^1(A, D, E) = 1/\|L\|) \tag{41}$$

where L is defined by (39) (resp. (40)).

This time domain interpretation of the complex stability radius is the starting point for various characterizations of $r_{\mathbb{C}}$ which can be used to compute $r_{\mathbb{C}}$. We first consider the continuous-time case.

If

$$J_\rho(x_0, v) = \int_0^\infty \left[\|v(t)\|^2 - \rho^2\|y(t)\|^2\right] dt \tag{42}$$

where

$$\begin{aligned} \dot{x}(t) &= Ax(t) + Dv(t), \quad t \in \mathbb{R}_+, \quad x(0) = x_0 \\ y(t) &= Ex(t) \end{aligned} \tag{43}$$

an alternative way of formulating Cor. 4.5 is that $J_\rho(0, v) \geq 0$ for all $v \in L^2$ iff $\rho \leq r_{\mathbb{C}}^-$. The Riccati equation associated with the problem of minimizing $J_\rho(x_0, v)$ subject to (43) is

$$A^*X + XA - \rho^2 E^*E - XDD^*X = 0 \tag{ARE$_\rho$}$$

Extending the work of Brockett [1] and Willems [23], the complex stability radius $r_{\mathbb{C}}^-$ can be characterized via the existence of Hermitian solutions of the parametrized algebraic Riccati equation (ARE$_\rho$).

Let $\mathcal{H}_n \subset \mathbb{C}^{n\times n}$ denote the real vector space of all Hermitian $n \times n$ matrices and $\mathcal{H}_n{}^+$ (resp. $\mathcal{H}_n{}^-$) the convex cones of positive semi-definite (resp. negative semi-definite) matrices in $\mathcal{H}_n$. We write $P \preceq Q$ if $Q - P \in \mathcal{H}_n{}^+$.

Theorem 4.6. *Suppose* $\sigma(A) \subset \mathbb{C}_-$, $\rho \geq 0$. *Then*

(i) *If* $\rho \in [0, r_{\mathbb{C}}^-(A; D, E))$, *there exists a solution* $P_\rho \in \mathcal{H}_n^-$ *of* (ARE_ρ) *such that* P_ρ *is stabilizing, i.e.* $\sigma(A - DD^*P_\rho) \subset \mathbb{C}_-$. P_ρ *is unique among all Hermitian solutions with this property.*

(ii) *If* $\rho = r_{\mathbb{C}}^-(A; D, E) < \infty$ *then there exists a solution* $P_\rho \in \mathcal{H}_n^-$ *of* (ARE_ρ) *satisfying* $\sigma(A - DD^*P_\rho) \subset \overline{\mathbb{C}}_-$ *and* $\sigma(A - DD^*P_\rho) \cap i\mathbb{R} \neq \emptyset$.

(iii) *If* $\rho > r_{\mathbb{C}}^-(A; D, E)$, *there does not exist a Hermitian solution of* (ARE_ρ).

If (A, E) *is observable then* $P_\rho \prec 0$ *for all* $\rho \in (0, r_{\mathbb{C}}^-]$.

The above theorem can be used to construct a Liapunov function of "maximal robustness". Indeed if P is any solution of (ARE_ρ) with $\rho = r_{\mathbb{C}}^-$ then it can be shown (see [15]) that $V(x) = -\langle x, Px\rangle$ is a joint Liapunov function for the set of systems

$$\Sigma_\Delta : \qquad \dot{x} = (A + D\Delta E)x, \quad \|\Delta\| < r_{\mathbb{C}}^-.$$

Clearly there does not exist a joint Liapunov function for a larger ball of systems Σ_Δ, $\|\Delta\| < \alpha$ where $\alpha > r_{\mathbb{C}}^-$.

This Liapunov function can also be used to consider nonlinear perturbations, see Section 6.

The parametrized Hamiltonian matrix associated with (ARE_ρ) is

$$H_\rho = \begin{bmatrix} A & -DD^* \\ \rho^2 E^*E & -A^* \end{bmatrix} \tag{44}$$

A reliable algorithm (`stabrad`) which computes $r_{\mathbb{C}}^-$ is based on the following characterization of $r_{\mathbb{C}}^-$ in terms of H_ρ [8].

Proposition 4.7. *If* H_ρ *is defined by (44) then*

$$\rho < r_{\mathbb{C}}^-(A; D, E) \quad \text{iff} \quad \sigma(H_\rho) \cap i\mathbb{R} = \emptyset. \tag{45}$$

Moreover,

$$i\omega_0 \in \sigma\left(H_{r_{\mathbb{C}}^-}\right) \quad \text{iff} \quad \|G(i\omega_0)\| = \max_{\omega \in \mathbb{R}} \|G(i\omega)\|. \tag{46}$$

The proposition is illustrated in Fig. 4.

The idea of `stabrad` is to determine $r_{\mathbb{C}}^-$ as that value of ρ for which the spectrum of H_ρ hits the imaginary axis for the first time. Starting from some lower and upper estimates

$$\rho_0^- < r_{\mathbb{C}}^- < \rho_0^+$$

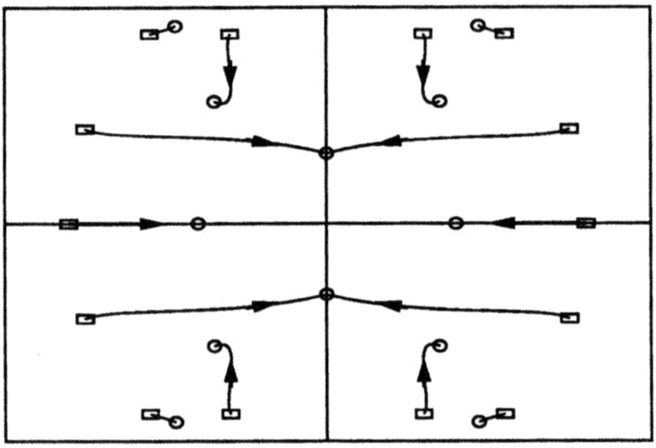

Figure 4. Eigenloci of H_ρ as $\rho \to r_{\mathbb{C}}^-$

successively better estimates ρ_k^-, ρ_k^+, $k = 0, 1, 2, \cdots$ are obtained by the following steps:
Take ρ_k such that $\rho_k^- < \rho_k < \rho_k^+$.
If H_{ρ_k} has eigenvalues on the imaginary axis set $\rho_{k+1}^- := \rho_k^-$ and $\rho_{k+1}^+ := \rho_k$.
Otherwise set $\rho_{k+1}^- := \rho_k$ and $\rho_{k+1}^+ := \rho_k^+$.

In `stabrad` I the strategy for the choice of ρ_k is that of bisection $\rho_k = (\rho_k^- + \rho_k^+)/2$. An acceleration can be achieved by using an extrapolation technique (`stabrad` II) which utilizes the fact that the graph of the function $\rho^2 \to \operatorname{dist}(\sigma(H_\rho), i\mathbb{R})$ has in most examples an almost parabolic shape, see [8] for details.

We conclude this section with an outline of the corresponding results for the discrete-time case which shows some special features. These features are mainly due to the fact that a difference system (7) may have nontrivial trajectories which end at the origin in finite time. To incorporate these discrete-time systems (with *singular* matrix A) one has to consider *symplectic pencils* instead of symplectic matrices (corresponding to the Hamiltonian matrices in the differentiable case), see [19].

With any real triple (A, D, E), $\sigma(A) \subset \mathbb{C}_1$ we associate the family of matrix pencils

$$W_\rho(\lambda) = M_\rho - \lambda N, \quad \lambda \in \mathbb{C} \quad (\rho \geq 0) \tag{47}$$

where

$$M_\rho = \begin{bmatrix} A & 0 \\ \rho^2 E^T E & I \end{bmatrix}, \quad N = \begin{bmatrix} I & DD^T \\ 0 & A^T \end{bmatrix}. \tag{48}$$

Given $\rho \geq 0$, a number $\lambda \in \mathbb{C}$ is called a *characteristic value* of the pencil W_ρ if $\det(M_\rho - \lambda N) = 0$. For each $\rho \geq 0$, the set of all characteristic values of the pencil W_ρ is denoted by $\sigma(W_\rho)$ and is called the *spectrum* of the pencil.

The following symmetry property of $\sigma(W_\rho)$ is well known: For every $\lambda \in \mathbb{C}$, $\lambda \neq 0$:

$$\lambda \in \sigma(W_\rho) \Longleftrightarrow 1/\bar{\lambda} \in \sigma(W_\rho) \tag{49}$$

Since

$$W_\rho(\lambda) = \begin{bmatrix} A - \lambda I & -\lambda DD^T \\ \rho^2 E^T E & I - \lambda A^T \end{bmatrix} \tag{50}$$

we see that for all $\rho \geq 0$

$$0 \in \sigma(W_\rho) \Longleftrightarrow 0 \in \sigma(A)$$

W_ρ is said to be *singular* at $\rho_0 \geq 0$ if the polynomial

$$p(\rho, \lambda) = \det(M_\rho - \lambda N) \tag{51}$$

is the zero polynomial (in λ) for $\rho = \rho_0$, otherwise it is said to be *regular* at ρ_0. If W_ρ is regular at ρ_0 then W_{ρ_0} has at most $2n$ characteristic values. W_ρ is singular at ρ_0 iff $\sigma(W_{\rho_0}) = \mathbb{C}$.

By (50), W_ρ is regular for sufficiently small $\rho \geq 0$. For every $\lambda \in \mathbb{C}$, $p(\rho, \lambda)$ is a nontrivial polynomial of degree $\leq n$ in ρ^2 so that W_ρ is singular for at most n positive values of the parameter ρ.

The following proposition is the discrete-time counterpart of Prop. 4.7 [17].

Proposition 4.8. *For any $\theta_0 \in [0, 2\pi]$ and any $\rho \in (0, \|G(e^{\imath\theta_0})\|^{-1}]$,*

$$\rho < r_{\mathbb{C}}^1 \Longleftrightarrow \sigma(W_\rho) \cap \partial\mathbb{C}_1 = \emptyset. \tag{52}$$

In particular, if $r_{\mathbb{C}}^1 < \infty$ then

$$r_{\mathbb{C}}^1 = \min\{\rho \in \mathbb{R}_+;\ \sigma(W_\rho) \cap \partial\mathbb{C}_1 \neq \emptyset\}. \tag{53}$$

Thus $r_{\mathbb{C}}^1$ is that value of ρ for which the spectrum of the matrix pencil W_ρ hits the unit circle for the first time as ρ increases from 0 to ∞. Again, a bisectional algorithm for the computation of $r_{\mathbb{C}}^1$ can be based on this characterization. Just how this algorithm may be accelerated is still an open problem.

Remark 4.9. Prop. 4.8 is not completely analogous to its continuous-time counterpart. In the continuous-time case, the parametrized Hamiltonian H_ρ has always at least one eigenvalue on the imaginary axis if $\rho \geq r_{\mathbb{C}}^-$. In the discrete-time case it is, however, *not* true that the symplectic pencil W_ρ always has at least one characteristic value on the unit circle if $\rho \geq r_{\mathbb{C}}^1$. In general, the set of all ρ for which $\sigma(W_\rho) \cap \partial\mathbb{C}_1 \neq \emptyset$ is a disjoint union of finitely many intervals, see Fig. 5. As a consequence the initial upper bound ρ_0^+ of the bisectional algorithm has to be chosen such that $\rho_0^+ \leq \|G(e^{\imath\theta})\|^{-1}$ for some $\theta \in [0, 2\pi]$. In terms of Fig. 5 this means that ρ_0^+ has to be chosen in the *first* interval (from the left) where the function $\rho \to \text{dist}(\sigma(W_\rho), \partial\mathbb{C}_1)$ vanishes. □

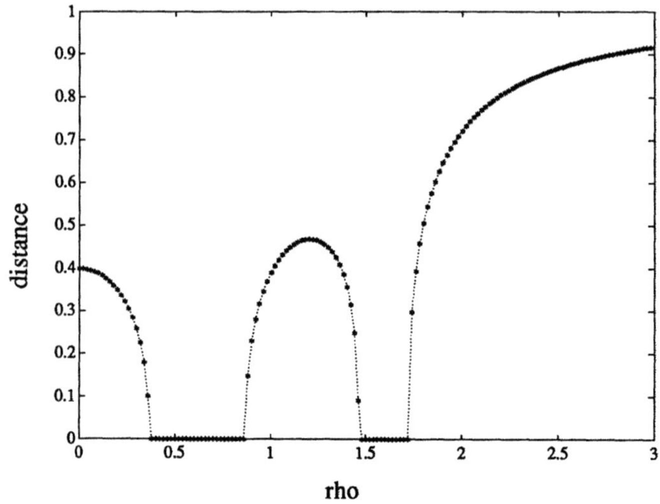

Figure 5. The distance of the spectrum of W_ρ from the unit circle as a function of ρ

5 The real stability radius

We now assume $\mathbb{K} = \mathbb{R}$ so that A, D, E are real matrices, $\sigma(A) \subset \mathbb{C}_-$. In this case we will, in general, have to consider perturbations of rank 2. Given $v_1, v_2 \in \mathbb{R}^\ell$, $y_1, y_2 \in \mathbb{R}^q$ we denote by $\mu(y_1, y_2; v_1, v_2)$ the smallest operator norm of all linear maps $\Delta : \mathbb{R}^q \to \mathbb{R}^\ell$ which take y_1, y_2 into v_1, v_2:

$$\mu = \mu(y_1, y_2; v_1, v_2) = \inf\left\{\|\Delta\|;\ \Delta \in \mathbb{R}^{\ell \times q}, \Delta y_1 = v_1 \text{ and } \Delta y_2 = v_2\right\} \tag{54}$$

Note that $\mu = \infty$ iff there is no such Δ and this will be the case iff $\alpha y_1 + \beta y_2 = 0$ and $\alpha v_1 + \beta v_2 \neq 0$ for some $\alpha, \beta \in \mathbb{R}$. If $\mu < \infty$ the infimum in (54) is attained.

The following theorem holds for arbitrary pairs of norms on $\mathbb{R}^\ell$, $\mathbb{R}^q$ and the corresponding operator norm on $\mathbb{R}^{\ell \times q}$ (see [10]).

Theorem 5.1. *If $G_R(s)$ and $G_I(s)$ are the real and imaginary parts of $G(s) = E(sI - A)^{-1}D$, then*

$$r_{\mathbb{R}}(A; D, E; \mathbb{C}_b) = \min_{\substack{s \in \partial\mathbb{C}_b \\ (v_1, v_2) \neq (0,0)}} \mu(G_R(s)v_1 - G_I(s)v_2, G_R(s)v_2 + G_I(s)v_1; v_1, v_2) \tag{55}$$

It is not difficult to show that for $s = \alpha + \imath\omega$, α, $\omega \in \mathbb{R}$

$$\begin{aligned} G_R(s) &= E[\omega^2 I + (\alpha I - A)^2]^{-1}(\alpha I - A)D, \\ G_I(s) &= -\omega E[\omega^2 I + (\alpha I - A)^2]^{-1}D \end{aligned} \tag{56}$$

If $\mathbb{R}^q$ and $\mathbb{R}^\ell$ are both provided with the usual Euclidean norm $\|\cdot\| = \|\cdot\|_2$, the function $\mu(y_1, y_2; v_1, v_2)$ can be determined explicitly. This is trivial if y_1, y_2 are linearly dependent. Otherwise one has the following

Proposition 5.2. *If y_1, $y_2 \in \mathbb{R}^q$ are linearly independent, then $\mu^2 = \mu(y_1, y_2; v_1, v_2)^2$ is the largest of the two real solutions of the quadratic equation in ρ:*

$$\begin{aligned} \rho^2 \left[\|y_1\|^2\|y_2\|^2 - \langle y_1, y_2\rangle^2\right] - \rho \left[\|y_1\|^2\|v_2\|^2 + \|y_2\|^2\|v_1\|^2\right. \\ \left. -2\langle y_1, y_2\rangle\langle v_1, v_2\rangle\right] + \|v_1\|^2\|v_2\|^2 - \langle v_1, v_2\rangle^2 = 0 \end{aligned} \tag{57}$$

The proof is based on Weierstrass' Theorem on quadratic forms and can be found in [10].

To construct a minimum norm destabilizing perturbation in the case $r_{\mathbb{R}} < \infty$ we note that by Theorem 5.1 there exist $s \in \partial\mathbb{C}_b$ and v_1, $v_2 \in \mathbb{R}^\ell$, $(v_1, v_2) \neq (0, 0)$ such that

$$y_1 = G_R(s)v_1 - G_I(s)v_2\,, \quad y_2 = G_R(s)v_2 + G_I(s)v_1$$

satisfy

$$r_{\mathbb{R}}(A; D, E; \mathbb{C}_b) = \mu(y_1, y_2; v_1, v_2).$$

It follows by straightforward calculations that (for $y_1 \neq 0$)

$$\Delta = \begin{cases} v_1 y_1^T / \|y_1\|^2 & \text{if } y_1,\ y_2 \text{ are linearly dependent} \\ [v_1\ v_2] \begin{bmatrix} \|y_1\|^2 & \langle y_1, y_2\rangle \\ \langle y_1, y_2\rangle & \|y_2\|^2 \end{bmatrix}^{-1} \begin{bmatrix} y_1^T \\ y_2^T \end{bmatrix} & \text{if } y_1, y_2 \text{ are linearly independent} \end{cases} \tag{58}$$

is a minimum norm destabilizing disturbance matrix $\Delta \in \mathbb{R}^{\ell \times q}$.

(58) illustrates that for an analysis of the real stability radius $r_{\mathbb{R}}$ it suffices to consider rank 2 matrix perturbations. Note, however, that in general there will not exist a minimum norm perturbation Δ of rank 1.

This is in contrast with the complex case and indicates why the complex stability radius is easier to analyze than the real one. In fact we have not been able to obtain a "good" algorithm for the computation of the real stability radius under general conditions. However, there is a considerable reduction in difficulty if only rank one perturbations need to be considered.

Suppose for example $\ell = 1$ and denote by $d(y, z\mathbb{R})$ the distance in the normed space $(\mathbb{R}^q, \|\cdot\|_{\mathbb{R}^q})$ of the point $y \in \mathbb{R}^q$ from the linear subspace $z\mathbb{R} = \{\alpha z; \alpha \in \mathbb{R}\}$ spanned by $z \in \mathbb{R}^q$:

$$d(y, z\mathbb{R}) = \min_{\alpha \in \mathbb{R}} \|y - \alpha z\|_{\mathbb{R}^q} \tag{59}$$

It is easy to see that in the Euclidean case ($\|\cdot\|_{\mathbb{R}^q} = \|\cdot\|_2$)

$$d^2(y, z\mathbb{R}) = \begin{cases} \|y\|_2^2 - \langle y, z\rangle^2 / \|z\|_2^2 & , \quad z \neq 0 \\ \|y\|_2^2 & , \quad z = 0 \end{cases}, \tag{60}$$

see Fig. 6.

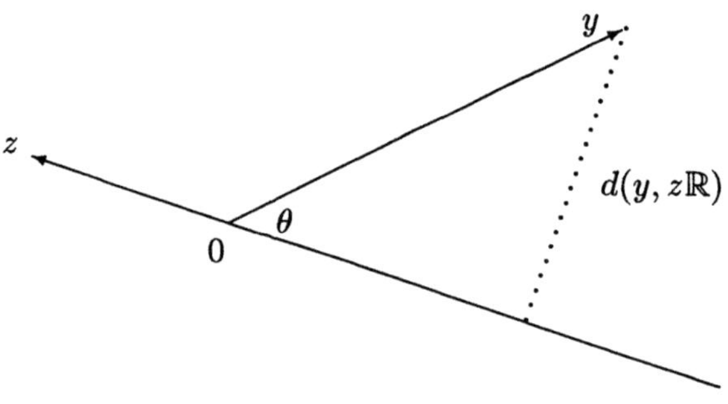

Figure 6. The distance $d(y, z\mathbb{R})$ in $(\mathbb{R}^q, \|\cdot\|_2)$ if $z \neq 0$

The following proposition holds for arbitrary norms on $\mathbb{R}^q$. Note that the disturbances Δ are *linear forms* on $\mathbb{R}^q$ if $\ell = 1$ and thus are represented by *row* vectors, i.e. $\Delta \in (\mathbb{R}^q)^* = \mathbb{R}^{1\times q}$. The corresponding operator norm is simply the dual norm of $\|\cdot\|_{\mathbb{R}^q}$ on $\mathbb{R}^{1\times q}$:

$$\|\Delta\| = \max\{|\Delta y|; y \in \mathbb{R}^q, \|y\|_{\mathbb{R}^q} = 1\}$$

Thus $\|\cdot\|_{\mathbb{R}^q}$ and $\|\cdot\|$ form a dual pair of norms on $\mathbb{R}^q = \mathbb{R}^{q\times 1}$ and $\mathbb{R}^{1\times q} = (\mathbb{R}^q)^*$, respectively.

Remark 5.3. Of particular interest are the 1, 2, ∞ norms on $\mathbb{R}^q$ and the corresponding dual norms on $\mathbb{R}^{1\times q} = (\mathbb{R}^q)^*$:

$$\|y\|_{\mathbb{R}^q} = \|y\|_\infty = \max_{i\in \underline{q}} |y_i|\,, \qquad \|\Delta\| = \|\Delta\|_1 = \sum_{i=1}^{q} |\Delta_i|$$

$$\|y\|_{\mathbb{R}^q} = \|y\|_2 = \left(\sum_{i=1}^{q} y_i^2\right)^{1/2}, \qquad \|\Delta\| = \|\Delta\|_2 = \left(\sum_{i=1}^{q} \Delta_i^2\right)^{1/2}$$

$$\|y\|_{\mathbb{R}^q} = \|y\|_1 = \sum_{i=1}^{q} |y_i|\,, \qquad \|\Delta\| = \|\Delta\|_\infty = \max_{i\in \underline{q}} |\Delta_i|$$

□

We have

Proposition 5.4. *If $\ell = 1$ then*

$$r_{\mathbb{R}}(A; D, E; \mathbb{C}_b) = \left[\max_{s\in\partial\mathbb{C}_b} d(G_R(s), G_I(s)\,\mathbb{R})\right]^{-1}. \tag{61}$$

In general the map $s \mapsto d(G_R(s), G_I(s)\,\mathbb{R})$ is *not* continuous at the zeros of $G_I(\cdot)$. The geometric reason for this can be seen from Fig. 6 by letting z tend to zero within a closed convex cone around y which does not contain orthogonal vectors to y. As a consequence, search methods cannot be applied directly in order to obtain the maximum of $s \mapsto d(G_R(s), G_I(s)\,\mathbb{R})$ on $\partial\mathbb{C}_b$. Instead one must investigate the jump points separately. Let

$$\Omega = \{s \in \partial\mathbb{C}_b; G_I(s) = 0\} \tag{62}$$

By (61) we obtain

$$r_{\mathbb{R}}(A; D, E; \mathbb{C}_b) = \min\left\{ \left[\max_{s\in\Omega} \|G_R(s)\|_{\mathbb{R}^q}\right]^{-1}, \left[\sup_{s\in\partial\mathbb{C}_b,\, s\notin\Omega} d(G_R(s), G_I(s)\,\mathbb{R})\right]^{-1} \right\} \tag{63}$$

Remark 5.5. In the case $\ell = q = 1$ the second term on the RHS of (63) is always infinite and so

$$r_{\mathbb{R}}(A; D, E; \mathbb{C}_b) = \left[\max_{s\in\Omega} |G_R(s)|\right]^{-1} \tag{64}$$

For the stability region $\mathbb{C}_g = \mathbb{C}_-$, this formula could have been easily deduced from the Nyquist criterion, see Fig. 7. □

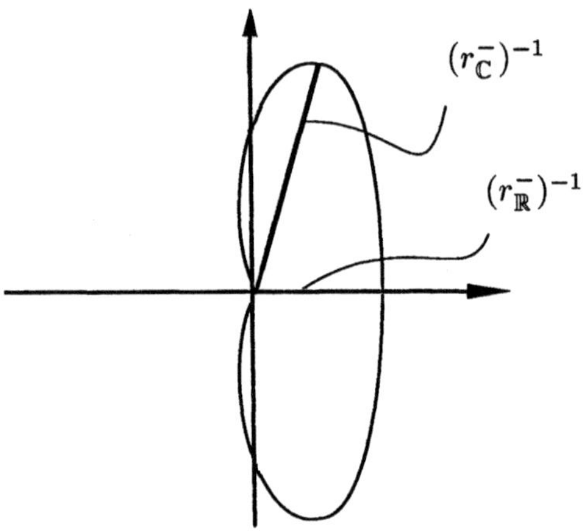

Figure 7. The real and complex stability radii in a Nyquist plot

The following example illustrates the application of formulae (64), (30) and presents a one parameter set of triples $(A_\varepsilon, D_\varepsilon, E)$ for which $r_{\mathbb{R}}^{-}(A_\varepsilon; D_\varepsilon, E) \to \infty$ and $r_{\mathbb{C}}^{-}(A_\varepsilon; D_\varepsilon, E) \to 1$ as $\varepsilon \to 0$.

Example 5.6. Let

$$A_\varepsilon = \begin{bmatrix} 0 & 1 \\ -1 & -\varepsilon \end{bmatrix}, \quad D_\varepsilon = \begin{bmatrix} 0 \\ -\varepsilon \end{bmatrix}, \quad E = \begin{bmatrix} 1 & 0 \end{bmatrix}.$$

This triple describes a linear oscillator with small damping coefficient $\varepsilon > 0$ and perturbed restoring force 1. The corresponding transfer function is

$$G_\varepsilon(\imath\omega) = E(\imath\omega I - A_\varepsilon)^{-1} D_\varepsilon = \frac{-\varepsilon}{1 - \omega^2 + \varepsilon\imath\omega}.$$

By (64),

$$r_{\mathbb{R}}^{-}(A_\varepsilon; D_\varepsilon, E) = 1/\varepsilon.$$

If $\varepsilon < \sqrt{2}$, a simple calculation shows that $|G_\varepsilon(\imath\omega)|^2 = \varepsilon^2/[(1-\omega^2)^2 + \varepsilon^2\omega^2]$ is maximized for $\omega^2 = 1 - \varepsilon^2/2$ so that by (30)

$$r_{\mathbb{C}}^{-}(A_\varepsilon; D_\varepsilon, E)^2 = \frac{\varepsilon^4/4 + \varepsilon^2(1 - \varepsilon^2/2)}{\varepsilon^2} = 1 - \varepsilon^2/4$$

□

Equation (63) gives rise to a computable formula for the real stability radius. In fact if $\mathbb{C}_g = \mathbb{C}_-$, so that $\partial\mathbb{C}_b = \imath\mathbb{R}$, we write

$$\Omega = \{\omega \in \mathbb{R}; G_I(\imath\omega) = 0\} = \{\omega \in \mathbb{R}; \omega = 0 \text{ or } E(\omega^2 I + A^2)^{-1}D = 0\}.$$

Ω is the zero set of a real rational function in ω^2 and hence can be computed.

With respect to the Hilbert norm we have

$$f(\omega^2) := d^2(G_R(\imath\omega), G_I(\imath\omega)\,\mathbb{R}) = \|G_R(\imath\omega)\|_2^2 - \frac{\langle G_R(\imath\omega), G_I(\imath\omega)\rangle^2}{\|G_I(\imath\omega)\|_2^2}, \quad \omega \notin \Omega$$

where

$$G_R(\imath\omega) = -E(\omega^2 I + A^2)^{-1}AD, \quad G_I(\imath\omega) = -\omega E(\omega^2 I + A^2)^{-1}D.$$

So $f(\omega^2)$ is a strictly proper real rational function of ω^2. The maximum of this function can be approximately determined via a computer plot of the function $f(\cdot)$ or, more precisely, by some global optimization procedure for real functions on $\mathbb{R}$.

An alternative method consists of the following three steps:

1. Compute a Cholesky factorization of $f(\omega^2)$, i.e. a strictly proper real rational function $g(s)$ which has its poles and zeros in $\mathbb{C}_-$ such that

$$f(\omega^2) = g(-\imath\omega)g(\imath\omega) = |g(\imath\omega)|^2 \tag{65}$$

2. Determine a realization $(\hat{A}, \hat{d}, \hat{e})$ of $g(s)$.

3. Compute $r_{\mathbb{C}}^-(\hat{A}; \hat{d}, \hat{e})$ by means of the algorithm `stabrad` II.

Then

$$r_{\mathbb{R}}^-(A; D, E) = \min\left\{\left[\max_{\omega\in\Omega}\|G_R(\imath\omega)\|_2\right]^{-1}, r_{\mathbb{C}}^-(\hat{A}; \hat{b}, \hat{c})\right\}.$$

Remark 5.7. It may seem that rank one perturbations are a very restrictive class, and of course this is true. However, they are sufficient to develop a general robustness theory for the root location of *polynomials* (under arbitrary affine perturbations of the coefficient vector and for arbitrary perturbation norms). The idea is simply to apply the above results to single input systems (A, D, E) in control canonical form where A is the companion matrix of the given polynomial and E describes the structure of the perturbations. In this way stability radii become useful for the robustness analysis of polynomials. Choosing e.g. the ∞-norm to measure the perturbation size, the above propositions can be applied to interval polynomials and establish a link to Kharitonov's Theorem. For details, see [16]. □

6 Other perturbation classes

If A is a real matrix it would seem that we should restrict considerations to real perturbations and real stability radii. However, with the exception of the cases where $\ell = 1$ or $q = 1$, we have seen that the formulae which characterize these radii are far more complicated than the corresponding complex ones. In fact the complex stability radius, $r_{\mathbb{C}}^{-}$, is computable and has nice control theoretic associations with Riccati equations and Hamiltonian matrices. It therefore seems reasonable to explore possible motives for using the complex stability radius in real problems.

One use of the complex stability radius is that in many cases it provides an acceptable lower estimate for the real stability radius. However, we have seen in Section 3 that there are cases where this estimate is too conservative to be useful, see Prop 3.6. On the other hand, the structured real stability radius may jump under slight changes of the data and hence, in certain cases, it may be an over-optimistic indicator of robustness whereas the complex stability radius depends continuously on (A, D, E). So far we have only considered parameter variations of the form $A \mapsto A + D\Delta E$. It is not a priori clear why perturbations of this form should play a particular role in robustness analysis. Since $r_{\mathbb{R}}$ and $r_{\mathbb{C}}$ are defined relative to this very restricted class one might expect that they are misleading indicators of robustness in the presence of more general perturbations of the system equation. In this section we will investigate how the real and the complex stability radii change if the perturbation class is extended.

We only consider the Hilbert norm case and the stability region $\mathbb{C}_g = \mathbb{C}_-$.

The notion of stability radius can be extended to include perturbations which are

- nonlinear
- linear but time-varying,
- nonlinear and time-varying,
- nonlinear, time-varying and dynamic.

We introduce these extensions for both the complex and the real case. In order to specify the last perturbation class, the following definition will be used.

Definition 6.1. A map $\mathcal{N} : L^2(0,\infty;\mathbb{K}^q) \to L^2(0,\infty;\mathbb{K}^\ell)$ is said to be weakly Lipschitz continuous if for every $y_0 \in \mathbb{K}^q$, $t_0 \in \mathbb{R}_+$ and $\varphi \in L^2(0,t_0;\mathbb{K}^q)$ there exist real numbers $k \geq 0$, $r > 0$ and $t_1 > t_0$, such that for all $y_1(\cdot)$, $y_2(\cdot) \in L^2(0,\infty;\mathbb{K}^q)$ with

$$y_1(t) = y_2(t) = \varphi(t), \quad 0 \le t < t_0, \quad \|y_i(\cdot) - y_0\|_{L^\infty(t_0,t_1;\mathbb{K}^q)} \le r, \ i = 1,2$$

we have

$$\|\mathcal{N}(y_1(\cdot)) - \mathcal{N}(y_2(\cdot))\|_{L^\infty(t_0,t_1;\mathbb{K}^\ell)} \quad < \quad k\,\|y_1(\cdot) - y_2(\cdot)\|_{L^\infty(t_0,t_1;\mathbb{K}^q)}$$

We consider the following perturbed system equations

$$\begin{aligned}
\Sigma_N: &\quad \dot{x}(t) = Ax(t) + DN(Ex(t)) \\
\Sigma_{\Delta(t)}: &\quad \dot{x}(t) = Ax(t) + D\Delta(t)Ex(t) \\
\Sigma_{N(t)}: &\quad \dot{x}(t) = Ax(t) + DN(Ex(t),t) \\
\Sigma_{\mathcal{N}}: &\quad \dot{x}(t) = Ax(t) + D\mathcal{N}(Ex(\cdot))(t)
\end{aligned}$$

where

(i) $N : \mathbb{K}^q \to \mathbb{K}^\ell$, $N(0) = 0$, N is differentiable, at 0, is Lipschitz continuous and of finite gain,

(ii) $\Delta(\cdot) \in L^\infty(\mathbb{R}_+, \mathbb{K}^{\ell\times q})$,

(iii) $N(\cdot,\cdot) : \mathbb{K}^q \times \mathbb{R}_+ \to \mathbb{K}^\ell$ is jointly continuous, $N(0,t) = 0$ for all $t \in \mathbb{R}_+$, it is Lipschitz continuous in $y \in \mathbb{K}^q$ and of finite gain (uniformly in t),

(iv) $\mathcal{N} : L^2(0,\infty;\mathbb{K}^q) \longrightarrow L^2(0,\infty;\mathbb{K}^\ell)$ is causal, weakly Lipschitz continuous, of finite gain and $N(0) = 0$.

As perturbation norms we choose:

(i) $\|N\| = \inf\{\gamma \ge 0; \forall y \in \mathbb{K}^q : \|N(y)\| \le \gamma\|y\|\}$,

(ii) $\|\Delta\| = \text{ess sup}_{t\in\mathbb{R}_+} \|\Delta(t)\|$,

(iii) $\|N\| = \inf\{\gamma \ge 0; \forall t \ge 0, y \in \mathbb{K}^q : \|N(y,t)\| \le \gamma\|y\|\}$,

(iv) $\|\mathcal{N}\| = \inf \big\{\gamma \ge 0; \forall y(\cdot) \in L^2(0,\infty;\mathbb{K}^q) : \|\mathcal{N}(y(\cdot))\|_{L^2(0,\infty;\mathbb{K}^\ell)} \le \gamma\|y(\cdot)\|_{L^2(0,\infty;\mathbb{K}^q)}\big\}$.

Before introducing stability radii for the above systems we have to clarify the existence of solutions. Standard theorems can be applied to the first three systems, so that for all $(t_0, x_0) \in \mathbb{R}_+ \times \mathbb{K}^n$ there exist unique solutions $x(\cdot) = x(\cdot; t_0, x_0)$ on $[t_0, \infty)$ satisfying $x(t_0) = x_0$. For the case of dynamic perturbations the situation is more complicated. This is not surprising since this is a very large perturbation class and contains operators which give rise to different classes of nonlinear infinite-dimensional systems, e.g. delay equations, integro-differential equations and other systems with unbounded memory. Since $\mathcal{N}(y(\cdot))(t)$ depends not only on $y(t)$ but on the whole "past" $y(\cdot)|_{[0,t]}$ of $y(\cdot)$, the initial value problem for $\Sigma_{\mathcal{N}}$ must include initial functions.

Definition 6.2. Suppose $(t_0, x_0, \varphi) \in \mathbb{R}_+ \times \mathbb{K}^n \times L^2(0,\infty;\mathbb{K}^q)$ are given and I an interval with $\min I = t_0$. A function $x(\cdot) : I \to \mathbb{K}^n$ is said to be a solution of $\Sigma_{\mathcal{N}}$ with initial data (t_0, x_0, φ) if it satisfies the following conditions

(i) $x(\cdot)$ is absolutely continuous on I

(ii) $x(\cdot)$ satisfies

$$\dot{x}(t) = Ax(t) + D\mathcal{N}(y_\varphi(\cdot))(t), \quad a.e.\ t \in I$$

where

$$y_\varphi(\tau) = \begin{cases} \varphi(\tau) & \text{if} \quad 0 \le \tau < t_0 \\ Ex(\tau) & \text{if} \quad \tau \in I \end{cases}$$

(iii) $x(t_0) = x_0$

Note that by causality $\mathcal{N}(Ex(\cdot))(t)$ only depends on φ and $Ex(\cdot)|_{[t_0,t]}$ for every $t \in I$. The following theorem is proved in [13].

Theorem 6.3. *For any initial data $(t_0, x_0, \varphi) \in \mathbb{R}_+ \times \mathbb{K}^n \times L^2(0,t_0;\mathbb{K}^q)$ there exists a (unique) solution $x(\cdot;t_0,x_0,\varphi)$ of the above initial value problem on $[t_0,\infty)$.*

The stability concept that we use is global asymptotic stability. This definition is the standard one for Σ_N, $\Sigma_{\Delta(t)}$, $\Sigma_{N(t)}$; its generalization to systems $\Sigma_{\mathcal{N}}$ is

Definition 6.4. $\Sigma_{\mathcal{N}}$ is said to be globally asymptotically stable (g. a. s.) if it satisfies the following two conditions

(i) $\Sigma_{\mathcal{N}}$ is *stable*, i.e. for any $t_0 \ge 0$ and any $\varepsilon > 0$ there exists $\delta = \delta(t_0,\varepsilon) > 0$ such that

$$\|x_0\| < \delta, \quad \|\varphi\|_{L^2(0,t_0;\mathbb{K}^q)} < \delta \Longrightarrow$$
$$\|x(t;t_0,x_0,\varphi)\|_{\mathbb{K}^n} < \varepsilon \quad \text{for all} \quad t \ge t_0$$

(ii) The origin $0 \in \mathbb{K}^n$ is *globally attractive*, i.e. for any $(t_0,x_0,\varphi) \in \mathbb{R}_+ \times \mathbb{K}^n \times L^2(0,t_0;\mathbb{K}^q)$,

$$\lim_{t\to\infty} x(t;t_0,x_0,\varphi) = 0\,.$$

Definition 6.5. The stability radius of A with respect to the structure (D, E) for dynamic perturbations is defined by

$$r^-_{\mathbb{K},d}(A; D, E) = \inf\{\|\mathcal{N}\|;\ \Sigma_{\mathcal{N}} \text{ is not g. a. s.}\} \qquad \mathbb{K} = \mathbb{R} \text{ or } \mathbb{C}.$$

The stability radii with respect to other perturbation classes are defined analogously and are denoted by $r^-_{\mathbb{K},n}(A; D, E)$, $r^-_{\mathbb{K},t}(A; D, E)$, $r^-_{\mathbb{K},nt}(A; D, E)$. It is shown in [13] that, in general, we have the following

$$\begin{aligned} r^-_{\mathbb{K}}(A; D, E) &\geq r^-_{\mathbb{K},n}(A; D, E) \geq r^-_{\mathbb{K},t}(A; D, E) \\ &\geq r^-_{\mathbb{K},nt}(A; D, E) \geq r^-_{\mathbb{K},d}(A; D, E). \end{aligned} \tag{66}$$

The complex stability radius, however, is invariant under the above extensions of the perturbation class so that we have equality in (66) if $\mathbb{K} = \mathbb{C}$.

Theorem 6.6.

$$r^-_{\mathbb{C}} = r^-_{\mathbb{C},n} = r^-_{\mathbb{C},t} = r^-_{\mathbb{C},nt} = r^-_{\mathbb{C},d} \tag{67}$$

Note that the stability result implied in the first equality means that the nonlinear differential system Σ_N will have an asymptotically stable equilibrium at zero provided there exists $\gamma < r^-_{\mathbb{C}}$ such that

$$\|N(y)\| \leq \gamma \|y\|, \quad y \in \mathbb{C}^q \tag{68}$$

It appears that the equality $r_{\mathbb{K}} = r_{\mathbb{K},nl}$ is closely related, albeit not equivalent, to a multivariable version of the famous *Aizermann conjecture* (see e.g. [22]). An extension of this conjecture to multivariable systems can be formulated over both the real and the complex fields:

Aizermann's conjecture: *For any $\rho > 0$ the linear systems*

$$\dot{x} = Ax + D\Delta Ex, \quad \Delta \in \mathbb{K}^{\ell \times q},\ \|\Delta\| < \rho$$

are asymptotically stable iff the origin of the nonlinear systems

$$\dot{x} = A + DN(Ex) \tag{69}$$

is globally asymptotically stable for all Lipschitz continuous $N : \mathbb{K}^q \to \mathbb{K}^\ell$ satisfying $N(0) = 0$ and

$$\|N(y)\| < \rho\|y\|, \quad y \in \mathbb{K}^q,\ y \neq 0, \tag{70}$$

Using the Liapunov function of maximal robustness described in Section 4, the following can be proved, see [13]:

Proposition 6.7. *The multivariable version of Aizermann's conjecture holds true over the field of complex numbers.*

If $r_{\mathbb{C}}^- < \infty$, this proposition implies that Σ_N will be globally asymptotically stable provided

$$\|N(y)\| < r_{\mathbb{C}}^- \|y\|, \quad y \in \mathbb{C}^q, y \neq 0 \tag{71}$$

which is an improvement on (68).

Remark 6.8. In [10] we have made the mistake of including *time-varying* nonlinearities $N(y,t)$ in Aizermann's conjecture. The corresponding statement in that article [10, Prop. 6.3] is not correct. The proof (using LaSalle's Principle) is only applicable to *time-invariant* nonlinearities. We apologize for this mistake. □

It is well known that Aizermann's conjecture does not hold over the field of real numbers. So one might expect that in general $r_{\mathbb{R}}^- > r_{\mathbb{R},n}^-$. Unfortunately , the usual counterexamples against the classical (scalar) Aizermann conjecture have the property that the nonlinear gain $|N(y)|$ approaches for large values of $|y|$ a gain for which the resulting linear system is not asymptotically stable. But, to prove $r_{\mathbb{R},n}^- < r_{\mathbb{R}}^-$ for a particular system one has to construct a destabilizing nonlinearity which remains inside a *strictly smaller sector* than that determined by the smallest destabilizing *linear* gain. This can be done by analyzing an example given by *Willems* in [22].

Example 6.9. Consider the fourth order scalar system

$$\begin{aligned} x^{(4)}(t) + 10\ddot{x}(t) + 9x(t) + \varepsilon \left[\alpha x^{(3)}(t) + \beta\ddot{x}(t) + \gamma\dot{x}(t) + \delta x(t)\right] \\ + \varepsilon N(\ddot{x}(t)) = 0 \end{aligned} \tag{72}$$

where $N : \mathbb{R} \to \mathbb{R}$ is continuously differentiable and $N(0) = 0$. If $x^T = [x_1, x_2, x_3, x_4] = [x, \dot{x}, \ddot{x}, x^{(3)}]$, an equivalent state space form is

$$\dot{x}(t) = Ax(t) + DN(y(t)), \quad y(t) = Ex(t) \tag{73}$$

where $A = A_0 - \varepsilon A_1$,

$$A_0 = \begin{bmatrix} 0 & 1 & 0 & 0 \\ 0 & 0 & 1 & 0 \\ 0 & 0 & 0 & 1 \\ -9 & 0 & -10 & 0 \end{bmatrix}, A_1 = \begin{bmatrix} 0 & 0 & 0 & 0 \\ 0 & 0 & 0 & 0 \\ 0 & 0 & 0 & 0 \\ \delta & \gamma & \beta & \alpha \end{bmatrix}, D = \begin{bmatrix} 0 \\ 0 \\ 0 \\ -\varepsilon \end{bmatrix} \tag{74}$$

and $E = \begin{bmatrix} 0 & 0 & 1 & 0 \end{bmatrix}$.

For $|\varepsilon|$ small

$$\sigma(A) \subset \mathbb{C}_- \;\Leftrightarrow (\gamma - \alpha)(\gamma - 9\alpha) < 0 \text{ and } \begin{array}{ll} \text{(i)} & \varepsilon > 0, \gamma > 0, \alpha > 0, \\ \text{(ii)} & \varepsilon < 0, \gamma < 0, \alpha < 0. \end{array} \tag{75}$$

Now

$$G(\imath\omega) = E(\imath\omega I - A)^{-1}D = \frac{\varepsilon\omega^2}{\omega^4 - 10\omega^2 + 9 + \varepsilon(\delta - \beta\omega^2) + \imath\varepsilon\omega(\gamma - \alpha\omega^2)}$$

so that (cf. Section 5) $\Omega = \{\omega; G_I(\imath\omega) = 0\} = \{0, |\gamma/\alpha|^{1/2}\}$. Hence if either (75) (i) or (ii) hold we obtain (see (64))

$$r_{\mathbb{R}}^-(A; D, E) = \left|\frac{(\gamma - 9\alpha)(\gamma - \alpha) + \alpha\varepsilon(\delta\alpha - \beta\gamma)}{\alpha\gamma\varepsilon}\right| \tag{76}$$

Note that $r_{\mathbb{R}}^-(A; D, E) \to \infty$ as $\varepsilon \to 0$.By choosing $\omega^2 = 1$ and $\omega^2 = 9$, it is easy to see that

$$r_{\mathbb{C}}^-(A; D, E) \leq$$
$$\min\left\{[(\delta - \beta)^2 + (\gamma - \alpha)^2]^{1/2}, 1/9\,[(\delta - 9\beta)^2 + 9(\gamma - 9\alpha)^2]^{1/2}\right\}$$

so that $r_{\mathbb{C}}^-(A; D, E)$ is uniformly bounded for bounded parameters ε, α, β, γ, δ. By elementary albeit cumbersome calculations it is even possible to show that

$$\lim_{\varepsilon \to 0} r_{\mathbb{C}}^-(A; D, E) = \min\{|\gamma - \alpha|, 1/3|\gamma - 9\alpha|\}.$$

Using *averaging theory*, Willems [22] shows that if e.g. $N(y) = \tanh y$ it is possible to choose initial states $x_0 = [x_{01}, x_{02}, x_{03}, x_{04}]^T$ and parameter values α_0, β_0, γ_0, δ_0 satisfying either (75) (i) or (ii) such that for ε sufficiently small there exist parameters α_ε, β_ε, γ_ε, δ_ε with

$$\lim_{\varepsilon \to 0}(\alpha_\varepsilon, \beta_\varepsilon, \gamma_\varepsilon, \delta_\varepsilon) = (\alpha_0, \beta_0, \gamma_0, \delta_0) \tag{77}$$

yielding a nonlinear differential equation

$$x^{(4)}(t) + 10\ddot{x}(t) + 9x(t) + \varepsilon\left[\alpha_\varepsilon x^{(3)}(t) + \beta_\varepsilon \ddot{x}(t) + \gamma_\varepsilon \dot{x}(t) + \delta_\varepsilon x(t)\right]$$
$$+\varepsilon \tanh \ddot{x}(t) = 0$$

with a periodic solution of period 2π starting at x_0. But $\|\tanh(\cdot)\| = 1$ and hence if we set $\alpha = \alpha_\varepsilon$, $\beta = \beta_\varepsilon$, $\gamma = \gamma_\varepsilon$, $\delta = \delta_\varepsilon$ in (74) and denote the corresponding matrices A, D (defined via (73)) by A_ε, D_ε then

$$r_{\mathbb{R},n}^-(A_\varepsilon; D_\varepsilon, E) \leq 1$$

On the other hand, we have by (76)

$$r_{\mathbb{R}}^{-}(A_\varepsilon; D_\varepsilon, E) = \left| \frac{(\gamma_\varepsilon - 9\alpha_\varepsilon)(\gamma_\varepsilon - \alpha_\varepsilon) + \alpha_\varepsilon \varepsilon(\delta_\varepsilon \alpha_\varepsilon - \beta_\varepsilon \gamma_\varepsilon)}{\alpha_\varepsilon \gamma_\varepsilon \varepsilon} \right|.$$

By (77) there exists $\varepsilon_0 > 0$ such that, for all ε, $|\varepsilon| \leq \varepsilon_0$, γ_ε is uniformly bounded away from zero and $(\gamma_\varepsilon - 9\alpha_\varepsilon)(\gamma_\varepsilon - \alpha_\varepsilon) < -\varepsilon_0$. Hence by choosing ε sufficiently small $r_{\mathbb{R}}^{-}(A_\varepsilon; D_\varepsilon, E)$ can be made as large as we please, yet $r_{\mathbb{R},n}^{-}(A_\varepsilon; D_\varepsilon, E) \leq 1$. In particular $r_{\mathbb{R}}^{-}/r_{\mathbb{R},n}^{-}$ is unbounded. □

We have not been able to characterize $r_{\mathbb{R},n}^{-}$ in general, and the situation for real time-varying linear perturbations seems to be equally difficult. The following example illustrates that $r_{\mathbb{R},t}^{-}$ may lie anywhere between $r_{\mathbb{C}}^{-}$ and $r_{\mathbb{R}}^{-}$, and $r_{\mathbb{R},n}^{-}$ may be greater than $r_{\mathbb{R},t}^{-}$

Example 6.10. Consider the linear oscillator

$$\dot{x} = \begin{bmatrix} 0 & 1 \\ -1 & -2b \end{bmatrix} x + \Delta(t) \begin{bmatrix} 0 & 0 \\ 1 & 0 \end{bmatrix} x \tag{78}$$

where the restoring force coefficient is subjected to time-varying perturbations $\Delta(t)$. For the corresponding triple

$$A = \begin{bmatrix} 0 & 1 \\ -1 & -2b \end{bmatrix}, \quad D = \begin{bmatrix} 0 \\ 1 \end{bmatrix}, \quad E = [1 \ 0]$$

it is easy to show that $r_{\mathbb{R}}^{-} = 1$ for all $b \in (0, 2)$, whereas $r_{\mathbb{C}}^{-} = 1$ for $b \in (0.708, 2)$ and $r_{\mathbb{C}}^{-} < r_{\mathbb{R}}^{-}$ for $b \in (0, 0.708)$. Moreover, $r_{\mathbb{C}}^{-} \to 0$ as $b \to 0$, see Fig. 8 For time-varying perturbations $\Delta(t)$, *Colonius and Kliemann* [3] determined analytically the maximal achievable Liapunov exponent for (78) when $\Delta(t)$ is restricted to a given interval $[\alpha, \beta] \subset \mathbb{R}$. Their analysis involves a separate discussion of many different cases even for this simple example. $r_{\mathbb{R},t}^{-}$ is the smallest positive real ρ for which there exists a perturbation $\Delta : \mathbb{R}_+ \to [-\rho, \rho]$ such that the maximum Liapunov exponent of (78) is greater than or equal zero. As a consequence of the analysis in [3] we obtain the dependence of $r_{\mathbb{R},t}^{-}$ on b as shown in Fig. 8. We find $r_{\mathbb{R},t}^{-} = 1$ for $b \in (0.26, 2)$, but more interestingly, we have this equality also for parameter values $b \in (0.26, 0.708)$ where $r_{\mathbb{C}}^{-} < r_{\mathbb{R}}^{-}$. Moreover Fig. 8 shows that as parameter b runs from 0 to 1, $r_{\mathbb{R},t}^{-}$ moves in the varying interval $(r_{\mathbb{C}}^{-}, r_{\mathbb{R}}^{-}]$ from one extreme to the other.

Now consider the nonlinear perturbed problem

$$\dot{x} = Ax + DN(Ex)$$

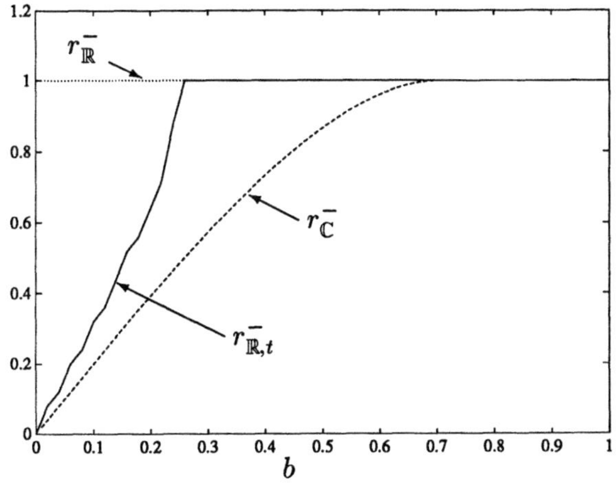

Figure 8. Stability radii $r_{\mathbb{C}}^-$, $r_{\mathbb{R},t}^-$ and $r_{\mathbb{R}}^-$ of the linear oscillator (78)

where $N : \mathbb{R} \to \mathbb{R}$ is differentiable, of gain $\|N\| < 1$ and $N(0) = 0$. It is easy to verify that along the flow of this system the real valued function

$$V(x) = x_1^2 + x_2^2 + 2\int_0^{x_1} N(s)\,ds + 2bx_1x_2 + 2b^2x_1^2 > 0\,, \quad x \in \mathbb{R}^2, x \neq 0$$

has the derivative

$$\dot{V}(x) = -2b[x_1^2 + x_2^2 + N(x_1)x_1]$$

Since $\|N\| < 1$ we conclude that V is a Liapunov function for the system, hence $r_{\mathbb{R},n}^- = 1$ for all $b \in (0,2)$. □

In contrast with the other real stability radii where not much is known, there is a complete characterization of the real stability radius with respect to dynamic perturbations [13].

Definition 6.11. A real state space system (F, G, H, K) is said to be a *multivariable oscillator* if its transfer matrix is of the form

$$H(sI - F)^{-1}G + K = \frac{1}{p(s)}\left[K_0 + K_1 s + K_2 s^2\right]$$

where the K_i are constant matrices and $p(s)$ is a quadratic polynomial.

If the oscillator is stable then the H^∞-norm of its transfer matrix is equal to the norm of the associated input-output operator $\mathcal{L}$, cf. (39).

Proposition 6.12. *Suppose (A, D, E) is a given $\mathbb{C}_-$-stable triple. Then, for any $\varepsilon > 0$, there exists a stable multivariable oscillator with input-output operator $\mathcal{L}$ of norm $\|\mathcal{L}\| < r_{\mathbb{C}}^- + \varepsilon$ which destabilizes (A, D, E) (i.e. $\Sigma_{\mathcal{L}}$ is not asymptotically stable). In particular,*

$$r_{\mathbb{R},d}^-(A; D, E) = r_{\mathbb{C}}^-(A; D, E) \tag{79}$$

The proof of the above proposition uses a similar technique to that in [2] (see [13]).

As a consequence of Proposition 6.12 one should take $r_{\mathbb{C}}^-$ rather than $r_{\mathbb{R}}^-$ as a robustness measure whenever neglected dynamics play an important role in the model uncertainty.

7 Multiple perturbation structures

In applications we have often found that the parameter uncertainties can be represented by a *single perturbation* structure of the form (1). However we have seen in Remark 2.1 that this cannot always be the case, so here we consider multiple perturbations which provide us with much more flexibility. Since we only have a few preliminary results the section is of a rather different tone than that of the others.

We assume that the perturbed system matrix is

$$A + \sum_{i=1}^{N} D_i \Delta_i E_i \qquad \text{where } D_i \in \mathbb{K}^{n \times \ell_i}, E_i \in \mathbb{K}^{q_i \times n}, \quad i \in \underline{N}$$

are given matrices defining the structure of the perturbation and $\Delta_i \in \mathbb{K}^{\ell_i \times q_i}$, $i \in \underline{N}$ are unknown disturbance matrices. If

$$\Delta = \begin{bmatrix} \Delta_1 & & 0 \\ & \ddots & \\ 0 & & \Delta_N \end{bmatrix} \tag{80}$$

the size of the perturbation matrix Δ is measured by

$$\begin{aligned} \|\Delta\| &= \max_{i \in \underline{N}} \|\Delta_i\| \\ \|\Delta_i\| &= \max\{\|\Delta_i y_i\|_{\mathbb{K}^{\ell_i}} : y_i \in \mathbb{K}^{q_i}, \|y_i\|_{\mathbb{K}^{q_i}} \leq 1\} \end{aligned} \tag{81}$$

where $\|\cdot\|_{\mathbb{K}^{\ell_i}}$, $\|\cdot\|_{\mathbb{K}^{q_i}}$ are arbitrary norms on $\mathbb{K}^{\ell_i}$, $\mathbb{K}^{q_i}$, $i \in \underline{N}$.

Example 7.1. Suppose $\left[\begin{smallmatrix} & * \\ * & * \end{smallmatrix}\right]$ is an uncertain 2×2 matrix where the elements marked $*$ are uncertain parameters. The perturbed system cannot be represented by a single perturbation structure, but instead a 2-block structure of the form $\begin{smallmatrix} & \left[\begin{smallmatrix} * \\ * \end{smallmatrix}\right] \\ [*] & \end{smallmatrix}$ (corresponding to $N = 2,\ \ell_1 = q_1 = 1,\ \ell_2 = 2, q_2 = 1$) or $\begin{smallmatrix} & [*] \\ [* & *] \end{smallmatrix}$ (corresponding to $N = 2,\ \ell_1 = q_1 = 1,\ \ell_2 = 1,\ q_2 = 2$), or a 3-block structure $\begin{smallmatrix} & & [*] \\ [*] & [*] & \end{smallmatrix}$ must be used (where $N = 3$ and $\ell_i = q_i = 1$ for $i = 1, 2, 3$). The corresponding Δ's take the form

$$\begin{bmatrix} * & 0 \\ 0 & * \\ 0 & * \end{bmatrix} \qquad \begin{bmatrix} * & 0 & 0 \\ 0 & * & * \end{bmatrix} \qquad \begin{bmatrix} * & 0 & 0 \\ 0 & * & 0 \\ 0 & 0 & * \end{bmatrix}$$

We see that different multiple perturbation structures can be used to represent the uncertainty in a model. It is not clear which one of these should be chosen. The choice of the perturbation structure and the norms on the individual blocks will be influenced by the precise nature of the uncertainty in the particular application. □

Definition 7.2. The stability radius of A with respect to the multi perturbation structure $(D_i, E_i)_{i \in \underline{N}}$ is defined by

$$\begin{aligned} r_{\mathbb{K},m} &= r_{\mathbb{K},m}(A; (D_i, E_i)_{i \in \underline{N}}; \mathbb{C}_b) \\ &= \inf\{\|\Delta\|; \Delta_i \in \mathbb{K}^{\ell_i \times q_i}, \sigma(A + \textstyle\sum_{i=1}^{N} D_i \Delta_i E_i) \cap \mathbb{C}_b \neq \emptyset\} \end{aligned} \tag{82}$$

$r_{\mathbb{K},m} = \infty$ iff there do not exist Δ_i, $i \in \underline{N}$ with $\sigma(A + \sum_{i=1}^{N} D_i \Delta_i E_i) \cap \mathbb{C}_b \neq \emptyset$. On the other hand just as in the single perturbation case if $r_{\mathbb{K},m} < \infty$ we may replace the infimum in (82) by a minimum. Moreover

$$r_{\mathbb{K},m}(A; (D_i, E_i)_{i \in \underline{N}}; \mathbb{C}_b) = r_{\mathbb{K},m}(A; (D_i, E_i)_{i \in \underline{N}}; \partial\mathbb{C}_b) \tag{83}$$

Note also for $\alpha_i \in \mathbb{R}_+ = (0, \infty)$, $i \in \underline{N}$

$$r_{\mathbb{K},m}(A; (\alpha_i D_i, \alpha_i^{-1} E_i)_{i \in \underline{N}}; \mathbb{C}_b) = r_{\mathbb{K},m}(A; (D_i, E_i)_{i \in \underline{N}}; \mathbb{C}_b) \tag{84}$$

An important role will be played by the following block transfer matrix $G_m(s) \in \mathbb{C}^{q \times \ell}$, $q = \sum_{i=1}^{N} q_i$, $\ell = \sum_{i=1}^{N} \ell_i$.

$$G_m(s) = \begin{bmatrix} G_{11}(s) & \cdots & G_{1N}(s) \\ \vdots & & \vdots \\ G_{N1}(s) & \cdots & G_{NN}(s) \end{bmatrix} \tag{85}$$

$$G_{ij}(s) = E_i(sI - A)^{-1} D_j \qquad i, j \in \underline{N}$$

Just as in the case of a single perturbation we have

Proposition 7.3. $\Delta = \operatorname{diag}(\Delta_1, \cdots, \Delta_N)$ *is destabilizing iff there exist* $y_i \in \mathbb{C}^{q_i}$, $i = 1, \ldots, N$ *not all zero and* $s_0 \in \mathbb{C}_b$, *such that* $y^T = [y_1^T, \cdots, y_N^T]$ *satisfies*

$$y = G_m(s_0)\Delta y. \tag{86}$$

In [4] Doyle introduced the concept of *structured singular values.*

Definition 7.4. The structured singular value $\mu_{\mathbb{K}}(M)$ of a matrix $M \in \mathbb{C}^{q\times\ell}$ with respect to the block structure Δ given by (80) is

$$\mu_{\mathbb{K}}(M) = [\min\{\|\Delta\|; \Delta \in \mathbb{K}^{\ell\times q} \text{ with structure } (80) : \det(I - M\Delta) = 0\}]^{-1} \tag{87}$$

As a corollary to Proposition 7.3, we have

Corollary 7.5.

$$r_{\mathbb{K},m}(A; (D_i, E_i)_{i\in\underline{N}}; \mathbb{C}_b) = \left[\sup_{s\in\partial\mathbb{C}_b} \mu_{\mathbb{K}}(G_m(s))\right]^{-1} \tag{88}$$

For Euclidean norms and $M \in \mathbb{C}^{q\times\ell}$, Doyle has given a characterization of $\mu_{\mathbb{C}}(M)$, but this has not as yet led to a computable algorithm for the RHS of (88).

Formally (86) looks the same as (15) for the single perturbation case. So at first sight one might think that it is possible to mirror the development of Section 4 here. Indeed since by (86) any destabilizing Δ must have the property $\|\Delta\| \geq \|G_m(s_0)\|^{-1}$ for some $s_0 \in \mathbb{C}_b$, we have

$$r_{\mathbb{K},m}(A; (D_i, E_i)_{i\in\underline{N}}; \mathbb{C}_b) \geq \left[\max_{s\in\partial\mathbb{C}_b} \|G_m(s)\|\right]^{-1} \tag{89}$$

But note that this lower bound can be improved by utilizing (84); which yields

$$r_{\mathbb{K},m}(A; (D_i, E_i)_{i\in\underline{N}}; \mathbb{C}_b) \geq \left[\sup_{s\in\partial\mathbb{C}_b} \inf_{\alpha\in\mathbb{R}_+^N} \|G_m(s,\alpha)\|\right]^{-1} \tag{90}$$

where $\alpha = [\alpha_1, \cdots, \alpha_N]^T$

$$G_m(s,\alpha) := (G_{ij}(s,\alpha)), \quad G_{ij}(s,\alpha) = \alpha_i^{-1} G_{ij}(s)\alpha_j \quad i,j \in \underline{N}. \tag{91}$$

Specializing to Euclidean norms and $\mathbb{K} = \mathbb{C}$ it has been shown (see [4]) that for many problems (in particular $N \leq 3$) equality holds in (90). Moreover since the substitution $\alpha_i = \log \alpha_i'$, $i \in \underline{N}$ transforms $\|G_m(s,\alpha)\|$ into a convex function of α', any minima of $\|G_m(s,\cdot)\|$ must be global. This helps considerably in the computation of the RHS of (90) and has led to the suggestion that it should be used as a safe lower bound.

Remark 7.6. For the case $\mathbb{C}_g = \mathbb{C}_-$ we have

$$r^-_{\mathbb{C},m}(A;(D_i,E_i)_{i\in\underline{N}}) := r_{\mathbb{C},m}(A;(D_i,E_i)_{i\in\underline{N}};\mathbb{C}_+) = \left[\max_{\omega\in\mathbb{R}} \mu_{\mathbb{C}}(G_m(\imath\omega))\right]^{-1}$$

$$\geq \left[\sup_{\omega\in\mathbb{R}} \inf_{\alpha\in\mathbb{R}^N_+} \|G_m(\imath\omega,\alpha)\|\right]^{-1} \geq \left[\inf_{\alpha\in\mathbb{R}^N_+} \sup_{\omega\in\mathbb{R}} \|G_m(\imath\omega,\alpha)\|\right]^{-1} =: r^\ell_{\mathbb{C},m} \quad (92)$$

A realization of $G(s,\alpha)$ is $G(s,\alpha) = E(\alpha)(sI-A)^{-1}D(\alpha)$, where

$$\begin{aligned} E(\alpha) &= \left[\alpha_1^{-1}E_1^T, \cdots, \alpha_N^{-1}E_N^T\right]^T \\ D(\alpha) &= \left[\alpha_1 D_1, \cdots, \alpha_N D_N\right] \end{aligned}$$

Thus $r^-_{\mathbb{C}}(A;D(\alpha),E(\alpha)) = [\sup_{\omega\in\mathbb{R}} \|G_m(\imath\omega,\alpha)\|]^{-1}$ can be associated with the Riccati equation

$$XA + A^*X - \rho^2 E^*(\alpha)E(\alpha) - XD(\alpha)D^*(\alpha)X = 0 \quad (93)$$

This suggests that it may be possible to develop an algorithm for the computation of the lower bound $r^\ell_{\mathbb{C},m}$ based on the corresponding Hamiltonian matrix

$$H_{\rho,\alpha} = \begin{bmatrix} A & -D(\alpha)D^*(\alpha) \\ \rho^2 E^*(\alpha)E(\alpha) & -A^* \end{bmatrix}.$$

Also it is easy to see that multiple nonlinear, time-varying and dynamic perturbations will not destabilize the system providing their norms are less than $r^\ell_{\mathbb{C},m}$ (see [13]). Thus a natural safe (but not optimal) candidate for a robustness measure of a linear system when subjected to very general multiple perturbations is $r^\ell_{\mathbb{C},m}$. □

8 Maximizing the complex stability radius

Throughout this section we assume that $(A,D,E) \in \mathbb{C}^{n\times n} \times \mathbb{C}^{n\times \ell} \times \mathbb{C}^{q\times n}$, $B \in \mathbb{C}^{n\times m}$ and consider the *controlled* nominal system

$$\dot{x} = Ax + Bu \quad (94)$$

where (A,B) is assumed to be *stabilizable*. The perturbed system is

$$\dot{x} = (A + D\Delta E)x + Bu \quad (95)$$

Our aim is to determine the supremum of the complex stability radii which can be achieved by linear state feedback $u = Fx$, $F \in \mathbb{C}^{m\times n}$:

$$\bar{r}_{\mathbb{C}}(A;D,E;B) = \sup\{r^-_{\mathbb{C}}(A+BF;D,E); F \in \mathbb{C}^{m\times n},\ \sigma(A+BF) \subset \mathbb{C}_-\} \quad (96)$$

The results which we present in this section have been proved in [12]. Since, by definition,

$$\bar{r}_{\mathbb{C}}(A+BF;D,E;B)=\bar{r}_{\mathbb{C}}(A;D,E;B)\,,\quad F\in\mathbb{C}^{m\times n}$$

we may assume without any restriction of generality that $\sigma(A)\subset\mathbb{C}_-$.

Remark 8.1. By Prop. 3.1 the achievable robustness is *infinite*, i.e. $\bar{r}_{\mathbb{C}}(A;D,E;B)=\infty$ if and only if, for any $\delta>0$, there exists a stabilizing state feedback $F\in\mathbb{C}^{m\times n}$ such that $G_F(s)=E(sI-A-BF)^{-1}D$ satisfies

$$\|G_F\|_{H_\infty}=\max_{\omega\in\mathbb{R}}\|G_F(i\omega)\|<\delta\,. \tag{97}$$

In the terminology of [24] this is equivalent to saying that the system

$$\begin{aligned}\dot{x}(t)&=Ax(t)+Bu(t)+Dv(t)\\ y(t)&=Ex(t)\end{aligned} \tag{98}$$

(with control $u(t)$ and disturbance $v(t)$) is L_2*-almost disturbance decouplable* by *stabilizing* state feedback.

Note that, even if $\bar{r}_{\mathbb{C}}<\infty$ there will, in general, *not* exist an optimal feedback matrix F. □

For any $F\in\mathbb{C}^{m\times n}$ and $v(\cdot)\in L^2(0,\infty;\mathbb{C}^\ell)$ let

$$J_\rho(x_0,v(\cdot),F)=\int_0^\infty[\|v(t)\|^2-\rho^2\|y_F(t)\|^2]\,dt \tag{99}$$

where

$$\begin{aligned}\dot{x}_F&=(A+BF)x_F+Dv\,,\quad x_F(0)=x_0\\ y_F&=Ex_F\end{aligned} \tag{100}$$

Then, using the results of Section 4 we have

$$\begin{aligned}\bar{r}_{\mathbb{C}}(A;D,C;B)=\sup\{\rho;\exists F\in\mathbb{C}^{m\times n}:\ &\sigma(A+BF)\subset\mathbb{C}_-\text{ and}\\ \forall v\in L^2(0,\infty;\mathbb{C}^\ell):\ &J_\rho(0,v(\cdot),F)\geq 0\}\end{aligned}$$

Setting $u(t)=Fx_F(t)$ in (100) this leads us to consider the following parametrized maximin problem:

$$\sup_{u(\cdot)\in L^2(0,\infty;\mathbb{C}^m)}\ \inf_{v(\cdot)\in L^2(0,\infty;\mathbb{C}^\ell)}\int_0^\infty[\|v(t)\|^2-\rho^2\|y_{u,v}(t)\|^2]\,dt \tag{101}$$

subject to

$$\begin{aligned}\dot{x}_{u,v}&=Ax_{u,v}+Dv+Bu\,,\quad x(0)=x_0\\ y_{u,v}&=Ex_{u,v}\end{aligned} \tag{102}$$

Since this problem will, in general, not admit an optimal solution we consider the regularized version where the control action is slightly penalized and the cost functional is replaced by

$$J_{\rho,\varepsilon}(x_0, v(\cdot), u(\cdot)) = \int_0^\infty [\|v(t)\|^2 - \varepsilon^2\|u(t)\|^2 - \rho^2\|y_{u,v}(t)\|^2]\, dt, \quad \rho, \varepsilon > 0 \tag{103}$$

Associated with this problem is the algebraic Riccati equation

$$A^*X + XA - \rho^2 E^*E - XDD^*X + \varepsilon^{-2}XBB^*X = 0 \qquad (\text{ARE}_{\rho,\varepsilon})$$

In Section 4 we saw that it is possible to characterize the complex stability radius $r_{\mathbb{C}}^-(A; D, E)$ via the associated Riccati equation (ARE_ρ). In a similar way one can characterize the supremal achievable radius $\bar{r}_{\mathbb{C}}(A; D, E; B)$ via ($\text{ARE}_{\rho,\varepsilon}$). For this, let $\mathcal{P}(A, B, D, E)$ be the set of all pairs $(\rho, \varepsilon) \in (0, \infty)^2$ such that ($\text{ARE}_{\rho,\varepsilon}$) has a solution $P_{\rho,\varepsilon} \in \mathcal{H}_n{}^-$ and set

$$\rho(\varepsilon) = \sup\{\rho \in \mathbb{R}_+;\ (\rho, \varepsilon) \in \mathcal{P}(A, B, D, E)\}, \quad \varepsilon > 0.$$

If $(\rho, \varepsilon) \in \mathcal{P}(A, B, D, E)$, the feedback generated by the solution $P_{\rho,\varepsilon}$

$$F_{\rho,\varepsilon} = \varepsilon^{-2} B^* P_{\rho,\varepsilon} \tag{104}$$

increases the robustness of the closed loop system (as measured by the complex stability radius) to at least ρ. This is a consequence of the following lemma.

Lemma 8.2. *Suppose $\sigma(A) \subset \mathbb{C}_-$ and $(\rho, \varepsilon) \in \mathcal{P}(A, B, D, E)$.*

(i) If $P_{\rho,\varepsilon} \in \mathcal{H}_n{}^-$ solves ($ARE_{\rho,\varepsilon}$) then $\sigma(A + \varepsilon^{-2}BB^*P_{\rho,\varepsilon}) \subset \mathbb{C}_-$ and $r_{\mathbb{C}}^-(A + \varepsilon^{-2}BB^*P_{\rho,\varepsilon}; D, C) \geq \rho$.

(ii) $0 < \varepsilon' < \varepsilon \implies \rho(\varepsilon') \geq \rho(\varepsilon)$

Roughly speaking, the following proposition shows that the robustness achievable by any state feedback $u = Fx$ can also be obtained by using Riccati generated feedback matrices of the form (104).

Proposition 8.3. *Suppose $\sigma(A) \subset \mathbb{C}_-$. Then*

$$\begin{aligned} \bar{r}_{\mathbb{C}} &= \sup\{\rho;\ (ARE_{\rho,\varepsilon}) \text{ has a solution } P \in \mathcal{H}_n{}^- \text{for some } \varepsilon > 0\} \\ &= \lim_{\varepsilon \to 0+} \rho(\varepsilon) \end{aligned} \tag{105}$$

Closely related results have been obtained in the context of H^∞-optimization, see [5], [20], [27].

So far we have not carried out a complete asymptotic analysis of what happens with the feedback operators $F_{\rho,\varepsilon}$ (104) as $\varepsilon \to 0$ and $\rho \to \bar{r}_{\mathbb{C}}$. A first step towards such an analysis is made in the next proposition.

Proposition 8.4. *Suppose* $(\rho, \varepsilon) \in \mathcal{P}(A, B, D, E)$ *and fix* ρ. *For any decreasing sequence* (ε_k) *in* $(0, \varepsilon)$ *with* $\lim_{k\to\infty} \varepsilon_k = 0$, *there exists an increasing sequence of solutions* P_{ρ,ε_k} *of* $(ARE_{\rho,\varepsilon_k})$ *in* $\mathcal{H}_n^-$ *such that* $\lim_{k\to\infty} P_{\rho,\varepsilon_k} = P_{\rho,0}$ *exists and satisfies:*

(i) $B^* P_{\rho,0} = 0$

$$\tag{106}$$

(ii) There exists $P_{\rho,1} \in \mathcal{H}_n$ *such that* $P_{\rho,1}[(\operatorname{Im} B)^\perp] \subset \operatorname{Im} B$ *and*

$$A^* P_{\rho,0} + P_{\rho,0} A - \rho^2 E^* E - P_{\rho,0} DD^* P_{\rho,0} + P_{\rho,1} BB^* P_{\rho,1} = 0 \tag{107}$$

This proposition allows us to compute upper bounds for $\bar{r}_{\mathbb{C}}$. In fact, for any $\rho < \bar{r}_{\mathbb{C}}$ the above equations must have a pair of solutions $P_{\rho,0} \in \mathcal{H}_n^-$, $P_{\rho,1} \in \mathcal{H}_n$. If such solutions cease to exist for a certain value of $\rho > 0$ we have found an upper bound for $\bar{r}$. This is illustrated in the following example.

Example 8.5. Consider

$$A = \begin{bmatrix} -4 & 2 \\ -3 & 1 \end{bmatrix}, \quad B = \begin{bmatrix} 1 \\ 0 \end{bmatrix}, \quad D = \begin{bmatrix} 0 \\ 1 \end{bmatrix}, \quad E = I$$

Let $P_{\rho,0} = \begin{bmatrix} p_1^0 & p_2^0 \\ p_2^0 & p_3^0 \end{bmatrix}$, then (106) is satisfied iff $p_1^0 = p_2^0 = 0$. Setting $B^* P_{\rho,1} = [\tilde{p}_1 \ \tilde{p}_2]$, (107) is equivalent to the following three equations

$$\tilde{p}_1^2 = \rho^2, \quad \tilde{p}_1 \tilde{p}_2 = 3p_3^0, \quad 2p_3^0 - \rho^2 - (p_3^0)^2 + \tilde{p}_2^2 = 0$$

Hence $\tilde{p}_1 = \pm\rho$, $\tilde{p}_2 = \pm 3p_3^0/\rho$ and it is easy to see that

$$\begin{aligned} p_3^0 &= -\frac{\rho^2}{9-\rho^2}[1 \pm \sqrt{10-\rho^2}] && \text{if} \quad \rho^2 < 9 \\ p_3^0 &= \rho^2/2 && \text{if} \quad \rho^2 = 9 \\ p_3^0 &= \frac{\rho^2}{9-\rho^2}[1 \pm \sqrt{10-\rho^2}] && \text{if} \quad 9 < \rho^2 \le 10 \end{aligned}$$

Since one can choose $P_{\rho,0} \in \mathcal{H}_n^-$ for $\rho < 3$ and $P_{\rho,0} \notin \mathcal{H}_n^-$ for $\rho \ge 3$ we see that $\bar{r}_{\mathbb{C}} \le 3$. In fact a more careful analysis shows that $\bar{r}_{\mathbb{C}} = 3$. □

We hope to be able to use these asymptotic results to construct an efficient algorithm for the computation of $\bar{r}_{\mathbb{C}}$.

The Riccati equation $(\text{ARE}_{\rho,\varepsilon})$ is associated with the Hamiltonian matrix

$$H = H_{\rho,\varepsilon} = \begin{bmatrix} A & \varepsilon^{-2} BB^* - DD^* \\ \rho^2 E^* E & -A^* \end{bmatrix} \tag{108}$$

The adjoint

$$H^* = H^*_{\rho,\varepsilon} = \begin{bmatrix} A^* & \rho^2 E^* E \\ \varepsilon^{-2} BB^* - DD^* & -A \end{bmatrix} \tag{109}$$

is associated with the *dual* Riccati equation

$$AX + XA^* + \rho^2 XE^*EX - \varepsilon^{-2}BB^* + DD^* = 0 \qquad (\mathrm{ARE}^*_{\rho,\varepsilon})$$

or, equivalently,

$$A(-X) + (-X)A^* - \rho^2(-X)E^*E(-X) + \varepsilon^{-2}BB^* - DD^* = 0.$$

The following lemma is easily verified.

Lemma 8.6. *(i) A non-singular matrix $P \in \mathcal{H}_n$ solves $(ARE_{\rho,\varepsilon})$ if and only if $-P^{-1}$ solves $(ARE^*_{\rho,\varepsilon})$.*

(ii) If (A,E) is observable, then every solution of $(ARE_{\rho,\varepsilon})$ is nonsingular.

Existence results for Riccati equations with semi-definite quadratic term are well-known [23], [25]. They imply, in particular, that $(\mathrm{ARE}^*_{\rho,\varepsilon})$ always has a Hermitian solution if

$$\sigma(H_{\rho,\varepsilon}) \cap i\mathbb{R} = \emptyset \tag{110}$$

In [12] it is shown that if $\varepsilon > 0$ and $0 \leq \rho < \rho(\varepsilon)$ then (110) holds. So it is interesting to determine the set of parameter values (ρ, ε) for which the Hamiltonian $H_{\rho,\varepsilon}$ does not have an eigenvalue on the imaginary axis. This condition can be expressed in terms of the transfer matrices

$$G_D(s) = E(sI - A)^{-1}D \quad , \quad G_B(s) = E(sI - A)^{-1}B.$$

Proposition 8.7. *Suppose that $\sigma(A) \cap i\mathbb{R} = \emptyset$. Then, for all ρ, $\varepsilon > 0$,*

$$\sigma(H_{\rho,\varepsilon}) \cap i\mathbb{R} = \emptyset \iff \rho^{-2} > \sup_{\omega \in \mathbb{R}} \lambda_{\max}(H_\varepsilon(i\omega)) \tag{111}$$

where $\lambda_{\max}(H_\varepsilon(i\omega))$ is the largest eigenvalue of the $q \times q$ Hermitian matrix

$$H_\varepsilon(i\omega) = G_D(i\omega)G_D(i\omega)^* - \varepsilon^{-2}G_B(i\omega)G_B(i\omega)^* \tag{112}$$

As a consequence we have

Proposition 8.8. *Suppose that $\sigma(A) \subset \mathbb{C}_-$. Then*

(a) for any $\varepsilon > 0$

$$\rho(\varepsilon)^{-2} \geq \sup_{\omega \in \mathbb{R}} \lambda_{\max}(H_\varepsilon(\imath\omega)) =: \lambda(\varepsilon) \tag{113}$$

where $H_\varepsilon(\imath\omega)$ is defined by (112).

(b) *If* $\lambda(\varepsilon) > 0$ *then for any* $\rho \in (\rho(\varepsilon), \lambda(\varepsilon)^{-1/2})$, *the dual Riccati equation* ($ARE^*_{\rho,\varepsilon}$) *does not have a positive definite solution.*

(c) *If* $\rho > \lambda(\varepsilon)^{-1/2}$ *then* ($ARE_{\rho,\varepsilon}$) *does not have Hermitian solutions.*

$\lambda(\varepsilon)$ decreases as $\varepsilon \to 0$ and tends to $-\infty$ once it becomes negative for some $\varepsilon > 0$. Let

$$\lambda_0 = \lim_{\varepsilon \to 0} \max\{\lambda(\varepsilon), 0\} \tag{114}$$

Then $\bar{r}_{\mathbb{C}} \leq \lambda_0^{-1/2}$ and by Prop. 8.7

$$\lambda_0^{-1/2} = \sup\{\rho; \rho \in \mathbb{R}, \exists \varepsilon > 0 : \sigma(H_{\rho,\varepsilon} \cap \imath\mathbb{R} = \emptyset\}$$

The quantity $\lambda_0^{-1/2}$ admits an interesting interpretation in terms of stabilizability.

Definition 8.9. The complex *stabilizability radius* of the system $\dot{x} = Ax + Bu$ with respect to structured perturbations $A \mapsto A + D\Delta E$ of the system matrix is defined by

$$\delta(A, B, D, E) = \inf\{\|\Delta\|;\ \Delta \in \mathbb{C}^{\ell \times q},\ (A + D\Delta E, B) \text{ is non-stabilizable}\} \tag{115}$$

We have proved in [12]:

Theorem 8.10. *Suppose that* $\sigma(A) \subset \mathbb{C}_-$. *Then*

$$\delta(A, B, D, E) = |\lambda_0|^{-1/2} \tag{116}$$

As a consequence of the previous proposition, two cases are possible:

1. $\bar{r}_{\mathbb{C}}(A; D, E; B) = \delta(A, B, D, E)$, i.e. the supreme complex structured stability radius achievable by static state feedback is equal to the structured stabilizability radius: This case occurs, roughly speaking, if, with increasing ρ, all Hermitian solutions of ($ARE_{\rho,\varepsilon}$) cease to exist (for small ε) because the spectrum of $H_{\rho,\varepsilon}$ hits the imaginary axis.

2. $\bar{r}_{\mathbb{C}}(A; D, E; B) < \delta(A, B, D, E)$: This case occurs if, with increasing ρ, all (decreasing) solutions of ($ARE_{\rho,\varepsilon}$) in $\mathcal{H}_n^-$ cease to exist, escaping to infinity. If (A, E) is observable, the corresponding positive definite solutions of ($ARE^*_{\rho,\varepsilon}$) all become singular.

Remark 8.11. It is an interesting open question, to determine criteria by which the first case can be distinguished from the second for a given set of data (A, B, D, E). □

We illustrate the second case by an example.

Example 8.12.
We again consider the matrices A, B, D, E given in Example 8.5. Then $\lambda \in \sigma(H_\varepsilon(\imath\omega))$ if and only if $\det\left[(\imath\omega I - A)(\imath\omega I - A)^* - \operatorname{diag}(\frac{-1}{\varepsilon^2\lambda}, \frac{1}{\lambda})\right] = 0$, i.e.

$$\det\begin{bmatrix} \omega^2 + 20 + \frac{1}{\varepsilon^2\lambda} & 5\imath\omega + 14 \\ -5\imath\omega + 14 & \omega^2 + 10 - \frac{1}{\lambda} \end{bmatrix} = 0.$$

Solving the quadratic equation we see that the positive eigenvalue is given by

$$\begin{aligned}\lambda &= \frac{2}{\omega^2+10-\varepsilon^2(\omega^2+20)+\left[(\omega^2+10-\varepsilon^2(\omega^2+20))^2+4\varepsilon^2(\omega^2+1)(\omega^2+4)\right]^{1/2}} \\ &= \frac{2}{\omega^2+10-\varepsilon^2(\omega^2+20)+\left[(\omega^2+10+\varepsilon^2(\omega^2+20))^2-4\varepsilon^2(25\omega^2+196)\right]^{1/2}}.\end{aligned} \tag{117}$$

The first equation gives an upper bound and the second gives a lower bound for λ if ε^2 is sufficiently small:

$$\frac{1}{\omega^2+10} \leq \lambda \leq \frac{1}{\omega^2+10-\varepsilon^2(\omega^2+20)}.$$

Hence

$$\frac{1}{10} \leq \lambda(\varepsilon) \leq \frac{1}{10-20\varepsilon^2}, \; 0 < \varepsilon \ll 1 \quad \text{and} \quad \lambda_0 = \lim_{\varepsilon\to 0}\lambda(\varepsilon) = 1/10.$$

Taking into account what we have shown in Example 8.5 we conclude that

$$\delta(A, B, D, E) = \sqrt{10} > 3 \geq \bar{r}_{\mathbb{C}}(A; D, E; B).$$

□

As a consequence of Theorem 8.10, we have a characterization of those systems whose stability radius cannot be improved by static linear state feedback.

Corollary 8.13. *The complex stability radius of a stable system (A, B, D, E) is optimal with respect to state feedback if and only if its stability radius is equal to its stabilizability radius. In other words, the stability radius $r_{\mathbb{C}}^-(A; D, E)$ cannot be improved by static linear state feedback if and only if for every $\rho > r_{\mathbb{C}}^-(A; D, E)$ and all $\varepsilon > 0$ $\sigma(H_{\rho,\varepsilon}) \cap \imath\mathbb{R} \neq \emptyset$.*

Remark 8.14. If for a given set of data (A, B, D, E) there exists a *bounded* optimizing sequence (F_k) in $\mathbb{C}^{m\times n}$ then an easy compactness argument combined with the continuity of the map $A \mapsto r_{\mathbb{C}}^-(A; D, E)$ proves that there exists an optimal solution $F \in \mathbb{C}^{m\times n}$ such that

$$r_{\mathbb{C}}^-(A + BF; D, E) = \lim_{k\to\infty} r_{\mathbb{C}}^-(A + BF_k; D, E) = \bar{r}_{\mathbb{C}}(A; D, E; B).$$

Since the stabilizability radius is invariant with respect to feedback action, it follows from the previous Cor.8.13 that in this case we necessarily have

$$\bar{r}_{\mathbb{C}}(A; D, E; B) = \delta(A, B, D, E)$$

Hence, if $\bar{r}_{\mathbb{C}}(A; D, E; B) < \delta(A, B, D, E)$ there does not exist a feedback matrix maximizing the stability radius: in other words, *a gap between* $\bar{r}_{\mathbb{C}}(A; D, E; B)$ *and* $\delta(A, B, D, E)$ *causes high gain feedback.* □

REFERENCES

[1] R. Brockett. *Finite Dimensional Linear Systems.* J. Wiley, 1970.

[2] M.J. Chen and C.A. Desoer. Necessary and sufficient condition for robust stability of linear distributed feedback systems. *International Journal of Control*, 35:255–267, 1982.

[3] F. Colonius and W. Kliemann. Infinite time optimal control and periodicity. *Applied Mathematics and Optimization*, 20:113–130, 1989.

[4] J. Doyle. Analysis of feedback systems with structured uncertainties. *Proc. IEE*, 129:242–250, 1982.

[5] J.C Doyle, B.A. Francis, K. Glover, and P.P. Khargonekar. State space solutions to the standard H_2 and H_∞ control problems. In *Proc. American Control Conference*, Atlanta. U.S.A., 1988. submitted to IEEE Trans. Aut. Contr.

[6] B.A. Francis. *A course in H_∞ control theory.* Lecture Notes in Control and Information Sciences 88. Springer Verlag, Berlin-Heidelberg-New York, 1987.

[7] D. Hinrichsen, A. Ilchmann, and A.J. Pritchard. Robustness of stability of time-varying linear systems. *Journal of Differential Equations*, 82(2):219–250, 1989.

[8] D. Hinrichsen, B. Kelb, and A. Linnemann. An algorithm for the computation of the complex stability radius. *Automatica*, 25:771–775, 1989.

[9] D. Hinrichsen and M. Motscha. Optimization problems in the robustness analysis of linear state space systems. In *Approximation and Optimization*, LN Mathematics 1354, pages 54–78, Berlin-Heidelberg-New York, 1988. Springer Verlag.

[10] D. Hinrichsen and A. J. Pritchard. Robustness measures for linear state space systems under complex and real parameter perturbations. In *Proc. Summer School Perspectives in Control Theory*, Sielpia, 1988. Birkhäuser (Boston). to appear.

[11] D. Hinrichsen and A. J. Pritchard. A note on some differences between real and complex stability radii. Report 215, Institut für Dynamische Systeme, Universität Bremen, 1989. To appear. in: Systems & Control Letters.

[12] D. Hinrichsen and A. J. Pritchard. Parametrized Riccati equations and the problem of maximizing the complex stability radius. In S. Bittanti, editor, *Proc. Workshop "The Riccati Equation in Control, Systems and Signals"*, pages 136–142. Pitagora Editrice Bologna, 1989. Enlarged version to appear in: International Journal of Control.

[13] D. Hinrichsen and A. J. Pritchard. Destabilization by output feedback. Report 218, Institut für Dynamische Systeme, Universität Bremen, 1990.

[14] D. Hinrichsen and A.J. Pritchard. Stability radii of linear systems. *Systems & Control Letters*, 7:1–10, 1986.

[15] D. Hinrichsen and A.J. Pritchard. Stability radius for structured perturbations and the algebraic Riccati equation. *Systems & Control Letters*, 8:105–113, 1986.

[16] D. Hinrichsen and A.J. Pritchard. An application of state space methods to obtain explicit formulae for robustness measures of polynomials. In M. Milanese et al., editor, *Robustness in Identification and Control*, pages 183–206. Birkhäuser, 1989.

[17] D. Hinrichsen and N.K. Son. The complex stability radius of discrete-time systems and symplectic pencils. In *Proc. 28th IEEE Conference on Decision and Control*, pages 2265–2270, Tampa, 1989.

[18] V.L. Kharitonov. Asymptotic stability of an equilibrium position of a family of systems of linear differential equations. *Diff. Uravn.*, 14:2086–2088, 1979.

[19] T. Pappas, A. J. Laub, and N. R. Sandell. On the numerical solution of the discrete-time algebraic Riccati equation. *IEEE Transactions on Automatic Control*, AC-25:631–641, 1980.

[20] I.R. Petersen. Disturbance attenuation and H_∞-optimization: a design method based on the algebraic Riccati equation. *IEEE Transactions on Automatic Control*, AC-32(5):427–429, 1987.

[21] A.J. Pritchard and S. Townley. Robustness of linear systems. *Journal of Differential Equations*, 77(2):254–286, 1989.

[22] J.C. Willems. *The Analysis of Feedback Systems.* Research Monograph No. 62. The M.I.T. Press, Massachusetts, 1971.

[23] J.C. Willems. Least squares optimal control and the algebraic Riccati equation. *IEEE Transactions on Automatic Control*, AC-16:621–634, 1971.

[24] J.C. Willems. Almost invariant subspaces: An approach to high gain feedback design - Part I: Almost controlled invariant subspaces. *IEEE Transactions on Automatic Control*, AC-26:235–252, 1981.

[25] H.K. Wimmer. The algebraic Riccati equation without complete controllability. *SIAM J. Alg. Disc. Meth.*, 3:1–12, 1982.

[26] G. Zames. Feedback and optimal sensitivity: Model reference transformations, multiplicative seminorms, and approximate inverses. *IEEE Transactions on Automatic Control*, AC-26:301–320, 1981.

[27] K. Zhou and P.P. Khargonekar. An algebraic Riccati equation approach to H^∞ optimization. *Systems & Control Letters*, 11(2):85–91, 1988.

D. Hinrichsen
Institut für Dynamische Systeme
Universität Bremen
D-2800 Bremen 33
F.R.G.
dh@mathematik.uni-Bremen.de

A. J. Pritchard
Control Theory Centre
University of Warwick
Coventry CV4 7AL
UK
ajp@maths.warwick.ac.uk

Stabilization of Uncertain Sampled-Data Systems

K.H. Kienitz*

In this contribution the problem of obtaining stabilizing controllers for a class of sampled-data uncertain systems is investigated. Particular attention is given to the choice of a suitable mathematical discrete-time model for the uncertain system. A Lyapunov-based approach is then used to propose stabilizing extended state-feedback controllers and investigate the stability of the controlled system.

1 Introduction

Recently some effort has been devoted to the problem of obtaining stabilizing controllers for discrete-time uncertain systems [10], [3], [4], [6], [9]. Such systems are described by difference equations containing time-varying uncertainties. No statistical description is assumed for them. A Lyapunov min-max approach has been used to solve the stabilization problem; it combines the construction of a quadratic Lyapunov function and the design of a stabilizing memoryless full-state feedback controller. In most of the cited references it was explicitly assumed, that the uncertainties satisfy the so-called "matching condition", i.e., that the range of the uncertainties lies within the range of the system input matrix. In [6] a more general matching condition was considered.

However, particularly in sampled-data systems, "matching conditions" generally are not satisfied [10]. Nevertheless the design of discrete-time stabilizing controllers was extended to sampled-data systems for the case of continuous-time matched uncertainty in an approximate way in [10].

*Supported by Instituto Tecnológico de Aeronáutica, Brazil.

In the present contribution the problem of obtaining stabilizing state-feedback controllers for a class of sampled-data uncertain systems is investigated. Particular attention is given to the choice of an adequate discrete-time description of these systems. To gain some information about the uncertainty's structural properties after the sampling process, mismatch measures are defined based on generalized Euclidean norms. Roughly speaking, mismatch measures give information about which "amount" of uncertainty is "matched", i.e. lies in the range of the nominal system input matrix. (For the use of the matching property in previous research see [2].) Further, the information available about the uncertainty through the chosen generalized Euclidean norm and the mismatch measures, as well as a nominal discrete-time model of the system, is used in the Lyapunov-based design of a digital stabilizing extended state-feedback controller for the uncertain system. This means that not only the state vector but also the value of the last control vector will be used to compute the new control value. It turns out that the choice of the generalized Euclidean norm corresponds to the choice of a quadratic Lyapunov function for the controlled uncertain system. In the choice of this Lyapunov function, knowledge about the uncertainty's structure may be considered using non-linear programming methods.

The text is divided into six sections and an appendix with some definitions, formulas and proofs. In section 2 the control problem is stated. Section 3 discusses the construction of an adequate mathematical model for the uncertain sampled-data system. The next section addresses controller design and stability issues via a Lyapunov approach. In section 5 the choice of adequate Lyapunov functions for the overall system is discussed. Conclusions are found in section 6.

2 Problem statement

Consider an uncertain continuous-time system Σ described adequately by (1) with zero-order sample-holds (sampling period T) on its inputs.

$$\Sigma \; : \quad \dot{x}(t) = A(r)x(t) + B(r)u(t) + w(t) \tag{1}$$

where

$$x(t) \in \Re^n \quad \text{is the state vector;}$$

$$u(t) \in \Re^m \quad \text{is the control vector;}$$

$$A(r) = A_0 + \Delta A(r) = A_0 + \sum_{i=1}^{p} [r_i(t) A_i], \;\; A_i \in \Re^{n \times n}, \;\; i = 0, \ldots, p;$$

$$B(r) = B_0 + \Delta B(r) = B_0 + \sum_{i=1}^{p}[r_i(t)B_i],\; B_i \in \Re^{n\times m},\; i = 0,\ldots,p;$$

$$r = r(t) = [r_1(t)\ldots r_p(t)]^T \in \Re^p \text{ and } w(t) \in \Re^n, \text{ where}$$

$$\underline{r} \leq r(t) \leq \bar{r} \text{ and } \underline{w} \leq w(t) \leq \bar{w} \text{ with}$$

$$\underline{r}, \bar{r} \in \Re^p \text{ and } \underline{w}, \bar{w} \in \Re^n.$$

$r(t)$ and $w(t)$ model the uncertainties affecting Σ and have components which are piecewise continuous, unknown, but bounded functions of time with known bounds $\underline{r}$, $\bar{r}$ and $\underline{w}$, $\bar{w}$ respectively.

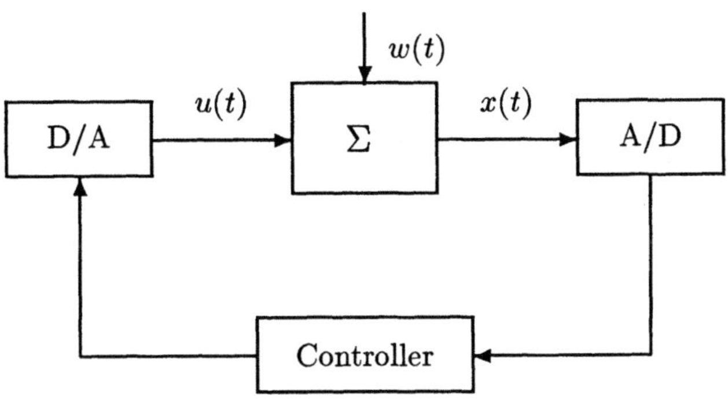

Figure 1. Uncertain sampled-data system

Figure 1 shows a diagram of the uncertain sampled data system. Strings D/A and A/D identify digital-analog and analog-digital converters respectively. The system states are supposed to be available for feedback. In the absence of uncertainty, i.e. for $r = 0$ and $w = 0$, the system is given by the matrix pair (A_0, B_0). This is called the nominal system.

Remark 1. Because of the sample-holds, $u(t)$ in (1) has the special form:

$$u(t) = u_k \quad \forall t \in [kT, (k+1)T), \quad k = 0, 1, \ldots$$

Note that this control function may be generated with a suitable sequence $\{v_k\}$ by the following system:

$$\dot{u}(t) = \sum_{k=0}^{\infty} v_k \delta[t - (k+1)T] = v(t), \quad u(0) = u_0 \tag{2}$$

where $\delta(.)$ is the Kronecker-delta function. From (2) it is seen that

$$u_{k+1} = u_k + v_k.$$

□

Recall the stability definition given in the appendix and consider (1) and (2). The problem treated henceforth consists of determining a control sequence $\{v_k\}$ and restrictions on the uncertainty, such that the controlled uncertain system is globally uniformly asymptotically stable (g.u.a.s.) about a bounded set B, which contains the origin (0). Additionally a condition assuring $B = \{0\}$ shall be given. The derivation of appropriate restrictions on the uncertainty is necessary because discrete-time systems generally cannot be stabilized for every uncertainty in (1) [6].

The stabilization task shall be accomplished using an extended static state-feedback; i.e., not only the state vector x, but also the value of the last control vector will be used to compute the new control. Thus define the extended state-vector $z = [x^T u^T]^T$ and consider (1) and (2). The extended continuous-time system equation is:

$$\dot{z}(t) = A_E(r)z(t) + B_E v(t) + D_E w(t) \tag{3}$$

where

$$A_E = \begin{bmatrix} A(r) & B(r) \\ 0 & 0 \end{bmatrix} = A_{E0} + \Delta A_E(r) = \begin{bmatrix} A_0 & B_0 \\ 0 & 0 \end{bmatrix} + \begin{bmatrix} \Delta A(r) & \Delta B(r) \\ 0 & 0 \end{bmatrix}$$

$$B_E = \begin{bmatrix} 0 \\ I_{m\times m} \end{bmatrix} \quad D_E = \begin{bmatrix} I_{n\times n} \\ 0 \end{bmatrix}$$

The nominal extended system is given by (4).

$$\dot{z}_n(t) = A_{E0}(r)z_n(t) + B_E v(t) \tag{4}$$

3 A description for the uncertain sampled-data system

In this section a discrete-time description of system (3) subject to (2) is derived to allow a discrete-time stabilizing controller design in section 4. The derivation of a discrete-time description is motivated by the next theorem.

Theorem 1. *Consider a difference equation which describes (3) subject to (2) at the sampling instants. If a controller stabilizes the origin for this discrete-time system in the sense of Lyapunov, then it also does so for (3) subject to (2). Furthermore, if in (3) $w \equiv 0$, and the controller asymptotically stabilizes the origin for the discrete-time system, then it also asymptotically stabilizes it for (3) subject to (2).*

PROOF. Consider (3) subject to (2). Choose any vector norm $\|.\|$ and denote the associated matrix measure by $\mu(.)$. Due to the special form of $v(t)$, the state trajectory between two consecutive but otherwise arbitrary sampling instants is bounded as follows (see [12])

$$\|z(t)\| \leq e^{\bar{\mu}(A_E)(t-kT)}\|z_k\| + \frac{e^{\bar{\mu}(A_E)(t-kT)} - 1}{\bar{\mu}(A_E)} \max_w \|D_E w\| \, ,$$

$$kT \leq t < (k+1)T \, ,$$

where $z_k = z(kT)$ and $\bar{\mu}(A_E) = \max_r \mu(A_E)$. Under the hypothesis of the theorem the state vector sequence $\{z_k\}$ and w are bounded. Thus the origin will be stable in the sense of Lyapunov for (3) subject to (2). If $w \equiv 0$ and $\{z_k\}$ converges to the origin, then the origin will be asymptotically stable for the overall uncertain system in question because

$$\lim_{k \to \infty} \|z_k\| = 0.$$

Q.E.D.

Before deriving a discrete-time description for (3) at the sampling instants, one should observe that the ordinary discretization procedure for time-invariant continuous-time systems (solution of the state equation) is not applicable here because of the possibly time-varying uncertain functions r and w . One could ask which kind of difference equations is being looked for. Some authors have considered interval parameter matrix descriptions such as (1) also in the discrete-time case (see the application example in [3]). While its usefulness in the treatment of continuous-time and *pure* discrete-time systems has been shown, its use in the sampled-data case with time-varying uncertainty is at least problematic, since there is no known analytic solution to the state equations. Examples may be easily constructed to show that in general the uncertainty of a sampled-data system does not depend linearly on the uncertain parameters of the continuous-time system, even when the uncertainty is not time-varying.

The two facts exposed above, namely that system uncertainty is possibly time-varying and that discrete-time interval parameter matrix descriptions for sampled-data systems are problematic even when uncertainty is constant during each sampling period, led to the choice of a different system representation. The discrete-time model to be used is of the form (5).

$$z(k+1) = F_0 z(k) + Gv(k) + \xi(k+1) \tag{5}$$

where z is the state vector of the continuous-time system, $v(k)$ is the control sequence and ξ stands for the uncertainty. F_0 and G are matrices of suitable dimensions. Expressions for ρ_0 and ρ_1 will be derived, such that for any vector norm $\|.\|$ and any $z(k)$ the following bound holds:

$$\|\xi(k+1)\| \leq \rho_0 + \rho_1 \|z(k)\|. \tag{6}$$

Structural information on the uncertainty will be considered using mismatch measures, which size up the uncertainty "portion" outside the range of the discrete-time system input matrix. With respect to the P-inner product defined by $\langle x, y\rangle_P = x^T P y$, the uncertainty (after sampling) is decomposed into two orthogonal components: a matched one and a mismatched one. This is done as in (7) using the projection GH, which is the orthogonal projection with respect to the P-inner product, onto the range of G.

$$\begin{gathered} \xi = (I - GH)\xi + GH\xi \\ (GH\xi)^T P (I - GH)\xi = 0, \quad P > 0 \end{gathered} \tag{7}$$

The mismatched part of the uncertainty is then described as follows:

$$\|(I - GH)\xi(k+1)\|_P \leq a\rho_0 + b\rho_1 \|z(k)\|_P, \quad a, b \leq 1 \tag{8}$$

where

$$\|\xi\|_P = \sqrt{\langle \xi, \xi \rangle_P}.$$

Expressions for a and b will be derived in addition to those for ρ_0 and ρ_1. Because the input $v(t)$ of the extended system is a sum of delta-functions, the discrete version of (3), although not known completely, may be written as

$$z(k+1) = F_0 z(k) + B_E v(k) + \xi(k+1) \tag{9}$$

where $F_0 = e^{A_{E0} \cdot T}$, and $\xi(k+1)$ stands for the uncertainty from (3). To derive expressions for ρ_0, ρ_1, a and b, let (4) and (9) have the same initial condition z_k at $t_0 = kT$. Then

$$z(t) = z_n(t) + [z(t) - z_n(t)] = z_n(t) + \xi(t), \quad t \geq t_0$$

and

$$\dot{\xi}(t) = \dot{z}(t) - \dot{z}_n(t) = A_E \xi(t) + \Delta A_E z_n(t) + D_E w(t), \quad \xi(t_0) = 0. \tag{10}$$

Theorem 2. *Let $\|.\|_P$ denote the generalized Euclidean vector norm defined in (8) and let it also denote its induced matrix norm. Let $\mu(.)$ be the associated matrix measure (for details on $\mu(.)$ see [5]). Then for ξ in (9) bounds (6) and (8) hold with*

$$\rho_0 = \frac{e^{\bar{\mu}(\tilde{A})T} - 1}{\bar{\mu}(\tilde{A})} \|D_E w\|_P^{\max}$$

$$\rho_1 = \frac{e^{\bar{\mu}(\tilde{A})T} - e^{\mu(A_{E0})T}}{\bar{\mu}(\tilde{A}) - \mu(A_{E0})} \max_r \|\Delta A_E(r)\|_P$$

$$a = \frac{\|HD_E w\|_P^{\max}}{\|D_E w\|_P^{\max}}$$

$$b = \frac{\max_r \|H\Delta A_E(r)\|_P}{\max_r \|\Delta A_E(r)\|_P}$$

where

$$H = -(B_E^T P B_E)^{-1} B_E^T P$$

$$\|D_E w\|_P^{\max} = \max_{t\in[kT;(k+1)T]} \|D_E w(t)\|_P$$

$$\bar{\mu}(\tilde{A}) = \max_r \mu\left(\begin{bmatrix} A(r) & 0 \\ 0 & 0_{m\times m}\end{bmatrix}\right)$$

For a proof of this theorem consult the appendix.

Remark 2. In the derivation the theorem, the following crucial fact was needed: the extended system and the nominal extended system have the same input matrix B_E. Further this matrix is not affected by uncertainty, and ξ does not depend (directly) on the control variable v. The desirability of this fact justified the system extension used herein and also gives an idea of the additional "difficulty" when trying to deal directly with (1), where it generally does not hold. □

The usefulness of the previous theorem would be limited, if the maximization operations needed to determine ρ_0, ρ_1, a and b would require extensive numerical optimization procedures. The next known lemma shows that these aren't necessary. Based on norm and matrix measure convexity, it guarantees that for the maximizations in r and w on the domains given in (1), only the polytope corners defined by their bounds have to be considered.

Lemma 3. *Consider a convex function $g(r_1, \ldots, r_p)$. Then the maximum of $g(.)$ on a domain*

$$D = \{r \in \Re^p \ / \ \underline{r} \le r(t) \le \bar{r} \text{ with } \underline{r}, \bar{r} \in \Re^p\}$$

lies on one of its corners, i.e., on a point r_0 where each vector component r_i assumes either its maximum or its minimum value.

4 The Lyapunov min-max controller design

A Lyapunov approach is followed to propose candidate stabilizing controllers for Σ . For this purpose consider some quadratic Lyapunov function candidate $V(z) = \|z\|_P^2 > 0$. The Lyapunov difference then is:

$$\Delta V = V[z(k+1)] - V[z(k)] = \|F_0 z(k) + B_E v(k) + \xi(k+1)\|_P^2 - \|z(k)\|_P^2 \quad (11)$$

Remembering that g.u.a. stability about the zero state is guaranteed if $\Delta V < 0$, $\xi = \xi_w$ which maximizes $\Delta V(\xi, v)$ with respect to all allowed ξ is regarded as the worst admissible uncertainty in the context of a Lyapunov approach. In the same context, a controller which minimizes $\Delta V(\xi_w, v)$ with respect to the control variable v is a "strongly recommended" candidate stabilizing controller. However, the difficulty of working directly with this min-max approach for ΔV was already mentioned in [10] and [4]. It showed to be more tractable to consider upper bounds on ΔV rather than ΔV itself. In fact, for the system class under consideration, analytical expressions are available only for upper bounds.

To develop a suitable upper bound on ΔV rewrite (11) as (12), recalling the definition of the matrix H given in theorem 2.

$$\begin{aligned} \Delta V = \; & \|F_0 z(k) + B_E v(k)\|_P^2 + \|\xi(k+1)\|_P^2 - \|z(k)\|_P^2 \\ & +2[(I + B_E H) F_0 z(k)]^T P \xi(k+1) \\ & +2\{B_E[v(k) - H F_0 z(k)]\}^T P \xi(k+1) \end{aligned} \quad (12)$$

Now arbitrate the following upper bound on ΔV:

$$\begin{aligned} \Delta V \le \Delta V_{\text{sup}} = \; & \|F_0 z(k) + B_E v(k)\|_P^2 + \|\xi(k+1)\|_P^2 - \|z(k)\|_P^2 \\ & +2[(I + B_E H) F_0 z(k)]^T P \xi(k+1) \\ & +2\|B_E[v(k) - H F_0 z(k)]\|_P \; \|\xi(k+1)\|_P. \end{aligned} \quad (13)$$

Other upper bounds could be defined for ΔV; this one was chosen because as a consequence of the next theorem it satisfies the property

$$\Delta V_{\text{sup}}(v^*) = \min_{v} \max_{\text{possible } \xi} \Delta V_{\text{sup}}(v, \xi) = \Delta V(v^*).$$

Further it will most naturally allow the use of mismatch measures in the stability analysis of the uncertain controlled system.

The control v^* which minimizes ΔV_{sup} with respect to v is chosen as a candidate stabilizing controller for the uncertain system.

Theorem 4. *The unique solution of the minimization problem*

$$\Delta V_{\text{sup}}(k, z, v^*) = \min_{v} \max_{\text{possible } \xi} \Delta V_{\text{sup}}(k, z, v, \xi)$$

is

$$v^*(k) = HF_0 z(k). \tag{14}$$

PROOF. Observe that, for all z, $v^*(k) = HF_0 z(k)$ is the unique minimizer of both, $\|F_0 z(k) + B_E v(k)\|_P^2$ and $\|B_E[v(k) - HF_0 z(k)]\|_P$. Q.E.D.

Considering ΔV (and not $\Delta V_{\sup}$) it was shown in [6] that (14) is the unique min-max Lyapunov controller if ξ is "balanced" (i.e., if ξ is an admissible uncertainty, then $-\xi$ also is an admissible uncertainty) and satisfies the "matching condition" (i.e., there exists a function $e(k, z)$ such that $\xi = B_E e(k, z)$).

Candidate stabilizing controllers parameterized in P were proposed for the uncertain system. However, system stability has not yet been guaranteed for any of these controllers. For an overall stability analysis reconsider the rate of increase of the chosen Lyapunov function candidate for the uncertain sampled-data system subject to control (14):

$$\begin{aligned}\Delta V_{\sup} = \Delta V = \|F_0 z(k) + B_E v^*(k)\|_P^2 + \|\xi(k+1)\|_P^2 - \|z(k)\|_P^2 \\ +2[(I + B_E H)F_0 z(k)]^T P\xi(k+1).\end{aligned}$$

$\Delta V < 0$ guarantees g.u.a.s. of the origin, but clearly ΔV may be positive for some imaginable ξ. Therefore the aim is to establish conditions on ξ such that controllers defined in (14) yield stable behavior for the uncertain controlled system. For the sake of conciseness the following notation is now introduced:

$$F_0 z(k) + B_E v^*(k) = F_c z(k)$$

i.e., F_c is the nominal controlled system matrix.

Theorem 5. *Consider system (9) subject to control (14) and with ξ satisfying (6) and (8). Then the given controlled system is g.u.a.s. about*

$$B = \{z \in \Re^n \;/\; \|z\|_P < d\}$$

with

$$d = \rho_0 \frac{\rho_1 + a\|F_c\|_P + \sqrt{1 - 2\rho_1(b-a)\|F_c\|_P - (1-a^2)\|F_c\|_P^2}}{1 - \rho_1^2 - 2b\rho_1\|F_c\|_P - \|F_c\|_P^2} \tag{15}$$

if

$$\rho_1 < \sqrt{1 - (1-b^2)\|F_c\|_P^2} - b\|F_c\|_P \quad (\textit{and } \|F_c\|_P < 1) \tag{16}$$

PROOF. Because of the special form of F_c, the Lyapunov difference of the controlled system may be rewritten as

$$\Delta V_{\text{sup}} = \Delta V = \|F_0 z(k) + B_E v^*(k)\|_P^2 + \|\xi(k+1)\|_P^2 - \|z(k)\|_P^2 \\ +2[F_c z(k)]^T P(I + BH)\xi(k+1).$$

Consider (6) and (8). Then ΔV clearly will be negative if

$$(\|F_c\|_P^2 + \rho_1^2 + 2b\rho_1\|F_c\|_P - 1)\|z\|_P^2 + 2(\rho_0\rho_1 + \rho_0 a\|F_c\|_P)\|z\|_P + \rho_0^2 < 0.$$

This inequality is satisfied iff (16) and $\|z\|_P \geq d$ hold, where d is given as in (15). Q.E.D.

If additionally to (16) one has $\rho_0 = 0$, then $B = \{0\}$, i.e., the controlled uncertain system is g.u.a.s. about the origin. In the hypothetical case of matched uncertainties $a = 0$, $b = 0$ and the result of the theorem would be identical to that in [4].

5 Choosing adequate Lyapunov function candidates

In the foregoing section a family of candidate stabilizing controllers (parameterized in P) was proposed, and the stability of the controlled uncertain system was investigated. For the selection of a single controller out of the proposed family, the choice of the real positive definite but otherwise still undefined matrix P is required. For a first choice of P, consider the controlled nominal system

$$z(k+1) = F_c z(k).$$

In theorem 5 it is required that $\|F_c\|_P < 1$. Therefore one may choose $P > 0$ such that this inequality is satisfied. This is equivalent to choosing P as the solution of the Riccati equation

$$F_c^T(P)PF_c(P) - P = -Q, \quad Q > 0.$$

However, because all information on the uncertainty is neglected, the resulting stability analysis with theorem 5 may be conservative.

Obviously it is possible to consider information on the uncertainty only if expressions are known for ρ_1 and b. As this is the case for the present class of sampled-data systems, an alternative method for determining Lyapunov matrices will be outlined.

According to theorem 5, system stability is guaranteed if (16) and $P > 0$ are satisfied. Thus, following an idea in [8], the choice of an adequate

Lyapunov function can be formulated as an unconstrained minimization problem by means of the Courant penalty function concept (see [13]).

$$f(P_g) = \min_P f(P) = \min_P \{\max{}^2(0; \|F_c\|_P^2 + \rho_1^2 + 2b\rho_1\|F_c\|_P - 1) + \max{}^2[0; \lambda_{\min}(P)]\} \quad (17)$$

Stability is then guaranteed if $f(P_g) = 0$. The first term of $f(P)$ is due to the stability constraint (16). The second one considers the requirement $P > 0$. The choice of (17) is not unique and other, constrained or even non-smooth minimization problems could be considered.

Anyway, the choice of P_g may be accomplished using gradient-based non-linear programming techniques. In this context it is desirable to determine gradients of the problem function analytically. Because ρ_1, b and $\|F_c\|_P$ are more or less complex functions of P, the derivation of analytical expressions for $\delta f/\delta P$ (P symmetric) may be lengthy and tedious. For $f(P)$ specified in (17) it may be accomplished with the use of some results compiled in [11]. Auxiliary derivatives which allow the straightforward construction of $\delta f/\delta P$ may be found in the appendix.

6 Conclusions

In this contribution a class of uncertain sampled-data systems was considered for stabilization via static extended state-feedback. By system extension one sampling period is allowed for control value computation. The uncertainties are of the time-varying bounded parameter type.

The most critical part of the stabilization procedure turned out to be the choice of an adequate Lyapunov function for the controlled system. Although the restriction to quadratic time-invariant Lyapunov functions is a limitation, it allowed for the specification of practical Lyapunov function choice procedures and a straightforward stability analysis with the help of established concepts such as vector and matrix norms, and matrix measures.

APPENDIX

A Stability definition

Definition 1. [4] Consider a non-empty subset B of $\Re^n$ and a discrete-time system described by

$$x(k+1) = F(k, x(k)). \quad (18)$$

Under a given norm $\|.\|$ system (18) is said to be globally uniformly asymptotically stable (g.u.a.s.) about B iff it has the following properties:

- *Global uniform boundedness of solutions:* For each $r \geq 0$, there exists $d(r) \geq 0$ such that $\|x(k)\| \leq d(r)$ for all $k \geq 0$ and for any solution $x(.)$ of (18) with $\|x(0)\| \leq r$.
- *Uniform stability about B:* For each neighborhood N of B , there exists a neighborhood M of B such that $x(k) \in M$ for $k \geq 0$ and for any solution $x(.)$ of (18) with $x(0) \in N$.
- *Global uniform attractiveness of B :* For each neighborhood N of B and for each $r \geq 0$, there exists $T(N,r) \geq 0$ such that $x(k) \in N$ for all $k \geq T(N,r)$ and for any solution $x(.)$ of (18) with $\|x(0)\| \leq r$.

B Proof of theorem 2

The proof closely follows the proofs of theorem 1, pg. 89, and theorem 32, pg. 93, of [12]. From the integral form of (10),

$$\xi(t+\delta) = \xi(t) + \delta[A_E(r)\xi(t) + \Delta A_E(r) z_n(t) + D_E w(t)] + o(\delta), \ \forall \delta > 0 \quad (19)$$

where $o(\delta)$ is used to denote an error item with the property that

$$\lim_{\delta \to 0+} \frac{\|o(\delta)\|}{\delta} = 0.$$

Rearranging (19) and noting from (10) that

$$\xi = \tilde{I}\xi \quad \text{where} \quad \tilde{I} = \begin{bmatrix} I_{n\times n} & 0_{n\times m} \\ 0_{m\times n} & 0_{m\times m} \end{bmatrix},$$

we get successively

$$\xi(t+\delta) = [I + \delta A_E(r)\tilde{I}]\xi(t) + \delta \Delta A_E(r) z_n(t) + \delta D_E w(t) + o(\delta)$$

$$\begin{aligned} \|\xi(t+\delta)\|_P - \|\xi(t)\|_P = {} & (\|[I + \delta A_E(r)\tilde{I}]\|_P - \|I\|_P)\|\xi(t)\|_P \\ & + \delta \|\Delta A_E(r)\|_P \|z_n(t)\|_P + \delta \|D_E w(t)\|_P + \|o(\delta)\|_P \end{aligned}$$

$$\begin{aligned} \frac{d^+}{dt}\|\xi(t)\|_P \leq {} & \left[\lim_{\delta \to 0^0} \frac{\|[I + \delta A_E(r)\tilde{I}]\|_P - \|I\|_P}{\delta}\right] \|\xi(t)\|_P \\ & + \|\Delta A_E(r)\|_P \|z_n(t)\|_P + \|D_E w(t)\|_P \end{aligned}$$

where

$$\lim_{\delta \to 0^0} \frac{\|[I + \delta A_E(r)\tilde{I}]\|_P - \|I\|_P}{\delta} \leq \bar{\mu}(A_E(r)\tilde{I}).$$

Recalling that $\xi(t_0 = kT) = 0$ and proceeding similarly as in the proof of theorem 32, pg. 93, of [12], the expressions given for ρ_0 and ρ_1 are derived.

Multiplying (10) by $H = [I - B_E(B_E^T P B_E)^{-1} B_E^T P]$ one gets

$$H\dot{\xi}(t) = HA_E(r)\xi(t) + H\Delta A_E(r) z_n(t) + HD_E w(t), \quad H\xi(t_0) = 0. \tag{20}$$

Because of the special forms of ξ and H, (20) may be rewritten as

$$[H\dot{\xi}(t)] = HA_E(r)\tilde{I}[H\xi(t)] + H\Delta A_E(r) z_n(t) + HD_E w(t) = 0, \; H\xi(t_0) = 0$$

Hence

$$H\xi(t+\delta) = H[I + \delta A_E(r)\tilde{I}]H\xi(t) + \delta[H\Delta A_E(r) z_n(t) + HD_E w(t)] + o(\delta)$$

where $o(d)$ is used to denote the error item. Noting that $\|H\|_P = 1$, it follows that

$$\frac{d^+}{dt}\|H\xi(t)\|_P \leq \left[\lim_{\delta \to 0^0} \frac{\|[I + \delta A_E(r)\tilde{I}]\|_P - \|I\|_P}{\delta}\right] \|H\xi(t)\|_P$$
$$+\|H\Delta A_E(r)\|_P \|z_n(t)\|_P + \|HD_E w(t)\|_P.$$

The expressions for $a\rho_0$ and $b\rho_1$ are now obtained in a similar way as those for ρ_0 and ρ_1.

C Auxiliary derivatives for the non-linear programming problem of section 5

The problem function given in (17) is

$$f(P_g) = \min_P f(P) = \min_P \{\max{}^2(0; \|F_c\|_P^2 + \rho_1^2 + 2b\rho_1\|F_c\|_P - 1)$$
$$+\max{}^2[0; \lambda_{\min}(P)]\}.$$

To readily construct $\delta f/\delta P$ the following expressions are needed (where not otherwise stated, symbols have the same meaning as in previous statements):

1)

$$\frac{\delta\|F_c\|_P}{\delta P} = \frac{F_c\hat{x}\hat{x}^T F_c^T - \hat{x}\hat{x}^T\|F_c\|_P^2}{2\|F_c\|_P \hat{x}^T P\hat{x}}$$

with $\hat{x}$ such that

$$\|F_c\|_P^2 = \max_x \frac{x_T F_c^T P F_c x}{x_T P x} = \frac{\hat{x}_T F_c^T P F_c \hat{x}}{\hat{x}_T P \hat{x}}.$$

2)

$$\frac{\delta\lambda_{\min}(P)}{\delta P} = x_{\min}\, x_{\min}^T$$

where $x_{\min}$ is the eigenvector associated to the minimum eigenvalue of P, and $\|x_{\min}\|_2 = 1$.

3)

$$\frac{\delta\rho_1}{\delta P} = \frac{n}{d}\frac{\delta m}{\delta P} + \frac{m}{d}\left[Te^{\bar{\mu}(\tilde{A})T} - \frac{n}{d}\right]\frac{\delta\bar{\mu}(\tilde{A})}{\delta P}$$
$$+\frac{m}{d}\left[\frac{n}{d} - Te^{\mu(A_{E0})T}\right]\frac{\delta\bar{\mu}(\tilde{A})}{\delta P}$$

with

$$n = n(P) = e^{\bar{\mu}(\tilde{A})T} - e^{\mu(A_{E0})T}$$
$$d = d(P) = \bar{\mu}(\tilde{A}) - \mu(A_{E0})$$
$$m = m(P) = \|\Delta A_E(r)\|_P.$$

4)

$$\frac{\delta b\rho_1}{\delta P} = \frac{n}{d}\frac{\delta m}{\delta P} + \frac{m}{d}\left[Te^{\bar{\mu}(\tilde{A})T} - \frac{n}{d}\right]\frac{\delta\bar{\mu}(\tilde{A})}{\delta P}$$
$$+\frac{m}{d}\left[\frac{n}{d} - Te^{\mu(A_{E0})T}\right]\frac{\delta\bar{\mu}(\tilde{A})}{\delta P}$$

with n and d as before and

$$m = m(P) = \|H\Delta A_E(r)\|_P.$$

5) For square matrices M

$$\frac{\delta\mu(M)}{\delta P} = \frac{1}{2}\frac{M\hat{x}\hat{x}^T + \hat{x}\hat{x}^T M - 2\mu(M)\hat{x}\hat{x}^T}{\hat{x}^T P\hat{x}}$$

where $\hat{x}$ is such that

$$\mu(M) = \frac{1}{2}\max_x \frac{x^T(M^TP + PM)x}{x^TPx} = \frac{1}{2}\frac{\hat{x}^T(M^TP + PM)\hat{x}}{\hat{x}^TP\hat{x}}.$$

The calculation of all quantities does not present any problem. In fact only routines for simple matrix algebra, singular value decomposition and maximum (or minimum) eigenvalue-eigenvector calculation are needed. All derivatives were extensively checked by comparison with numerical gradients, and were used in several examples.

Remark 3. All expressions listed above are intended for symmetric matrices P. In the non-symmetric case several modifications are necessary. □

REFERENCES

[1] B. R. Barmish; M. Corless; G. Leitmann. "A New Class of Stabilizing Controllers for Uncertain Dynamical Systems", *SIAM J. Contr. Optimiz.* , v. 21, pp. 246-255, 1983.

[2] Y. H. Chen. "On the Robustness of Mismatched Uncertain Dynamical Systems", *J. Dyn. Syst., Meas., and Control* , v. 109, pp.29-35, 1987.

[3] M. Corless. "Stabilization of Uncertain Discrete-Time Systems", *Proc. IFAC Workshop on Model Error Concepts and Compensation* , Boston, USA, 1985.

[4] M. Corless; J. Manela. "Control of Uncertain Discrete-Time Systems", *Proc. American Control Conference* , v. 1, pp. 515-520, 1986.

[5] C. A. Desoer; M. Vidyasagar. *Feedback Systems: Input-Output Properties*, Academic Press, 1975.

[6] C. V. Hollot; M. Arabacioglu. "l-th Step Lyapunov Min-Max Controllers: Stabilizing Discrete-Time Systems Under Real Parameter Variations", *Proc. American Control Conference* , v. 1, pp. 496-501, 1987.

[7] R. Kalman; J. E. Bertram. "Control System Analysis and Design Via the 'Second Method' of Lyapunov - part II: Discrete-Time Systems", *J. Basic Eng.* , v. 82, pp. 394-400, 1960.

[8] K. H. Kienitz; M. Mansour. "Output Feedback Stabilization of Uncertain Linear Systems", *Proc. 28th Conference on Decision and Control,* pp. 1682-1687, 1989.

[9] M. E. Magaña; S. H. Zak. "Robust State Feedback Stabilization of Discrete-Time Uncertain Dynamical Systems", *Proc. 26th Conference on Decision and Control* , v. 2, pp.1530-1534, 1987.

[10] J. Manela. *Deterministic Control of Uncertain Linear Discrete and Sampled-Data Systems*, Ph.D. Dissertation, U.C. Berkeley, 1985.

[11] G. S. Rogers. *Matrix Derivatives*, Lecture Notes in Statistics, vol. 2, Marcel Dekker, New York, 1980.

[12] M. Vidyasagar. *Nonlinear Systems Analysis*, Prentice-Hall, Englewood Cliffs, N.J., 1978.

[13] R. Fletcher. *Practical Optimization Methods*, 2nd. Edition, John Wiley & Sons, New York, 1987.

Karl Heinz Kienitz
Institut für Automatik und Industrielle Elektronik
Eidgenössische Technische Hochschule Zürich
ETH-Zentrum / ETL
CH-8092 Zürich
Switzerland
automatic@nimbus.ethz.ch

Inverse Radial Matrices and Maximal Stability Robustness

Izchak Lewkowicz Raphael Sivan

In this work properties of inverse radial matrices and their relations to generalized stability radii are discussed. A characterization of an inverse radial matrix is introduced. The connection between these matrices and the stability radii is presented. The case of equality between the complex and the real stability radii is characterized.

1 Introduction

The problem of generalized stability robustness for unstructured perturbations is studied in this work. Roughly, the problem is as follows: for a given matrix A and a domain Ω in the complex plane, in which the eigenvalues of A are contained, find a *minimal norm* perturbation matrix P such that at least one of the eigenvalues of $A - P$ "leaves" Ω. The norm of P is called the *generalized stability radius of A with respect to Ω.*

Several properties of *inverse radial* matrices are discussed. It is shown that the notion of an inverse radial matrix is strongly related to maximal stability robustness.

Then, the special case where for $A \in \mathbb{R}^{n\times n}$, the *complex* and the *real* generalized stability radii are equal is characterized.

Finally, these results are applied to the special case where Ω is the open left half of $\mathbb{C}$. This is of interest in the investigation of the stability robustness of linear, time invariant, finite dimensional, autonomous, dynamical system described by the following state equations,

$$\dot{x}(t) = Ax(t),$$

where the system matrix $A \in \mathbb{R}^{n\times n}$.

2 Notations

Let $A \in \mathbb{C}^{n\times n}$ be a matrix with the following singular value decomposition, (SVD),

$$A = USV^* = \sum_{k=1}^{n} \sigma_k u_k v_k^*. \tag{1}$$

where, $(\cdot)^*$ denotes the complex conjugate transpose of $(\cdot)$. The singular values of A are ordered as follows,

$$\sigma_1 \geq \cdots \geq \sigma_{n-m} > \sigma_{n-m+1} = \cdots = \sigma_n, \tag{2}$$

for some $m, 1 \leq m \leq n$. The matrix $S \triangleq \text{diag}\{\sigma_1, \ldots, \sigma_n\}$ and U is a matrix whose columns, $u_k \in \mathbb{C}^n$, $1 \leq k \leq n$, are normalized eigenvectors of the matrix AA^*. Similarly V is a matrix whose columns, $v_k \in \mathbb{C}^n$, $1 \leq k \leq n$, are normalized eigenvectors of the matrix A^*A.

Note that the matrices U and V are not unique. One way of achieving a decomposition (1) is to choose the columns of U as normalized eigenvectors of AA^*, which then uniquely determines, V, e.g. [1, 7.3.5]; in the case that A is nonsingular V is simply given by $V = A^*US^{-1}$. In order to obtain other SVD's, one can replace U and V by $\hat{U}, \hat{V}$ given by,

$$\hat{U} = U \begin{bmatrix} e^{i\theta_1} & & \\ & e^{i\theta_{n-m}} & \\ & & U_o \end{bmatrix} \text{ and } \hat{V}^* = \begin{bmatrix} e^{i\theta_1} & & \\ & e^{i\theta_{n-m}} & \\ & & U_o^* \end{bmatrix} V^*, \tag{3}$$

where, θ_k, $1 \leq k \leq n-m$ are *arbitrary* in $[0, 2\pi)$ and $U_o \in \mathbb{C}^{m\times m}$ is *any* unitary matrix.

We shall order, up to section 7, the eigenvalues of the matrix $A \in \mathbb{C}^{n\times n}$ as follows

$$|\lambda_1| \geq \cdots \geq |\lambda_{n-l}| > |\lambda_{n-l+1}| = \cdots = |\lambda_n|, \tag{4}$$

for some l, $1 \leq l \leq n$. The algebraic and geometric multiplicity of each eigenvalue is, for our purposes, of no importance.

3 Perturbations in the Spectrum of a Matrix

In this section we review some facts concerning perturbations of the spectrum of a given complex matrix. Let the set of the eigenvalues of $A \in \mathbb{C}^{n\times n}$, be denoted by spect$(A)$. For $q \in \mathbb{C}$, $q \notin$ spect(A), if and only if, $\sigma_n(A - qI) > 0$. The following nomenclature will be used in the sequel.

DEFINITION. Given a matrix $A \in \mathbb{C}^{n\times n}$, and a scalar $q \in \mathbb{C}$, such that $q \notin$ spect(A). We denote by $\mathcal{P}(A, q)$ the set of *minimal norm perturbation matrices* given by,

$$\mathcal{P}(A, q) \triangleq \{P \in \mathbb{C}^{n\times n} \mid q \in \operatorname{spect}(A-P), \|P\| \text{ is minimal}\}.$$

Hereafter $\|\cdot\| = \|\cdot\|_2 = \sigma_1(\cdot)$, i.e. the spectral norm.

The following Lemma 3.1 characterizes the set $\mathcal{P}(A, q)$.

Lemma 3.1. (2, pp. 18–19) *Let $A \in \mathbb{C}^{n\times n}$ be a matrix, and $q \in \mathbb{C}$ be a scalar such that $q \notin$ spect(A). Then,* every *$P \in \mathcal{P}(A, q)$ is of the following form,*

$$P = \sigma_n u_n v_n^* + \tilde{P},$$

where,

$$A - qI = \sum_{k=1}^{n} \sigma_k u_k v_k^*, \tag{5}$$

is some SVD of the matrix $A - qI$, and $\tilde{P}$ is any *matrix satisfying*

$$\|\tilde{P}\| \le \sigma_n, \quad u_n^* \tilde{P} = 0, \quad \tilde{P} v_n = 0.$$

Every minimal norm perturbation matrix $P \in \mathcal{P}(A, q)$ may be constructed as follows:

a. Take any SVD of the matrix $A - qI$, (5) .

b. Choose an arbitrary vector u such that $u \in \operatorname{span}\{u_{n-m+1}, \dots, u_n\}$.

c. Normalize u so that $\|u\| = 1$.

d. Select v so that $u^*(A - qI) = \sigma_n v^*$.

e. The matrix $P \in \mathcal{P}(A, q)$ is given by,

$$P = \sigma_n u v^* + \tilde{P},$$

where $\tilde{P}$ is any matrix satisfying

$$\|\tilde{P}\| \le \sigma_n, \quad u^* \tilde{P} = 0, \quad \tilde{P} v = 0.$$

Next we characterize the distance in $\mathbb{C}^{n\times n}$ of the spectrum of a matrix from the boundary of a given domain Ω containing that spectrum. In order to be more precise we introduce the following notation.

DEFINITION. Given a matrix $A \in \mathbb{C}^{n\times n}$ and an open domain $\Omega \subset \mathbb{C}$, such that spect$(A) \subset \Omega$. We denote by $\mu_{\mathbb{C}}(A, \Omega)$ the *generalized complex stability radius of A with respect to Ω* given by,

$$\mu_{\mathbb{C}}(A, \Omega) \triangleq \min_{P \in \mathbb{C}^{n\times n}} \{\|P\| \mid \operatorname{spect}(A - P) \not\subset \Omega\}.$$

Using the previous notation,

$$\mu_{\mathbb{C}}(A,\Omega) = \min_{q \notin \Omega} \{\|P\| \mid P \in \mathcal{P}(A,q)\}.$$

In the special case where Ω is the open left half of $\mathbb{C}$, the attribute *general* is omitted, and we consider simply the complex stability radius, see Section 7.

The following Proposition 3.2 provides us with an explicit expression for $\mu_{\mathbb{C}}(A,\Omega)$.

Proposition 3.2. **([3])** *Let $A \in \mathbb{C}^{n\times n}$ be a matrix, $\Omega \subset \mathbb{C}$ an open domain such that* $\operatorname{spect}(A) \subset \Omega$, *and $\partial\Omega$ the boundary of Ω in $\mathbb{C}$. Then,*

$$\mu_{\mathbb{C}}(A,\Omega) = \min_{q \in \partial\Omega}[\sigma_n(A - qI)].$$

This Proposition is a special case of Theorem 10.20 in [8], where the Banach algebra in question, is the set of matrices over $\mathbb{C}^{n\times n}$, together with the following fact,

$$\min_{q \in \partial\Omega}[\sigma_n(A - qI)] = \frac{1}{\max_{q \in \partial\Omega} \|(A - qI)^{-1}\|} = \frac{1}{\max_{q \in \Omega} \|(A - qI)^{-1}\|}.$$

The first equality stems from the fact, e.g. [1, 7.4.1], that for every nonsingular matrix A, $\sigma_n(A) = \|A^{-1}\|^{-1}$. The second equality is based upon exercise 11 in [9, p. 264]. It is interesting to note that the domain Ω need not be either bounded or connected.

4 Inverse Radial Matrices

In this section the notions of *radial* and *inverse radial* matrices are introduced and discussed.

DEFINITION. ([10]) A matrix $A \in \mathbb{C}^{n\times n}$ is said to be *radial* if its spectral radius is equal to its spectral norm, i.e. $|\lambda_1| = \sigma_1$.

The following Proposition reveals that this is not a "pointwise" property.

Proposition 4.1. *A matrix $A \in \mathbb{C}^{n\times n}$ is radial, if and only if, there exists an $\varepsilon_R > 0$, such that $\forall q \in \mathbb{C}$, the fact $|q| \le \varepsilon_R$ implies $|\lambda_1(A - qI)| = \sigma_1(A - qI)$.*

The proof is given in Appendix 1.

We proceed by introducing matrices with the "complement" property, i.e. inverse radial.

DEFINITION. A matrix $A \in \mathbb{C}^{n\times n}$ is said to be *inverse radial* if there exists an $\varepsilon_I > 0$, such that, $\forall q \in \mathbb{C}$, the fact $|q| \leq \varepsilon_I$ implies $|\lambda_n(A-qI)| = \sigma_n(A-qI)$.

From the proof of Proposition 4.1, it can be deduced that if A is inverse radial with $l < n$, (4), then one can always take $\varepsilon_I = \frac{1}{2}(|\lambda_l| - |\lambda_{l+1}|)$.

If A is inverse radial then $\sigma_n = |\lambda_n|$ and $\sigma_n^2(A) = \sigma_n(A^2)$, however, the converse is always true only for nonsingular matrices. For a singular matrix, of course, $\sigma_n = \lambda_n = 0$, but the matrix is not necessarily inverse radial. For example,

$$A = \begin{bmatrix} 0 & 1 \\ 0 & 0 \end{bmatrix},$$

is singular, $\sigma_2 = \lambda_2 = 0$, but simple computation reveals that $\forall q \neq 0$, $q \in \mathbb{C}$, $\sigma_2(A-qI) < |\lambda_2(A-qI)|$, and hence this is not an inverse radial matrix.

Normal matrices are both radial and inverse radial, the converse is, of course, not necessarily true. Roughly, one can think of inverse radial matrices, as matrices whose "tail" is normal.

Nonsingular inverse radial matrices are important since, as we shall see, in the next section, they are in a sense, maximally robust. We proceed by reviewing some of the known characterizations of inverse radial matrices.

Proposition 4.2. ([10], [11]) *The following statements are all equivalent characterizations of $A \in \mathbb{C}^{n\times n}$ as an* inverse radial *matrix :*

(I) The Schur triangular form of A is

$$A_{\mathsf{Schur}} = \begin{bmatrix} T & 0 \\ 0 & \Lambda \end{bmatrix}$$

where $T \in \mathbb{C}^{(n-l)\times(n-l)}$ is a triangular matrix, $\Lambda = \operatorname{diag}\{\lambda_{n-l+1}, \ldots, \lambda_n\}$ and $\sigma_{n-l}(T) \geq |\lambda_{n-l+1}|$.

(II) $\sigma_{n-l} \geq |\lambda_{n-l+1}|$, and for each k, $n-l+1 \leq k \leq n$, if z_k is a right eigenvector, i.e. it satisfies $Az_k = \lambda_k z_k$ then necessarily, it is also a left eigenvector, i.e. $z_k^* A = z_k^* \lambda_k$.

(III) $\sigma_{n-l} \geq |\lambda_{n-l+1}|$, and there is a SVD, (1), so that $u_k = e^{i\psi_k} v_k$, for some $0 \leq \psi_k < 2\pi$, and $\forall k$, $n-l+1 \leq k \leq n$. [In fact, this implies that also $Au_k = \lambda_k u_k$, $n-l+1 \leq k \leq n$.]

Note that if A is inverse radial then necessarily, $m \geq l$, see (2), (4). This follows, say, from characterization 4.2(II).

The key idea in the proofs of Proposition 4.2 is to take known results for radial matrices, and to apply them to A^{-1}, whenever A is nonsingular, using the facts,

$$\frac{1}{\sigma_n(A)} = \|A^{-1}\| = \sigma_1\left(A^{-1}\right), \quad \lambda_1\left(A^{-1}\right) = \frac{1}{\lambda_n(A)}.$$

We introduce here an additional characterization of an inverse radial matrix.

Proposition 4.3. *Let $A \in \mathbb{C}^{n\times n}$ be a matrix with a SVD as in (1). A is inverse radial, if and only if, $\sigma_{n-l} \geq |\lambda_{n-l+1}|$, and there exist a $q \in \mathbb{C}$, $q \neq 0$ and θ_k, $0 \leq \theta_k < 2\pi$, $n-l+1 \leq k \leq n$, so that, for each k, $n-l+1 \leq k \leq n$, $(u_k, e^{i\theta_k}v_k)$ is some pair of singular vectors of the matrix $A - qI$.*

In fact, for an inverse radial matrix A, for every *$q \in \mathbb{C}$, there exist θ_k, $0 \leq \theta_k < 2\pi$, $n-l+1 \leq k \leq n$, [where $\theta_k = \theta_k(q)$], so that, for each k, $n-l+1 \leq k \leq n$, $(u_k, e^{i\theta_k}v_k)$, is some pair of singular vectors of the matrix $A - qI$.*

The proof is given in Appendix 2 .

In the next section we shall apply the notion of inverse radial matrices in order to characterize maximally robust matrices.

5 Characterization of Maximally Robust Matrices

In this section, we consider the problem of characterizing among all matrices having the same spectrum contained within an open domain $\Omega \subset \mathbb{C}$, the maximally robust ones.

For a formal presentation, we need further notations. The Euclidean distance in $\mathbb{C}$ between spect(A) and Ω^c, the complement of Ω is given by,

$$d[\text{spect}(A), \Omega^c] \stackrel{\Delta}{=} \min_{q\in\partial\Omega} \min_{1\leq k\leq n} |\lambda_k - q| = \min_{q\in\partial\Omega} |\lambda_n(A - qI)|.$$

Note that, there always exists a scalar ϕ, $0 \leq \phi < 2\pi$, such that spect$\left\{A - d[\text{spect}(A), \Omega^c]e^{i\phi}I\right\} \not\subset \Omega$, hence, in general, $\mu_{\mathbb{C}}(A, \Omega) \leq d[\text{spect}(A), \Omega^c]$. A matrix for which $\mu_{\mathbb{C}}(A, \Omega) = d[\text{spect}(A), \Omega^c]$ is said to be *maximally robust with respect to* Ω. The following Proposition 5.1 characterizes these matrices, but first we introduce additional nomenclature:

Let Q_d denote the set of all points in which the above $d[\text{spect}(A), \Omega^c]$ is attained, i.e.

$$Q_d \stackrel{\Delta}{=} \left\{q \in \partial\Omega \mid |\lambda_n(A - qI)| = d[\text{spect}(A), \Omega^c]\right\}. \tag{6}$$

Let Q_μ denote the set of all points where the minimum of $\sigma_n(A-qI)$ is attained, i.e.

$$Q_\mu \triangleq \{q \in \partial\Omega \mid \sigma_n(A-qI) = \mu_{\mathbb{C}}(A,\Omega)\}\,. \tag{7}$$

Thus, due to Proposition 3.2, the following relations hold,

$$\sigma_n(A-qI)\mid_{q\in Q_\mu} \leq \sigma_n(A-qI)\mid_{q\in Q_d} \leq |\lambda_n(A-qI)|\mid_{q\in Q_d} \leq |\lambda_n(A-qI)|\mid_{q\in Q_\mu}\,. \tag{8}$$

It should be remarked that in some special cases Q_d and Q_μ, may be uncountable sets.

Proposition 5.1. *Given a matrix* $A \in \mathbb{C}^{n\times n}$ *and an open domain* $\Omega \subset \mathbb{C}$, *such that* spect$(A) \subset \Omega$. *The matrix* A *is* maximally robust with respect to Ω, *if and only if, (i)* $Q_d \subset Q_\mu$, *and (ii)* each *of the matrices* $A - qI$, *where* $q \in Q_d$, *is inverse radial.*

PROOF. First, recall that by assumption $\forall q \in \partial\Omega$, the matrix $A - qI$ is nonsingular. Then, the proof amounts to noting that the leftmost inequality in (8), holds with equality, if and only if, $Q_d \subset Q_\mu$, and the center inequality in (8) holds with equality, if and only if, each of the matrices $A-qI$, where $q \in Q_d$, is inverse radial. Finally, these two equalities are equivalent to $\mu_{\mathbb{C}}(A,\Omega) = d[\text{spect}(A), \Omega^c]$. Q.E.D.

In the special case where all three relations in (8) hold with equality, we have the following,

Corollary 5.2. (to Proposition 5.1) *Given a matrix* $A \in \mathbb{C}^{n\times n}$ *and an open domain* $\Omega \subset \mathbb{C}$, *such that* spect$(A) \subset \Omega$. *The matrix* A *is* maximally robust with respect to Ω *and* $Q_d = Q_\mu$, *if and only if,* each *of the matrices* $A - qI$, *where* $q \in Q_\mu$, *is inverse radial.*

6 Characterization of the Case where the Minimal Norm Perturbation Matrix is Real

In this section we characterize the case where for a given matrix $A \in \mathbb{R}^{n\times n}$ and a scalar $q \notin \mathbb{R}$, the set of minimal norm perturbation matrices $\mathcal{P}(A,q)$, considered in Section 3, contains a *real* matrix. Note that if $q \in \mathbb{R}$ the case is trivial since then, there is always a real matrix in $\mathcal{P}(A,q)$. This is due to Lemma 3.1 and the fact that the SVD of a real matrix may always be taken to be real, e.g. [1, 1.4.7] .

Lemma 6.1. *Let* $A \in \mathbb{R}^{n\times n}$ *be a matrix,* $q \notin \mathbb{R}$ *is a scalar so that* $q \notin$ spect(A). *Then, there exists a* real *matrix* $P \in \mathcal{P}(A,q)$, *if and only if, there exists a SVD of* $A - qI$, *(5), so that,*

$$u_n^T u_n = v_n^T v_n = 0.$$

Furthermore, for that SVD, one choice of P is,

$$P = 2\sigma_n[\mathrm{Re}(u_n v_n^*)].$$

The proof is given in Appendix 3 .

In order to gain insight for the condition in Lemma 6.1 the following Corollary to Proposition 4.2 describes a special case where this condition holds.

Corollary 6.2. (to Proposition 4.2) *Let $A \in \mathbb{R}^{n\times n}$ be a matrix with a SVD as in (1), where $m = 2$, (2), and $\lambda_n \notin \mathbb{R}$. If A is inverse radial, then,*

$$u_n^T u_n = v_n^T v_n = 0$$

The proof is given in Appendix 4 .

Suppose $A \in \mathbb{C}^{n\times n}$ is a matrix such that $\mathrm{spect}(A) \subset \Omega$, where $\Omega \subset \mathbb{C}$ is an open domain. The problem of finding explicitly the the minimal norm perturbation matrix $P \in \mathbb{C}^{n\times n}$, such that $\mathrm{spect}(A-P) \not\subset \Omega$, is solved due to Lemma 3.1 and Proposition 3.2 . Now, if we confine the matrices A and P to be real, then the problem is well defined only if Ω is a domain symmetric with respect to the real axis, i.e. $q \in \Omega \Leftrightarrow \bar{q} \in \Omega$, where $(\bar{\cdot})$ denotes the complex conjugate of $(\cdot)$. Such a domain is the unit disc, for example. In order to elaborate on this problem quantitatively, we introduce the following notation.

DEFINITION. Given a matrix $A \in \mathbb{R}^{n\times n}$ and an open domain $\Omega \subset \mathbb{C}$, symmetric with respect to the real axis, such that $\mathrm{spect}(A) \subset \Omega$. We denote by $\mu_{\mathbb{R}}(A,\Omega)$ the *generalized real stability radius of A with respect to* Ω given by,

$$\mu_{\mathbb{R}}(A,\Omega) \triangleq \min_{P\in\mathbb{R}^{n\times n}} \{\|P\| \mid \mathrm{spect}(A-P) \not\subset \Omega\}.$$

As before, in the special case where Ω is the open left half of $\mathbb{C}$, the attribute *general* is omitted, and we consider simply the real stability radius, see Section 8.

The problem of finding an explicit expression for $\mu_{\mathbb{R}}(A,\Omega)$, in the spirit of Proposition 3.2, is apparently, in general, yet unsolved. Still, the following Proposition 6.3 solves a special case of it.

Proposition 6.3. *Given a matrix $A \in \mathbb{R}^{n\times n}$ and an open domain $\Omega \subset \mathbb{C}$, symmetric with respect to the real axis, such that* $\mathrm{spect}(A) \subset \Omega$. *Then, the generalized complex and real stability radii are equal, namely, $\mu_{\mathbb{C}}(A,\Omega) = \mu_{\mathbb{R}}(A,\Omega)$, if and only if, one (or both) of the following conditions holds: (i) There exists* a real *scalar $q \in Q_\mu$, (7). (ii) There exists a SVD of $A - qI$, (5), for a $q \in Q_\mu$, $q \notin \mathbb{R}$, such that*

$$u_n^T u_n = v_n^T v_n.$$

The proof for a *real* scalar $q \in Q_\mu$, follows from Lemma 3.1 recalling that the SVD of a real matrix may always be taken to be real, e.g. [1, 1.4.7]. For a *not real* scalar $q \in Q_\mu$, the proof is immediate from the definition of Q_μ, (7), and Lemma 6.1.

It should be emphasized that in order to have $\mu_{\mathbb{C}}(A,\Omega) = \mu_{\mathbb{R}}(A,\Omega)$, it is sufficient that condition (ii) in Proposition 6.3 holds for *one* $q \in Q_\mu$, where q is not real.

We conclude this section by presenting a known special case of Proposition 6.3. It is given here in a form slightly different from the original formulation.

Proposition 6.4. ([12]) *Given a matrix $A \in \mathbb{R}^{n\times n}$ and an open domain $\Omega \subset \mathbb{C}$, symmetric with respect to the real axis, such that* $\mathrm{spect}(A) \subset \Omega$. *Let $B \triangleq A \oplus A$, i.e. $B \in \mathbb{R}^{2n\times 2n}$, then,*

$$\mu_{\mathbb{C}}(A,\Omega) = \mu_{\mathbb{C}}(B,\Omega) = \mu_{\mathbb{R}}(B,\Omega).$$

Furthermore, a minimal norm matrix $P \in \mathbb{R}^{2n\times 2n}$ such that $\mathrm{spect}(B-P) \not\subset \Omega$*, can be constructed in the following way. Take any SVD of $A - qI$, (5), for arbitrary $q \in Q_\mu(A)$, (7). If we denote, $u_n v_n^* \triangleq X + iY$, $X, Y \in \mathbb{R}^{n\times n}$, then,*

$$P = \sigma_n \begin{bmatrix} X & Y \\ -Y & X \end{bmatrix}$$

Proof. The proof amounts to noting that, for any SVD of $A - qI$, (5), where q is arbitrary, $q \in Q_\mu(A)$, (7); considering the SVD of $B - qI$ for the same q, its U_o, (3), can always be chosen so that $u_{2n}(B-qI)$, $v_{2n}(B-qI)$, the last singular vectors of $B - qI$ are of the following form,

$$u_{2n}(B-qI) = \frac{1}{\sqrt{2}}\begin{bmatrix} u_n \\ iu_n \end{bmatrix} \quad v_{2n}(B-qI) = \frac{1}{\sqrt{2}}\begin{bmatrix} v_n \\ iv_n \end{bmatrix}.$$

Consequently, $[u_{2n}(B-qI)]^T[u_{2n}(B-qI)] = [v_{2n}(B-qI)]^T[v_{2n}(B-qI)] = 0$. Then, Lemma 6.1 can be applied, and the resulting P, is identical to the one appearing in this Proposition. Q.E.D.

In the rest of the work we apply the results presented so far to the special case where Ω is the open left half of $\mathbb{C}$.

7 First Application: The Complex Stability Radius

In this section various aspects of the complex stability radius defined for the state space model of a linear time invariant system are studied, using the results of the previous sections for general matrices.

In the rest of the work, we shall order the eigenvalues of a matrix $A \in \mathbb{C}^{n\times n}$, differently from (4), i.e. according to their real part,

$$\mathrm{Re}(\tilde{\lambda}_1) \leq \cdots \leq \mathrm{Re}(\tilde{\lambda}_{n-p}) < \mathrm{Re}(\tilde{\lambda}_{n-p+1}) = \cdots = \mathrm{Re}(\tilde{\lambda}_n) \stackrel{\Delta}{=} -\lambda_o(A), \quad (9)$$

for some $p, 1 \leq p \leq n$. The matrix A is said to be *stable* if $\lambda_o(A) > 0$.

Roughly, the question of stability robustness is: how large the perturbation P can be, for a given stable A, without losing the stability of $A - P$?

In order to analyze, quantitatively, the effect of perturbations, we consider the *complex stability radius*, $\mu_{\mathbb{C}}(A)$, of a stable matrix, given by, e.g. [3] – [7] ,

$$\mu_{\mathbb{C}}(A) = \mu_{\mathbb{C}}(A, \Omega) \Big|_{\Omega=\{q\in\mathbb{C}|\mathrm{Re}(q)<0\}} .$$

Since, $\mu_{\mathbb{C}}(A)$ is a special case of $\mu_{\mathbb{C}}(A, \Omega)$, where Ω is the open left half of $\mathbb{C}$, and $\partial\Omega = i\mathbb{R}$, from Proposition 3.2, we have the following, e.g. [3] – [7],

$$\mu_{\mathbb{C}}(A) = \min_{\Omega\in\mathbb{R}}[\sigma_n(A - i\omega I)].$$

Now, as a motivation, we mention that the complex stability radius is of importance, also, in the study of the stability robustness of the following linear *time varying* systems:

Theorem 7.1. **([13]–[15])** *Given the system,*

$$\dot{x} = [A + F(t)]x,$$

where, $A \in \mathbb{R}^{n\times n}$ is a known fixed stable matrix, and $F(t) \in \mathbb{R}^{n\times n}$ is any time varying matrix such that the solution $x(t)$ exists. If there exists an $\varepsilon > 0$, such that for each "frozen" t

$$\|F(t)\| \leq \mu_{\mathbb{C}}(A) - \varepsilon,$$

then the system is uniformly exponentially stable.

Next, we use the result of Section 5 to characterize the stable complex matrices with *maximal stability robustness.* First, note that for a stable matrix $A \in \mathbb{C}^{n\times n}$, $d = \lambda_o(A)$ and using (6) and (9), $Q_d = \{i[\mathrm{Im}(\tilde{\lambda}_k)], n - p + 1 \leq k \leq n\}$. So we can state the following Corollary to Proposition 5.1.

Corollary 7.2. **(to Proposition 5.1)** *Let $A \in \mathbb{C}^{n\times n}$ be a stable matrix, then A is maximally robust, if and only if, (i) $\forall k$, $n - p + 1 \leq k \leq n$, (9), $i[\mathrm{Im}(\tilde{\lambda}_k)] \in Q_\mu$, (7), and (ii)* each *of the matrices $A - i[\mathrm{Im}(\tilde{\lambda}_k)]I, n - p + 1 \leq k \leq n$, (9), is inverse radial.*

We conclude this section by presenting a known equivalent characterization of $A \in \mathbb{C}^{n\times n}$ with maximal stability robustness, which follows from Corollary 7.2 and characterization 4.2(I) of inverse radial matrices.

Proposition 7.3. **([16])** *Let $A \in \mathbb{C}^{n\times n}$ be a stable matrix with eigenvalues as ordered in (9), then A is maximally robust, if and only if, its Schur triangular form is,*

$$A_{\mathsf{Schur}} = \begin{bmatrix} T & 0 \\ 0 & \Lambda \end{bmatrix},$$

where $T \in \mathbb{C}^{(n-p)\times(n-p)}$ is a triangular matrix, $\Lambda = \operatorname{diag}\left\{\tilde{\lambda}_{n-p+1}, \ldots, \tilde{\lambda}_n\right\}$ and $\mu_{\mathbb{C}}(T) \geq \lambda_o(A)$.

8 Second Application: The Real Stability Radius

In this section we consider a stable *real* matrix $A \in \mathbb{R}^{n\times n}$ for which the *real stability radius* $\mu_{\mathbb{R}}(A)$, has been introduced in [3], given by

$$\mu_{\mathbb{R}}(A) = \mu_{\mathbb{R}}(A, \Omega)|_{\Omega=\{q\in\mathbb{C}|\mathrm{Re}(q)<0\}}.$$

By definition, for a stable matrix $A \in \mathbb{R}^{n\times n}$,

$$\mu_{\mathbb{C}}(A) \leq \mu_{\mathbb{R}}(A). \tag{10}$$

In the rest of the work we shall examine the case of equality in (10) . Unfortunately the exact computation of $\mu_{\mathbb{R}}(A)$ for the general case is apparently yet unknown. Exact formulas for two special cases are given in [5], [17]. A lower bound for $\mu_{\mathbb{R}}(A)$ was introduced in [18].

The following corollary to Proposition 6.3, answers the problem of characterizing the case of equality in (10), raised in [17, p. 60].

Corollary 8.1. **(to Proposition 6.3)** *Let $A \in \mathbb{R}^{n\times n}$ be a stable matrix, Q_μ as defined in (7). Then, the complex and the real stability radii are equal, if and only if, one (or both) of the following conditions holds: (i) $0 \in Q_\mu$, and (ii) There exists a SVD of $A - i\omega I$, (5), for an $i\omega \in Q_\mu$, $\omega > 0$, such that,*

$$u_n^T u_n = v_n^T v_n = 0.$$

Corollary 8.1 is a special case of Proposition 6.3 with $q = i\omega$, noting that $\sigma_n(A - i\omega I)$ is an *even* function of ω.

Expectedly, this result coincides with the 2×2 case described in [19, p. 66] .

As was already mentioned, a matrix A may have several $\omega's > 0$, where $i\omega \in Q_\mu$. It is sufficient that the orthogonality condition for the singular vectors of $A - i\omega I$ holds only for *one* of these $\omega's$ in order to ensure $\mu_{\mathbb{C}}(A) = \mu_{\mathbb{R}}(A)$.

Although typically $\mu_{\mathbb{C}}(A) < \mu_{\mathbb{R}}(A)$, the special case presented in Corollary 8.1, i.e. $\mu_{\mathbb{C}}(A) = \mu_{\mathbb{R}}(A)$ is of interest for the following reasons:

a. $\mu_{\mathbb{R}}(A)$ is apparently in general, very difficult to compute.

b. In view of Theorem 7.1 and the definition of $\mu_{\mathbb{R}}(A)$, this is a characterization of the linear dynamical systems with the stable nominal part such that for each time varying destabilizing perturbation $F(t)$, there exists a time invariant destabilizing perturbation ΔA with the same size, i.e. $\|F(t)\| = \|\Delta A\|$.

9 Conclusions

In this work, some properties of inverse radial matrices were studied. It was shown that these matrices are closely related to the set of matrices with maximal complex stability robustness. In addition, the case where the complex and the real stability radii are equal, was characterized. The problems of finding an explicit expression for $\mu_{\mathbb{R}}(A, \Omega)$, and characterizing the matrices with maximal *real* generalized stability robustness, are apparently, yet, unanswered.

Appendix 1 — Proof of Proposition 4.1

Clearly, if there exists an $\varepsilon_R > 0$, such that $\forall |q| \leq \varepsilon_R$ the matrix $A - qI$ is radial, then in particular A is radial. So, all we have to prove is that if A is radial, then some neighborhood of it, $A - qI$, $|q| \leq \varepsilon_R$, is also radial.

In this appendix, we shall order the eigenvalues of a matrix $A \in \mathbb{C}^{n \times n}$ as follows,

$$|\lambda_1| = \ldots = |\lambda_s| > |\lambda_{s+1}| \geq \cdots \geq |\lambda_n|,$$

for some s, $1 \leq s \leq n$.

It is known, [10], that if a matrix $A \in \mathbb{C}^{n \times n}$ is radial, then A_{Schur}, its Schur triangular form is the following,

$$A_{\text{Schur}} = \begin{bmatrix} \Lambda & 0 \\ 0 & T \end{bmatrix},$$

where $\Lambda = \text{diag}\{\lambda_1, \ldots, \lambda_s\}$, $T \in \mathbb{C}^{(n-s)\times(n-s)}$ is a triangular matrix, and $\sigma_{s+1} \leq |\lambda_s|$.

Thus, if A is a radial matrix, for any $q \in \mathbb{C}$,

$$(A - qI)_{\text{Schur}} = \begin{bmatrix} \Lambda - qI_s & 0 \\ 0 & T - qI_{n-s} \end{bmatrix},$$

where I_s is the $s \times s$ identity matrix. Hence, $\forall q \in \mathbb{C}$ and $\forall k$, $1 \leq k \leq s$, $|\lambda_k + q|$ is *some* singular value of the matrix $A - qI$.

Next, for every $A \in \mathbb{C}^{n\times n}$ and $\forall q \in \mathbb{C}$, the eigenvalues of $A - qI$ are $\lambda_k - q$, $1 \le k \le n$. Now, either $s < n$ or $s = n$. We assume first, $s < n$, then, if we take $\varepsilon_R = \frac{1}{2}(|\lambda_s| - |\lambda_{s+1}|)$, the following relations hold, $\forall |q| \le \varepsilon_R$,

$$|\lambda_s(A-qI)| \ge |\lambda_s| - |q| \ge \frac{1}{2}\left(|\lambda_s| + |\lambda_{s+1}|\right) \ge |\lambda_{s+1}| + |q| \ge |\lambda_{s+1}(A-qI)|.$$

Consequently, $\forall |q| \le \varepsilon_R$ each of the elements on the diagonal part of $(A - qI)_{\text{Schur}}$ is larger or equal then the eigenvalues of $(T - qI_{n-s})$. Namely, $\forall |q| \le \varepsilon_R$, $\sigma_1(A - qI) = |\lambda_1(A - qI)|$, so the proof is complete for $s < n$.

If $s = n$, then $(A - qI)_{\text{Schur}}$ is diagonal so $\forall q \in \mathbb{C}$ and $\forall k$, $1 \le k \le n$, $|\lambda_k - q|$ are some singular values of the matrix $A - qI$. Hence, in this case, the matrix $A - qI$ is inverse radial $\forall q \in \mathbb{C}$, which completes the proof.

Q.E.D.

Appendix 2 — Proof of Proposition 4.3

(i) We first show that if A is inverse radial it implies that for *every* $q \in \mathbb{C}$, and for each k, $n - l + 1 \le k \le n$, $(u_k, e^{i\theta_k} v_k)$ is some pair of singular vectors of the matrix $A - qI$.

Note that without loss of generality we can consider A_{Schur}, the Schur triangular form of A. Now, since A is inverse radial, using the notation in characterization 4.2(I), the Schur triangular form of $A - qI$ for any $q \in \mathbb{C}$ is

$$(A - qI)_{\text{Schur}} = \begin{bmatrix} T + qI_{n-l} & 0 \\ 0 & \Lambda - I_l \end{bmatrix}.$$

Next, recall that if a matrix $D \in \mathbb{C}^{l\times l}$ is diagonal, i.e. $D = \operatorname{diag}\{d_1, \ldots, d_l\}$, then a SVD of D is,

$$D = \sum_{k=1}^{n} |d_k| E_k e^{i\theta_k} E_k^T,$$

where, E_k is the k-th standard basis vector, e.g. $E_1^T = (1, 0, \ldots, 0)$, and

$$e^{i\theta_k} = \begin{cases} d_k/|d_k| & d_k \ne 0 \\ e^{i\theta}, \text{ arbitrary, } 0 \le \theta < \pi & d_k = 0 \end{cases}$$

Hence, for each k, $n - l + 1$, $(E_k, e^{i\theta_k} E_k^T)$ is a pair of singular vectors of the Schur triangular form of the matrix $A - qI$. So the first part of the proof is established.

(ii) Here we show if $\sigma_{n-l} \ge |\lambda_{n-l+1}|$ and there exist: a $q \in \mathbb{C}$, $q \ne 0$ and θ_k, $0 \le \theta_k < 2\pi$, $n - l + 1 \le k \le n$, so that, for each k, $n - l + 1 \le k \le n$, $(u_k, e^{i\theta_k} v_k)$ is some pair of singular vectors of the matrix $A - qI$; then A satisfies characterization 4.2(III) of inverse radial matrices.

It is known that there exists a one to one correspondence between each $k, n-l+1 \leq k \leq n$, and an integer j, $1 \leq j \leq n$ so that,

$$\sigma_j^2(A-qI)u_k = (A-qI)(A-qI)^* u_k = (AA^* + |q|^2 I)u_k - (qA^* + \bar{q}A)u_k.$$

Consequently,

$$(qA^* + \bar{q}A)u_k = [\sigma_k^2 + |q|^2 - \sigma_j^2(A-qI)]u_k.$$

Similarly, for the same k and j,

$$(qA^* + \bar{q}A)v_k = [\sigma_k^2 + |q|^2 - \sigma_j^2(A-qI)]v_k,$$

i.e. each pair, u_k, v_k are eigenvectors of the same hermitian matrix, which correspond to the same eigenvalue. Hence, the sets $\{u_k\}$, $\{v_k\}$, where $n-l+1 \leq k \leq n$; are two orthonormal bases to the same subspace. Let us denote by $\hat{U}, \hat{V} \in \mathbb{C}^{n \times l}$ the submatrices obtained by taking the last l columns of the matrices U, V in (1), respectively. Due to the discussion above, there exists a unitary matrix $U_1 \in \mathbb{C}^{l \times l}$ so that,

$$\hat{U} = \hat{V} U_1.$$

Next, the matrix U_1 is decomposed to,

$$U_1 = U_2^* D U_2,$$

where, the columns of $U_2 \in \mathbb{C}^{l \times l}$ are normalized eigenvectors of U_1, and D is the matrix of the eigenvalues of U_1, i.e. $D = \operatorname{diag}\left\{e^{i\psi_{n-l+1}}, \ldots, e^{i\psi_n}\right\}$ for some $0 \leq \psi_k < 2\pi$, $n-l+1 \leq k \leq n$. Combining the last two equations results in,

$$(\hat{U}U_2^*) = (\hat{V}U_2^*)D.$$

Denoting the columns of $(\hat{U}U_2^*)$ and $(\hat{V}U_2^*)$ by $\tilde{u}_k$, $\tilde{v}_k$, $n-l+1 \leq k \leq n$, respectively, yields for each k, $n-l+1 \leq k \leq n$,

$$\tilde{u}_k = e^{i\psi_k}\tilde{v}_k,$$

and since by assumption $\sigma_{n-l} \geq |\lambda_{n-l+1}|$, characterization 4.2(III) of A as an inverse radial matrix is completed. Q.E.D.

Appendix 3 — Proof of Lemma 6.1

We use the usual notation of the SVD (5), for the given A and q,

$$A - qI = \sum_{k=1}^{n} \sigma_k u_k v_k^* \tag{C1a}$$

$$A - \bar{q}I = \sum_{k=1}^{n} \sigma_k \bar{u}_k v_k^T \tag{C1b}$$

(i) To prove sufficiency, we assume that there exists a SVD of $A - qI$, such that $u_n^T u_n = v_n^T v_n = 0$.

Let $P = \sigma_n[u_n v_n^* + \bar{u}_n v_n^T] = 2\sigma_n[\text{Re}(u_n v_n^*)]$, then, clearly, $P \in \mathbb{R}^{n \times n}$. If $u_n^T u_n = 0$ and $v_n^T v_n = 0$ then simple calculation reveals that $\|P\| = \sigma_n$. Finally,

$$[A - P - qI]v_n = [(A - qI) - P]v_n = \sigma_n u_n - \sigma_n u_n = 0 \Rightarrow q \in \text{spect}(A - P),$$

hence, this implies the existence of a *real* $P \in \mathcal{P}(A, q)$, which completes the first part of the proof.

(ii) To prove necessity, we assume that there exists a *real* matrix $P \in \mathcal{P}(A, q)$.

Since $P \in \mathcal{P}(A, q)$, due to Lemma 3.1,

$$P = \sigma_n u_n v_n^* + \tilde{P},$$

where $\tilde{P}$ is a matrix satisfying,

$$u_n^* \tilde{P} = 0, \quad \tilde{P} v_n = 0, \quad \|\tilde{P}\| \le \sigma_n.$$

Due to this construction, $u_n^T(A - P) = \bar{q}u_n$, and, $(A - P)v_n = qv_n$, i.e. u_n^T, v_n are correspondingly left and right eigenvectors of the matrix $A - P$, for *different* eigenvalues, hence, e.g. [1, 1.4.7],

$$u_n^T v_n = 0. \tag{C2}$$

Next, using (C1a) and (C2) yields

$$(A - qI)v_n = \sigma_n u_n \Rightarrow u_n^T(A - qI)v_n = \sigma_n(u_n^T u_n) \Rightarrow u_n^T A v_n = \sigma_n(u_n^T u_n).$$

Similarly from (C1b) and (C2) we have

$$u_n^T(A - \bar{q}I) = \sigma_n v_n^T \Rightarrow u_n^T(A - \bar{q}I)v_n = \sigma_n(v_n^T v_n) \Rightarrow u_n^T A v_n = \sigma_n(v_n^T v_n).$$

Consequently, since $\sigma_n > 0$

$$u_n^T u_n = v_n^T v_n. \tag{C3}$$

Now, note the SVD in (C1a) can be replaced by,

$$A - qI = \sum_{k=1}^{n} \sigma_k (e^{i\theta_k} u_k)(e^{-i\theta_k} v_k^*),$$

where θ_k can be chosen arbitrarily $0 \leq \theta_k < 2\pi$, $1 \leq k \leq n$, and still yields the same P. Hence in (C3), *for all* θ_n,

$$v_n^T v_n e^{-2i\theta_n} = u_n^T u_n e^{+2i\theta_n},$$

which implies that,

$$u_n^T u_n = v_n^T v_n = 0,$$

so the proof is complete. Q.E.D.

Appendix 4 — Proof of Corollary 6.2

Due to characterization 4.2(III) of inverse radial matrices, and the given A and λ_n,

$$u_n^T A = \bar{\lambda}_n u_n^T, \quad A u_n = \lambda_n u_n,$$

i.e. u_n^T, v_n are correspondingly, left and right eigenvectors of *different* eigenvalues of the matrix A. Hence, e.g. [1, 1.4.7], $u_n^T u_n = 0$. Using 4.2(III) again, results in $u_n = e^{i\psi} v_n$ for some $0 \leq \psi < 2\pi$. Consequently, also $v_n^T v_n = 0$, so the proof is complete. Q.E.D.

Acknowledgement: The general infinite dimensional form of Proposition 3.2 was presented to us by Dr. G. Weiss from the Department of Mathematics in Virginia Polytechnic Institute, USA.

REFERENCES

[1] R. A. Horn, C. R. Johnson, *Matrix Analysis,* Cambridge University Press, New York, 1985.

[2] G. H. Golub and C. F. Van - Loan, *Matrix Computations,* The Johns Hopkins University Press, Baltimore Maryland, 1983.

[3] D. Hinrichsen, A. J. Pritchard, "Stability Radii of Linear Systems", *System & Control Letters,* Vol. 8, pp. 1–10, 1986.

[4] J. Kautsky, N. K. Nichols, P. Van Dooren, "Robust Pole Assignment in Linear State Feedback", *Int. J. Contr.,* Vol. 41, pp. 1129–1155, 1985.

[5] C. F. Van - Loan, "How Near is a Stable Matrix to an Unstable Matrix" in *Linear Algebra and its Role in Systems Theory,* in *Proc. AMS-IMS-SIAM Conf.* held July 29 - Aug. 4, 1984. R. A. Brualdi et al. Eds., *Contemporary Mathematics, American Mathematical Society,* Vol. 47, 1985.

[6] L. Qiu, E. J. Davison, "New Perturbation Bounds for the Robust Stability of Linear State Space Models", *Proc. C.D.C. Athens,* pp. 751–755, 1986.

[7] J. M. Martin, "State Space Measures for Stability Robustness", *IEEE Trans. Auto. Cont.,* Vol. AC-32, pp. 509–512, June 1987.

[8] W. Rudin, *Functional Analysis,* McGraw - Hill, 1973.

[9] W. Rudin, *Real and Complex Analysis,* McGraw - Hill, third edition, 1987.

[10] M. Goldberg, G. Zwass, "On Matrices Having Equal Spectral Radius and Spectral Norm", *Linear Algebra and its Applications,* Vol. 8, pp. 427–434, 1974.

[11] I. Lewkowicz, "Remarks on Inequality in Johnson's Lower Bound for the Smallest Singular Value", *Linear Algebra and its Applications,* Vol. 120, pp. 39–46, Aug. 1989.

[12] D. Hinrichsen, A. J. Pritchard, "A note on some differences between real and complex stability radii", Report Nr. 215, Inst. fuer Dynamische Systeme, Universität Bremen 1989, to appear in *Systems and Control Letters.*

[13] A. J. Pritchard, S. Townley, "Robustness of Linear Systems", *Journal of Differential Equations,* Vol 77, No. 2, pp. 254–286, February 1989.

[14] D. Hinrichsen, A. Ilchmann, A. J. Pritchard, "Robustness of Stability of Time Varying Linear Systems", *Journal of Differential Equations,* Vol. 82, No 2, pp. 219–250, December 1989.

[15] P. P. Khargonekar, I. R. Petersen, K. Zhou, "Robust Stabilization of Uncertain Linear Systems: Quadratic Stability and H_∞ Theory", to appear in *IEEE Trans. Auto. Control,*

[16] I. Lewkowicz, R. Sivan, "Maximal Stability Robustness for State Equations", *IEEE Trans. Auto. Control,* Vol. AC - 33, pp. 297–300, March 1988.

[17] D. Hinrichsen, A. J. Pritchard, "New Robustness Results for Linear Systems under Real Perturbations", *Proc. 27th C.D.C.,* Austin Texas, 1988, pp. 1375–1379.

[18] L. Qiu and E. J. Davison, "A New Method for the Stability Robustness Determination of State Space Models with Real Perturbations", *Proc. 27th C.D.C.,* Austin Texas, 1988, pp. 538–543.

[19] D. Hinrichsen, M. Motscha, "Optimization Problems in Robustness Analysis of Linear State Space Systems", in *Approximation and Optimization Proceedings,* Havana 1987, *Lecture Notes in Mathematics,* pp. 54–77 , No. 1354, Springer Verlag 1988.

Izchak Lewkowicz and Raphael Sivan
Department of Electrical Engineering
Technion — Israel Institute of Technology
Haifa 32000
Israel

Adaptive Control of a Family of Plants*

D.E. Miller E.J. Davison

Consider a linear time-invariant (LTI) plant which is not completely specified, but instead belongs to a finite set of known plants, say $\{\mathbf{P}_i : i \in \mathbf{p}\}$. Our objective is to design a controller which provides "good" tracking and disturbance rejection, in a sufficiently well-defined sense, for this partially known plant. We first design a high-performance LTI controller $\mathbf{K}_i$ for each possible $\mathbf{P}_i$; if the pair $(\mathbf{P}_i, \mathbf{K}_j)$ is stable iff $i = j$ and has no eigenvalues on the imaginary axis for any $i, j \in \mathbf{p}$, and if an upper bound on the magnitude of the unmeasurable disturbance signal is available, then it is shown that a switching mechanism can be used to find the correct LTI controller; furthermore, each LTI controller need only be tried once. This kind of problem often arises in an industrial setting, and is often approached using heuristic "gain-scheduling" techniques.

1 Introduction

The control of a partially known plant has received considerable attention in the adaptive control literature. Recently it was shown in Mårtensson [1] that in order to stabilize a controllable and observable, linear time-invariant (LTI) plant, it is sufficient to know the order of a LTI stabilizing compensator; by stability we mean that the state tends to zero as $t \to \infty$. A more noise tolerant controller which requires the same plant information was given in Miller and Davison [2]. It was subsequently shown in Fu and Barmish [3] that in order to get a stronger form of stability, namely

*This work has been supported by the Natural Sciences and Engineering Research Council of Canada under Grant No. A4396.

exponential Lyapunov stability, it is enough to know a compact set of controllable and observable plants in which the plant lies; it was later shown in Miller and Davison [4] that this can be achieved even if the prior plant information is the same as in [1] and [2]. Although these controllers all solve the stabilization problem, if the plant is square, they can be modified to track a finite number of sinusoidal signals of known frequencies by extending the modification described in [5] for modifying the controller of [1] to track constant signals; an alternative method is presented in [6] for extending the controller of [2] to this setup. However, when considering a wider range of control problems such as wide-band tracking and/or disturbance rejection, none of these controllers are suitable.

The problem is simplified if we assume that the plant belongs to a finite set of known plants $\{\mathbf{P}_i : i \in \mathbf{p}\}$. We first design a suitable high-performance LTI controller $\mathbf{K}_i$ for each of the possible plants $\mathbf{P}_i$, and then use a "switching mechanism" to switch between these controllers at appropriate points in time. Here we present two switching mechanisms: to design the first one we borrow ideas from [2] and [4] but exploit the greater amount of information available in this setup, while the second one uses observers to provide a potentially better transient response than the first mechanism. Assuming that (i) each plant-controller combination $(\mathbf{P}_i, \mathbf{K}_j)$ is stable iff $i = j$ and has no poles on the imaginary axis for any $i, j \in \mathbf{p}$, and (ii) a bound on the unmeasurable disturbance signal is known *a priori*, we show that it is only necessary to try each LTI controller at most once, and that the final controller applied is the correct one.

2 Notation

Let $\mathbf{R}$ denote the set of real numbers, $\mathbf{C}$ denote the set of complex numbers, N denote the set of natural numbers, and $\mathbf{I}$ denote the set of integers. $\mathbf{R}^+$ is the set of non-negative real numbers, $\mathbf{R}^n$ is the n-dimensional real vector space, and $\mathbf{R}^{n\times m}$ is the set of $n \times m$ real matrices.

With $x \in \mathbf{C}$, we use $\mathrm{Re}(x)$ to denote its real part and $\mathrm{Im}(x)$ to denote its imaginary part. With $p \in \mathbf{N}$, we define $\mathbf{p}$ to be the set of integers $\{1, 2, \ldots, p\}$.

The *norm* of $x \in \mathbf{R}^n$, which we denote by $\|x\|$, is the Holder 2-norm; with $A \in \mathbf{R}^{n\times m}$, $\|A\|$ denotes the corresponding induced norm of A.

With $A \in \mathbf{R}^{n\times n}$, we say that A is *stable* if for every eigenvalue λ of A, $\mathrm{Re}(\lambda) < 0$; otherwise we say that A is *unstable*.

Let PC denote the set of real vector-valued functions defined on $t \geq 0$ whose elements are piecewise continuous, and let PC_∞ denote the subset of functions which are bounded.

3 Changing the Dynamic Feedback Problem into a Static Feedback Problem

Consider the strictly proper, controllable and observable, LTI system $\mathbf{P}_i$, $i \in \mathbf{p}$, described by

$$\begin{aligned} \dot{x} &= A_i x + B_i u + E_i w && \text{(1a)} \\ y &= C_i x + F_i w && \text{(1b)} \\ e &= y_{\text{ref}} - y && \text{(1c)} \end{aligned}$$

where $x(t) \in \mathbf{R}^{n_i}$ is the state, $u(t) \in \mathbf{R}^m$ is the control input, $y(t) \in \mathbf{R}^r$ is the output to be regulated, $w(t) \in \mathbf{R}^v$ is the disturbance, and $e(t) \in \mathbf{R}^r$ is the error, which is the difference between the specified reference input y_{ref} and the output y. It is assumed that the parameters of each $\mathbf{P}_i$ are known exactly. The actual plant is one of $\mathbf{P}_i$, $i \in \mathbf{p}$, only y and y_{ref} are measurable, and only u can be excited.

For each $i \in \mathbf{p}$, design a high-performance LTI controller $\mathbf{K}_i$ (by any means at your disposal) for $\mathbf{P}_i$ of the form

$$\begin{aligned} \dot{z} &= G_i z + H_i y + J_i y_{\text{ref}} && \text{(2a)} \\ u &= K_i z + L_i y + M_i y_{\text{ref}}, && \text{(2b)} \end{aligned}$$

where $z(t) \in \mathbf{R}^{l_i}$; without loss of generality we can assume that $l_i = l_j = l$ for $i, j \in \mathbf{p}$, by adding unobservable stable modes if necessary. In particular, assume that each $\mathbf{K}_i$ has been designed so that it stabilizes the closed loop system corresponding to $\mathbf{P}_i$, *i.e.* the closed loop system matrix has all of its eigenvalues in the open left half of the complex plane; more importantly, it performs satisfactory disturbance regulation and/or tracking for $\mathbf{P}_i$, depending on the control objective.

Finding a controller (2) which provides stabilization of the closed loop system when the plant is $\mathbf{P}_i$ is equivalent to finding a matrix $\tilde{K}_i$ so that the static output feedback controller

$$\tilde{u} = \tilde{K}_i \tilde{y} \qquad (3)$$

stabilizes the following augmented controllable and observable system:

$$\begin{aligned} \dot{\tilde{x}} &= \tilde{A}_i \tilde{x} + \tilde{B}_i \tilde{u} + \tilde{E}_i w && \text{(4a)} \\ \tilde{y} &= \tilde{C}_i \tilde{x} + \tilde{D}_i y_{\text{ref}} + \tilde{F}_i w && \text{(4b)} \end{aligned}$$

where

$$\tilde{x} = \begin{bmatrix} x \\ z \end{bmatrix}, \quad \tilde{u} = \begin{bmatrix} u \\ \dot{z} \end{bmatrix}, \quad \tilde{y} = \begin{bmatrix} y \\ z \\ y_{\text{ref}} \end{bmatrix},$$

and

$$\tilde{A}_i = \begin{bmatrix} A_i & 0 \\ 0 & 0 \end{bmatrix}, \quad \tilde{B}_i = \begin{bmatrix} B_i & 0 \\ 0 & I \end{bmatrix}, \quad \tilde{C}_i = \begin{bmatrix} C_i & 0 \\ 0 & I \\ 0 & 0 \end{bmatrix},$$

$$\tilde{D}_i = \begin{bmatrix} 0 \\ 0 \\ I \end{bmatrix}, \quad \tilde{E}_i = \begin{bmatrix} E_i \\ 0 \end{bmatrix}, \quad \tilde{F}_i = \begin{bmatrix} F_i \\ 0 \\ 0 \end{bmatrix}, \quad \tilde{K}_i = \begin{bmatrix} L_i & K_i & M_i \\ H_i & G_i & J_i \end{bmatrix}.$$

So for each $i \in \mathbf{p}$, we have a gain $\tilde{K}_i$ so that with the controller given by (3) and the augmented system given by (4), the closed loop performance is acceptable. The question is: from this, how do we get a controller which will perform acceptably when the plant is not known exactly, but instead it is known only that the plant is one of $\{\mathbf{P}_i : i \in \mathbf{p}\}$. What we will do is switch between the feedback gains $\{\tilde{K}_i : i \in \mathbf{p}\}$ at appropriate points in time; this is discussed in detail in the next section.

4 The Main Result

One of the potential difficulties in deciding which gain $\{\tilde{K}_i : i \in \mathbf{p}\}$ should be used is that the initial condition and the disturbance are unknown, which can cause a "switching mechanism" to switch unnecessarily. Assuming that a bound on the disturbance is known in advance, the following preliminary result provides a way to get a bound on the initial condition.

Lemma 1. *Suppose that $u(\cdot) = 0$ in (1). Then for every $T > 0$, there exist constants α_{i_1} and α_{i_2}, so that for every initial condition $x(0)$ and every disturbance $w \in PC_\infty$, we have*

$$\|x(0)\|^2 \le \alpha_{i_1} \int_0^T \|y(\tau)\|^2 \, d\tau + \alpha_{i_2} \sup_{t \ge 0} \|w(t)\|^2.$$

PROOF. Fix $T > 0$, and let $x(0)$ and $w \in PC_\infty$ be arbitrary.

Since $u(\cdot) = 0$, it follows that

$$y(t) = C_i e^{A_i t} x(0) + \left[\int_0^t C_i e^{A_i(t-\tau)} E_i w(\tau) d\tau + F_i w(t) \right] =: y_1(t) + y_2(t).$$

So

$$y_1(t) = y(t) - y_2(t),$$

which means that

$$\beta := \int_0^T \|y_1(\tau)\|^2 \, d\tau \le 2 \int_0^T \|y(\tau)\|^2 \, d\tau + 2 \int_0^T \|y_2(\tau)\|^2 \, d\tau.$$

Define

$$W_i := \int_0^T e^{A_i'\tau} C' C e^{A_i \tau}\, d\tau;$$

then

$$\beta = x(0)' W_i x(0).$$

Let α_{i_3} = the smallest singular value of W_i; then

$$\beta - \alpha_{i_3} \|x(0)\|^2 = x(0)'[W_i - \alpha_{i_3} I] x(0) \geq 0,$$

so

$$\beta \geq \alpha_{i_3} \|x(0)\|^2.$$

Hence,

$$\|x(0)\|^2 \leq (2/\alpha_{i_3}) \int_0^T \|y(\tau)\|^2\, d\tau + (2/\alpha_{i_3}) \int_0^T \|y_2(\tau)\|^2\, d\tau.$$

Now

$$\|y_2(t)\| \leq \left(\int_0^t \|C_i e^{A_i(t-\tau)} E_i\|\, d\tau + \|F_i\| \right) \sup_{\tau \in [0,t]} \|w(\tau)\|,$$

so if we define $\alpha_{i_1} := 2/\alpha_{i_3}$ and

$$\alpha_{i_2} := (2/\alpha_{i_3}) \int_0^T \left[\int_0^t \|C_i e^{A_i(t-\tau)} E_i\|\, d\tau + \|F_i\| \right]^2 dt,$$

then the result follows. Q.E.D.

Hence, if we have a bound of $\bar{b}$ on $\|w(t)\|$, then we can compute a bound on the norm of the initial condition by setting the input u to zero, computing the square of the norm of the output for a period of time, and then applying Lemma 1. At this point we take two routes. The first approach is to only use a small amount of information about the possible plants $\mathbf{P}_i$, $i \in \mathbf{p}$, at the expense of poor performance but with the advantage of being impervious to small modeling errors. The second approach is to use all of the plant information to result in a potentially better transient response at the expense of extreme sensitivity to even small modeling errors.

4.1 A Robust Approach

The ideas used to design the controller of [2] and [4] can be combined to design a switching controller which allows utilization of the greater amount of prior plant information which is available here.

First, we need the following result, which provides a way to get a running upper bound on the state.

Lemma 2. *There exist constants* γ_{i_1}, γ_{i_2}, γ_{i_3}, *and* $\lambda_i < 0$, *so that for every* $\tilde{u}, y_{\rm ref}, w \in PC$ *and initial condition* $\tilde{x}(0)$, *the solution of (4) satisfies*

$$\|\tilde{x}(t)\| \leq \gamma_{i_1}\|\tilde{x}(0)\|e^{\lambda_i t} + \int_0^t e^{\lambda_i(t-\tau)}\left[\gamma_{i_2}\|\tilde{u}(\tau) - \tilde{K}_i(\tilde{y}(\tau) - \tilde{D}_i y_{\rm ref}(\tau))\| + \gamma_{i_3}\|w(\tau)\|\right] d\tau$$

for $t \geq 0$.

PROOF. We can write (4a) as

$$\dot{\tilde{x}}(t) = (\tilde{A}_i + \tilde{B}_i\tilde{K}_i\tilde{C}_i)\tilde{x}(t) + \tilde{B}_i\tilde{u}(t) + \tilde{E}_i w(t) - \tilde{B}_i\tilde{K}_i(\tilde{y}(t) - \tilde{D}_i y_{\rm ref}(t) - \tilde{F}_i w(t))$$

$$= (\tilde{A}_i + \tilde{B}_i\tilde{K}_i\tilde{C}_i)\tilde{x}(t) + \tilde{B}_i[\tilde{u}(t) - \tilde{K}_i(\tilde{y}(t) - \tilde{D}_i y_{\rm ref}(t))] + (\tilde{E}_i + \tilde{B}_i\tilde{K}_i\tilde{F}_i)w(t).$$

Since $\tilde{A}_i + \tilde{B}_i\tilde{K}_i\tilde{C}_i$ is stable, there exist constants $\gamma_{i_1} > 0$ and $\lambda_i < 0$ so that $\|e^{(\tilde{A}_i + \tilde{B}_i\tilde{K}_i\tilde{C}_i)t}\| \leq \gamma_{i_1}e^{\lambda_i t}$ for $t \geq 0$. Hence,

$$\|\tilde{x}(t)\| \leq \gamma_{i_1}e^{\lambda_i t}\|\tilde{x}(0)\| + \gamma_{i_1}\int_0^t e^{\lambda_i(t-\tau)} \cdot \left[\|\tilde{B}_i\| \cdot \|\tilde{u}(\tau) - \tilde{K}_i(\tilde{y}(\tau) - \tilde{D}_i y_{\rm ref}(\tau))\| + \|\tilde{E}_i + \tilde{B}_i\tilde{K}_i\tilde{F}_i\| \cdot \|w(\tau)\|\right] d\tau, \quad t \geq 0.$$

If we define $\gamma_{i_2} := \gamma_{i_1}\|\tilde{B}_i\|$, and $\gamma_{i_3} := \gamma_{i_1}\|\tilde{E}_i + \tilde{B}_i\tilde{K}_i\tilde{F}_i\|$, then the result holds. Q.E.D.

So we can use Lemma 1 together with knowledge of an upper bound on the disturbance to construct a bound on the initial condition; we combine this with Lemma 2 to construct a signal r_i which is an upper bound on the state of the augmented system if the plant is $\mathbf{P}_i$. At this point, we could compare $\tilde{y} - \tilde{D}_i y_{\rm ref}$ with r_i to see if the former is too large for the plant to be $\mathbf{P}_i$. However, $\tilde{y} - \tilde{D}_i y_{\rm ref}$ is discontinuous if Fw is, and we prefer to deal with continuous signals, so we will filter $\tilde{y} - \tilde{D}_i y_{\rm ref}$ first. We will use the feedback gain $\tilde{K}_i$ until the filtered $\tilde{y} - \tilde{D}_i y_{\rm ref}$ hits the bound on the output imposed by r_i, at which point we will switch gains. If the disturbance satisfies the prespecified bound, then the gain will switch only if the present gain is incorrect. Notice that $\tilde{D}_i = \tilde{D}_j$ for all $i, j \in \mathbf{p}$, so define $\tilde{D} := \tilde{D}_1$.

We define Controller 1 as follows. Fix $T > 0$, $\bar{b} > 0$, and $\varepsilon > 0$, choose α_{i_1}, α_{i_2}, $i \in \mathbf{p}$, so that the inequality given in Lemma 1 holds, and choose $\gamma_{i_1}, \gamma_{i_2}, \gamma_{i_3}, \lambda_i$, $i \in \mathbf{p}$, so that the inequality given in Lemma 2 holds. Then define

$$\lambda := \min\{\lambda_i : i \in \mathbf{p}\},$$

and fix $\tilde{\lambda} < \lambda$.

The controller is effected in two phases.

Phase 1: Finding a bound on the initial condition.
With $\tilde{u}(t) = 0$ for $t \in [0, T]$ and $z(0) = 0$, find

$$\theta := \int_0^T \|y(\tau)\|^2 \, d\tau$$

and define the following p auxiliary signals:

$$\dot{r}_j(t) = \lambda_j r_j(t) + \gamma_{j_2} \left\| \tilde{K}_j(\tilde{y}(t) - \tilde{D}_j y_{\text{ref}}(t)) \right\| + \gamma_{j_3}\bar{b}, r_j(0) = 0, \;\; t \in [0, T].$$

Define

$$\mu_j := [\alpha_{j_1}\theta + \alpha_{j_2}\bar{b}^2]^{\frac{1}{2}}, \quad j \in \mathbf{p};$$

if $\|w(t)\| \leq \bar{b}$ for $t \in [0, T]$ and the plant is $\mathbf{P}_j$, then it follows from Lemma 1 that $\|\tilde{x}(0)\| \leq \mu_j$.

Phase 2: Searching the gains.
Now we apply the control action. Let the control input be

$$\tilde{u}(t) = \tilde{K}_i \tilde{y}(t), t \in (t_i, t_{i+1}],$$

let the p auxiliary signals be

$$\dot{r}_j(t) = \lambda_j r_j(t) + \gamma_{j_2} \|\tilde{u}(t) - \tilde{K}_j(\tilde{y}(t) - \tilde{D}_j y_{\text{ref}}(t))\| + \gamma_{j_3}\bar{b}, j \in \mathbf{p},$$

with $r_j(T^+) = r_j(T) + \gamma_{j_1} e^{\lambda_j T} \mu_j$[1], and let the filtering signal be

$$\dot{\tilde{r}}(t) = \tilde{\lambda}\tilde{r}(t) + (\lambda - \tilde{\lambda}) \left\| \tilde{y}(t) - \tilde{D} y_{ref}(t) \right\|, \tilde{r}(T) = 0.$$

So $r_j(t)$ generates an upper bound on the norm of the state vector for $t \geq T$ in the event that the plant is $\mathbf{P}_j$, while $\tilde{r}$ filters $\tilde{y} - \tilde{D} y_{\text{ref}}$. We define the switching times recursively as follows: set $t_1 := T$, and for every $i \in \{2, \ldots, p+1\}$ for which $t_{i-1} \neq \infty$, define t_i by

$$\begin{aligned} \min \{t \geq t_{i-1} : \;\; & \text{there exists a time } \tilde{t} \in [T, t] \text{ for which} \\ & \tilde{r}(\tilde{t}) = \|\tilde{C}_{i-1}\| r_{i-1}(\tilde{t}) + \|\tilde{F}_{i-1}\|\bar{b} + \varepsilon\} \end{aligned}$$

if the set is nonempty (the minimum exists if it is nonempty), and ∞ otherwise; if t_{p+1} is defined and finite, then define $t_{p+2} = \infty$ and set $\tilde{K}_{p+1} = 0$. Observe that it could be that $t_i = t_{i-1}$ for some i; if this occurs then it means that by the time that we are ready to try the gain $\tilde{K}_i$, we have already ruled it out.

---

[1] $r_j(T^+) := \lim_{t \to T, t > T} r_j(t)$

Theorem 1. *Suppose that $y_{\text{ref}}, w \in PC_\infty$, and that $\|w(t)\| \le \bar{b}$ for $t \ge 0$. For every initial condition $x(0)$, when Controller 1 is applied to the plant, the closed loop system has the following properties:*

(i) the gain eventually remains constant at an element of $\{\tilde{K}_i : i \in \mathbf{p}\}$, and

(ii) the state $\tilde{x}(t)$ is bounded.

To prove this we need the following preliminary technical result on filtering.

Lemma 3. *Suppose that $\tilde{\lambda} < \lambda \le \lambda_j < 0$, $\gamma_1 \ge 0$, $\gamma_2 \ge 0$, $\phi_1(\cdot) \ge 0$, $\phi_2(\cdot) \ge 0$,*

$$\begin{aligned}\dot{v}_1(t) &= \lambda_j v_1(t) + \phi_1(t), v_1(0) \ge 0,\\ \dot{v}_2(t) &= \tilde{\lambda} v_2(t) + (\lambda - \tilde{\lambda})\phi_2(t), v_2(0) = 0.\end{aligned}$$

If $\phi_2(t) \le \gamma_1 v_1(t) + \gamma_2$ for $t \ge 0$, then $v_2(t) \le \gamma_1 v_1(t) + \gamma_2$ for $t \ge 0$ as well.

PROOF. Set

$$\hat{e}(t) := \gamma_1 v_1(t) + \gamma_2 - v_2(t);$$

then

$$\dot{\hat{e}}(t) = \tilde{\lambda}\hat{e}(t) + (\lambda - \tilde{\lambda})[-\phi_2(t) + \gamma_1 v_1(t) + \gamma_2] + (\lambda_j - \lambda)\gamma_1 v_1(t) + \gamma_1 \phi_1(t) - \lambda\gamma_2,$$

with $\hat{e}(0) = \gamma_1 v_1(0) + \gamma_2$. But all of the last four terms on the RHS are non-negative. Since $\hat{e}(0) \ge 0$ as well, it follows that $\hat{e}(t) \ge 0$ for $t \ge 0$, as required. Q.E.D.

PROOF OF THEOREM 1. Let $y_{\text{ref}}, w \in PC_\infty$ be such that $\|w(t)\| \le \bar{b}$ for $t \ge 0$, and let $x(0)$ be arbitrary. It follows from the definition of the controller that $\tilde{x}(t)$ is defined for $t \ge 0$.

Choose $j \in \mathbf{p}$ so that the plant is $\mathbf{P}_j$. It follows from Lemma 1 that

$$\|\tilde{x}(0)\|^2 \le \alpha_{j_1}\theta + \alpha_{j_2}\bar{b}^2,$$

so $\|\tilde{x}(0)\| \le \mu_j$. From Lemma 2 it follows that

$$\begin{aligned}\|\tilde{x}(t)\| \le\ & \gamma_{j_1} e^{\lambda_j t}\|\tilde{x}(0)\| \\ & + \int_0^t e^{\lambda_j(t-\tau)}\left[\gamma_{j_2}\|\tilde{u}(\tau) - \tilde{K}_j(\tilde{y}(\tau) - \tilde{D}_j y_{\text{ref}}(\tau))\| + \gamma_{j_3}\|w(\tau)\|\right] d\tau \\ \le\ & \gamma_{j_1}\mu_j e^{\lambda_j t} \\ & + \int_0^t e^{\lambda_j(t-\tau)}\left[\gamma_{j_2}\|\tilde{u}(\tau) - \tilde{K}_j(\tilde{y}(\tau) - \tilde{D}_j y_{\text{ref}}(\tau))\| + \gamma_{j_3}\bar{b}\right] d\tau \\ =\ & r_j(t), t \ge T. \end{aligned} \tag{5}$$

Hence,

$$\|\tilde{y}(t) - \tilde{D}_j y_{\text{ref}}(t)\| \leq \|\tilde{C}_j\| \cdot r_j(t) + \|\tilde{F}_j\|\bar{b}, \quad t \geq T,$$

so it follows from Lemma 3 that

$$\tilde{r}(t) \leq \|\tilde{C}_j\| \cdot r_j(t) + \|\tilde{F}_j\|\bar{b}, \quad t \geq T.$$

Therefore, if t_{j+1} is defined, then it must be ∞. Hence, property (i) holds.

It remains to show that $\tilde{x}(t)$ is bounded; from (5) it follows that it is enough to show that $r_j(t)$ is bounded. Let the final gain be $\tilde{K}_{\hat{i}}$. Now $r_{\hat{i}}(t)$ is bounded since $y_{\text{ref}}, w \in PC_\infty$. Hence, by definition of $t_{\hat{i}+1}$ it follows that $\tilde{r}(t)$ is as well. But

$$\int_T^t e^{\lambda_j(t-\tau)}(\lambda - \tilde{\lambda}) \left\|\tilde{y}(\tau) - \tilde{D}y_{\text{ref}}(\tau)\right\| d\tau$$
$$= \tilde{r}(t) + (\lambda_j - \tilde{\lambda}) \int_T^t e^{\lambda_j(t-\tau)}\tilde{r}(\tau)\, d\tau], \quad t \geq T,$$

so

$$\sup_{t>T} \int_T^t e^{\lambda_j(t-\tau)} \|\tilde{y}(\tau)\| \, d\tau < \infty.$$

Therefore, it follows that $r_j(t)$ is bounded. Q.E.D.

Notice that Controller 1 does not guarantee that the final feedback gain is the correct one, i.e. the one designed for the actual plant, so the closed loop system performance may be poor. In a particular case, however, more can be said.

Suppose that the control objective is to provide asymptotic error regulation for all reference and disturbance signals whose elements are described by the known differential equation

$$(\cdot)^{(\nu)} + \rho_{\nu-1}(\cdot)^{(\nu-1)} + \cdots + \rho_1(\cdot)^{(1)} + \rho_0(\cdot) = 0, \tag{6}$$

with independent initial conditions. It is assumed that the zeros of

$$s^\nu + \rho_{\nu-1}s^{\nu-1} + \cdots + \rho_1 s + \rho_0 \tag{7}$$

are simple and lie on the imaginary axis in the complex plane. Furthermore, suppose that we would like the error regulation to be maintained for all perturbations in the plant model which maintains closed loop stability. This problem is denoted the *robust servomechanism problem* in [7], and it is shown there how to solve this problem for a known plant: first design a servocompensator corresponding to (6), and then design a stabilizing controller for the plant-servocompensator combination; the details are omitted. Of course, in order for this procedure to work, it is necessary that the plant satisfy several conditions: (i) it must be detectable, (ii) stabilizable, (iii) have as many inputs as outputs, and (iv) have no transmission zeros (in the sense of [8]) at the zeros of (7), and we assume that each of the p plants satisfy these conditions. Such a controller configuration has the property that the transfer function from the reference and the disturbance to the error is identically zero at each zero of (7), as long as the stabilizing part of the controller is such that there is no closed loop system pole there.

Now suppose that the following assumption holds:

Assumption 1. For every $i, j \in \mathbf{p}$, $\tilde{A}_i + \tilde{B}_i \tilde{K}_j \tilde{C}_i$ has no eigenvalues on the imaginary axis.

This means that no plant-controller combination has a pole on the imaginary axis; this can always be arranged by perturbing the gains of the stabilizing part of the compensator if necessary. This leads to the following result.

Corollary 1. *Suppose that each element of y_{ref} and w satisfies (6), that $\|w(t)\| \leq \bar{b}$ for $t \geq 0$, that $\mathbf{K}_i$ solves the robust servomechanism problem for $\mathbf{P}_i$ for each $i \in \mathbf{p}$, and that Assumption 1 holds. Then for every initial condition $x(0)$, when Controller 1 is applied to the plant, $e(t) \to 0$ as $t \to \infty$.*

PROOF. Let y_{ref} and w be such that each element satisfies (6) and so that $\|w(t)\| \leq \bar{b}$ for $t \geq 0$, and let $x(0)$ be arbitrary. It follows from Theorem 1 that the gain eventually remains constant and $\tilde{x}(t)$ is bounded.

By Assumption 1, the final gain, say $K_{\bar{i}}$, is such that the final LTI closed loop system has no poles on the imaginary axis. Hence, the transfer function from y_{ref} and w to e is identically zero at each zero of (7). Therefore, the error is a sum of weighted exponentials corresponding to poles of the closed loop system, none of which lie on the imaginary axis, so it follows from the boundedness of e (which follows from the boundedness of $\tilde{x}$) that only the exponentials which decay to zero have non-zero weights. Hence, $e(t) \to 0$ as $t \to \infty$, as required. Q.E.D.

One of the drawbacks of Controller 1 is that we cannot guarantee that we end up with the correct gain. So suppose that the following assumption is made as well as Assumption 1.

Assumption 2. For each $i, j \in \mathbf{p}$, $\tilde{A}_i + \tilde{B}_i \tilde{K}_j \tilde{C}_j$ is stable iff $i = j$.

This says that each possible plant is sufficiently different from the others that its associated high-performance controller will not even stabilize any of them. If both Assumptions 1 and 2 hold, then in order for the final gain to be incorrect and for $\tilde{x}$ to remain bounded as well, there must be an unstable mode which is not excited. From a practical point of view, of course, this is virtually impossible. Nevertheless, in principle it could occur, so we will modify the controller slightly.

Suppose that the plant is $\mathbf{P}_i$ and the controller is $\mathbf{K}_j$. Suppose we add an auxiliary input $\hat{u}$ to give $\tilde{u} = \tilde{K}_j\tilde{y} + \hat{u}$. Then the closed loop system for this feedback is described by

$$\dot{\tilde{x}} = (\tilde{A}_i + \tilde{B}_i\tilde{K}_j\tilde{C}_i)\tilde{x} + \tilde{B}_i\hat{u} + \tilde{B}_i\tilde{K}_j\tilde{D}_i y_{\text{ref}} + (\tilde{B}_i\tilde{K}_j\tilde{F}_i + \tilde{E}_i)w.$$

Assume that $i \neq j$, so by Assumption 2 there are some poles in the closed right half plane, and by Assumption 1 there are none on the imaginary axis. We can do a similarity transformation S_{ij}, i.e. $\begin{bmatrix}\tilde{x}_1' & \tilde{x}_2'\end{bmatrix}' = S_{ij}\tilde{x}$, to transform the system into its stable and unstable parts:

$$\begin{bmatrix}\dot{\tilde{x}}_1 \\ \dot{\tilde{x}}_2\end{bmatrix} = \begin{bmatrix}A_{ij}^+ & 0 \\ 0 & A_{ij}^-\end{bmatrix}\begin{bmatrix}\tilde{x}_1 \\ \tilde{x}_2\end{bmatrix} + \begin{bmatrix}B_{ij}^+ \\ B_{ij}^-\end{bmatrix}\hat{u} + \begin{bmatrix}D_{ij}^+ \\ D_{ij}^-\end{bmatrix}y_{\text{ref}} + \begin{bmatrix}E_{ij}^+ \\ E_{ij}^-\end{bmatrix}w,$$

where A_{ij}^- and $-A_{ij}^+$ are both stable.

Now suppose that $\tilde{x}(t)$ is bounded. Since

$$\tilde{x}_1(t) = e^{A_{ij}^+ t}\left\{\tilde{x}_1(0) + \int_0^t e^{-A_{ij}^+\tau}\left[B_{ij}^+\hat{u}(\tau) + D_{ij}^+ y_{\text{ref}}(\tau) + E_{ij}^+ w(\tau)\right]\,d\tau\right\},$$

if $\tilde{x}_1(t)$ is bounded then it must be that

$$\tilde{x}_1(0) = -\int_0^\infty e^{-A_{ij}^+\tau}\left[B_{ij}^+\hat{u}(\tau) + D_{ij}^+ y_{\text{ref}}(\tau) + E_{ij}^+ w(\tau)\right]\,d\tau,$$

or equivalently that

$$\int_0^\infty e^{-A_{ij}^+\tau}B_{ij}^+\hat{u}(\tau)\,d\tau = -\tilde{x}_1(0) - \int_0^\infty e^{-A_{ij}^+\tau}\left[D_{ij}^+ y_{\text{ref}}(\tau) + E_{ij}^+ w(\tau)\right]\,d\tau. \tag{8}$$

We can measure an upper bound on $\|\tilde{x}(0)\|$, we have a bound on w, and we can measure y_{ref}. Each time that we switch gains, we will define $\hat{u}$ so that (8) cannot be satisfied if the gain is not correct. If y_{ref} is bounded and w satisfies its bound $\bar{b}$, then eventually $\hat{u}(t)$ will be zero, so that its effect on the closed loop system will tend to zero.

Since $-A_{ij}^+$ is stable, there exist constants $\bar{\lambda}_{ij} < 0$ and $\bar{\gamma}_{ij} > 0$ so that $\|e^{-A_{ij}^+ t}\| \leq \bar{\gamma}_{ij}e^{\bar{\lambda}_{ij}t}$, $t \geq 0$. So the RHS of (8) has a norm of at most

$$\|\tilde{x}_1(0)\| + \int_0^\infty \bar{\gamma}_{ij}e^{\bar{\lambda}_{ij}\tau}\|D_{ij}^+\| \cdot \|y_{\text{ref}}(\tau)\|\,d\tau + \bar{\gamma}_{ij}\|E_{ij}^+\|\bar{b}/|\bar{\lambda}_{ij}|. \tag{9}$$

Suppose that we fix $\hat{T} > 0$ and set $\hat{u}(t) = \bar{u}$ for $t \in [0, \hat{T}]$ and $\hat{u}(t) = 0$ for $t > \hat{T}$. Then the LHS of (8) is exactly

$$\left(I - e^{-A^+_{ij}\hat{T}}\right)\left(A^+_{ij}\right)^{-1} B^+_{ij}\bar{u}, \quad t \in [\hat{T}, \infty). \tag{10}$$

The idea is to guess a bound on (9) using the measured bound on $\|\tilde{x}(0)\|$ and the assumed bound on $\|w\|$, and an estimated bound on $\int_0^\infty e^{\bar{\lambda}_{ij}\tau}\|y_{\mathrm{ref}}(\tau)\|\,d\tau$; we then choose $\bar{u}$ so that the norm of (10) is greater than the anticipated norm of (9), apply the $\hat{u}$ as defined above, and compute the integral (from 0 to t) appearing in (9) as we go. If the value of the integral becomes larger than anticipated, then we start over with a new estimation of a bound on $\int_0^\infty e^{\bar{\lambda}_{ij}\tau}\|y_{\mathrm{ref}}(\tau)\|\,d\tau$. Each time that we switch gains we have a new closed loop LTI system, so we will restart the procedure. This is made precise below.

We define Controller 1′ as follows. Fix $T > 0$, $\hat{T} > 0$, $\bar{b} > 0$, $\hat{b} > 0$, and $\varepsilon > 0$, and let the constants α_{i_1}, α_{i_2}, γ_{i_1}, γ_{i_2}, γ_{i_3}, λ_i, $i \in \mathbf{p}$, and the constants λ and $\tilde{\lambda}$ be chosen the same way as for Controller 1. For each $i, j \in \mathbf{p}$ for which $i \neq j$, choose a similarity transformation S_{ij} as above and define the associated matrices and constants accordingly. Now define

$$\bar{s}_j := \max\{\|S_{ij}\| : i \in \mathbf{p}, i \neq j\}, \quad j \in \mathbf{p},$$

$$\bar{\gamma}_j := \max\{\bar{\gamma}_{ij} : i \in \mathbf{p}, i \neq j\}, \quad j \in \mathbf{p},$$

$$\bar{\lambda}_j := \max\{\bar{\lambda}_{ij} : i \in \mathbf{p}, i \neq j\}, \quad j \in \mathbf{p},$$

$$\bar{d}_j := \max\{\|D^+_{ij}\| : i \in \mathbf{p}, i \neq j\}, \quad j \in \mathbf{p},$$

$$\bar{e}_j := \max\{\|E^+_{ij}\| : i \in \mathbf{p}, i \neq j\}, \quad j \in \mathbf{p},$$

$$\bar{\delta}_j := \bar{\gamma}_j \bar{e}_j / |\bar{\lambda}_j|, \quad j \in \mathbf{p}.$$

For each $j \in \mathbf{p}$, choose $\bar{u}_j$ so that

$$\min\left\{\left\|\left(I - e^{-A^+_{ij}\hat{T}}\right)\left(A^+_{ij}\right)^{-1} B^+_{ij}\bar{u}_j\right\| : i \in \mathbf{p}, i \neq j\right\} = 1.$$

So if

(i) the controller is $\mathbf{K}_j$,

(ii) the plant is an arbitrary element of $\{\mathbf{P}_i : i \in \mathbf{p}, i \neq j\}$,

(iii) $\|w(t)\| \leq \bar{b}$ for $t \geq 0$,

(iv) $\int_0^\infty \bar{\gamma}_j \bar{d}_j e^{\bar{\lambda}_j \tau} \|y_{\text{ref}}(\tau)\| \, d\tau \le \tilde{b}$, and

(v) $\hat{u}(t) = \gamma \bar{u}_j$ for $t \in [0, \hat{T}]$, $\hat{u}(t) = 0$ for $t > \hat{T}$, and $\gamma \ge \tilde{b} + \bar{\delta}_j \bar{b} + \bar{s}_j \|\tilde{x}(0)\| + \varepsilon$,

then (8) will not be satisfied.

Controller $1'$ can be effected in two phases.

Phase 1: Finding a bound on the initial condition.

We proceed in exactly the same manner as for Controller 1.

Phase 2: Searching the gains.

We proceed as for Controller 1, but with two changes. First, we redefine the control input as

$$\tilde{u}(t) = \tilde{K}_i \tilde{y}(t) + \hat{u}(t), t \in (t_i, t_{i+1}].$$

Second, in the background we define $\hat{u}(t)$ so that it excites the system when necessary to ensure that some of the unstable modes (if they exist) are excited. For $t \in (t_i, t_{i+1}]$ we define $\hat{u}(t)$ as follows: set $t_{i_0} := t_i$, and set

$$\begin{aligned} \dot{\bar{r}}(t) &= \bar{\gamma}_i \bar{d}_i e^{\bar{\lambda}_i (t - t_{i_j})} \|\bar{y}_{\text{ref}}(t)\|, \bar{r}(t_{i_j}^+) = 0, t \in (t_{i_j}, \quad t_{i_{j+1}}], \\ \hat{u}(t) &= \left[(j+1)\hat{b} + \bar{\delta}_i \bar{b} + \bar{s}_i \max_{k \in \mathbf{p}} r_k(t_{i_j}) + \varepsilon \right] \bar{u}_i, \\ &\quad t \in \left(t_{i_j}, \min\left\{ t_{i_{j+1}}, t_{i_j} + \hat{T} \right\} \right] \end{aligned}$$

(set $\hat{u}(t) = 0$ where it is not defined by the above). For each $j > 0$ for which $t_{i_{j-1}} \ne t_{i+1}$, we define

$$t_{i_j} := \min\left\{ t > t_{i_{j-1}} : \bar{r}(t) = j\hat{b} \text{ or } t = t_{i+1} \right\}$$

if the set is nonempty, and ∞ otherwise. So while we are using $\bar{K}_i$, we define $\hat{u}$ to ensure that all modes are excited. Whenever the gain changes, we redefine $\hat{u}$ accordingly.

Theorem 2. *Suppose that $y_{\text{ref}}, w \in PC_\infty$, that $\|w(t)\| \le \bar{b}$ for $t \ge 0$, and that Assumptions 1 and 2 hold. For every initial condition $x(0)$, when Controller $1'$ is applied to the plant, the closed loop system has the following properties:*

(i) the gain eventually remains constant at an element of $\{\tilde{K}_i : i \in \mathbf{p}\}$; in fact, if the plant is $\mathbf{P}_i$, then the final gain is $\tilde{K}_i$, and

(ii) the state $\tilde{x}(t)$ remains bounded.

PROOF. Let $y_{\rm ref}, w \in PC_\infty$ be such that $\|w(t)\| \leq \bar{b}$ for $t \geq 0$, and let $x(0)$ be arbitrary.

Since $y_{\rm ref} \in PC_\infty$, it follows that for every $t_i < \infty$ which is defined,

$$\bar{r}(t) \leq \bar{\gamma}_i \bar{d}_i \|y_{\rm ref}\|_\infty / |\bar{\lambda}_i|, t \in (t_i, t_{i+1}],$$

so for every $t_i < \infty$ which is defined, there are only a finite number of j's for which t_{i_j} is defined. It follows that $\tilde{x}(t)$ is defined and that $\hat{u}(t)$ is bounded for $t \geq 0$.

Using an argument almost identical to that used to prove Theorem 1, we conclude that the first part of property (i) and property (ii) both hold; it remains to prove the last part of property (i). Suppose that the plant is $\mathbf{P}_i$ but that the final gain is $\tilde{K}_j$ with $j \neq i$. Now choose $k \geq 0$ so that $t_{j_{k+1}} = \infty$. Then it follows that

$$\int_{t_{j_k}}^\infty \bar{\gamma}_j \bar{d}_j e^{\bar{\lambda}_j \tau} \|y_{\rm ref}(\tau)\| \, d\tau \leq (k+1)\hat{b},$$

and since $\|\tilde{x}(t_{j_k})\| \leq \max_{l \in \mathbf{p}} r_l(t_{j_k})$, it follows from the discussion preceding the controller definition that

$$\int_{t_{j_k}}^\infty e^{-A_{ij}^+ \tau} B_{ij}^+ \hat{u}(\tau) \, d\tau \neq -\tilde{x}_1(t_{j_k}) - \int_{t_{j_k}}^\infty e^{-A_{ij}^+ \tau} [D_{ij}^+ y_{\rm ref}(\tau) + E_{ij}^+ w(\tau)] \, d\tau.$$

Hence, $\tilde{x}(t)$ is unbounded, which is a contradiction; we conclude that $i = j$.

Q.E.D.

Remark 1. Both of these controllers have tolerance against noisy measurements of the output; simply write the noise term as part of w. This means, of course, that a bound on the noise must be available in advance. □

Remark 2. Now suppose that the disturbance is not bounded by $\bar{b}$, but it is still bounded. If Controller 1 or Controller 1′ is applied and the gain stops at $\tilde{K}_i$ for some $i \in \mathbf{p}$, then it can be shown that $\tilde{x}(t)$ is bounded. However, it may be that the gain stops at $\tilde{K}_{p+1} = 0$, in which case instability may result. This can be remedied simply by restarting the controller with a new value of $\bar{b}$ double the previous one, whenever we reach t_{p+1}. Eventually, the disturbance will be bounded by $\bar{b}$, so this process will stop. □

Remark 3. The approach that we have taken to solve the problem only needs gross information about the plant, at least for Controller 1. This means that we do not need to know each $\mathbf{P}_i$ exactly; it would be enough if each plant parameter is in a sufficiently small known ball; of course, now the associated constants have to be defined so that they hold for every possible plant in the ball. □

4.2 A "Faster" Approach

We now present an alternative approach to the problem which potentially has a better transient response than for the previous controller; this is borne out in a simulation to follow, but at this time we have no theoretical justification that this is always the case.

Suppose that for each plant $\mathbf{P}_i$, we design a Luenberger-like observer $\hat{\mathbf{P}}_i$ described by

$$\dot{\hat{x}}_i(t) = \left(\tilde{A}_i + \tilde{H}_i\tilde{C}_i\right)\hat{x}_i(t) + \tilde{B}_i\tilde{u}(t) - \tilde{H}_i\tilde{y}(t) + \tilde{H}_i\tilde{D}_i y_{\text{ref}}(t)$$

$$\hat{y}_i(t) = \tilde{C}_i\hat{x}_i(t) + \tilde{D}_i y_{\text{ref}}(t).$$

If we let ϕ_i be the difference between the state of the observer $\hat{\mathbf{P}}_i$ and the state of the augmented plant corresponding to $\mathbf{P}_i$, then we get

$$\dot{\phi}_i(t) = \left(\tilde{A}_i + \tilde{H}_i\tilde{C}_i\right)\phi_i(t) - \left(\tilde{E}_i + \tilde{H}_i\tilde{F}_i\right)w(t).$$

Now choose $\gamma_i > 0$ and $\lambda_i < 0$ so that $\left\| e^{(\tilde{A}_i + \tilde{H}_i\tilde{C}_i)t} \right\| \leq \gamma_i e^{\lambda_i t}$ for $t \geq 0$; then

$$\phi_i(t) \leq \gamma_i e^{\lambda_i t}\|\phi_i(0)\| + \int_0^t \gamma_i e^{\lambda_i(t-\tau)} \left\|\tilde{E}_i + \tilde{H}_i\tilde{F}_i\right\| \cdot \|w(\tau)\|\, d\tau. \qquad (11)$$

If we have an upper bound on w and on $\|\phi_i(0)\|$, then we can get a bound on $\|\phi_i(t)\|$.

So what we will do is as follows. We implement all of the observers simultaneously, and try the feedback gains one at a time. While we are using the i-th gain $\tilde{K}_i$, we consider the i-th observer $\hat{\mathbf{P}}_i$. If we choose $\tilde{H}_i := \tilde{B}_i\tilde{K}_i$, then

$$\dot{\hat{x}}_i = \left(\tilde{A}_i + \tilde{B}_i\tilde{K}_i\tilde{C}_i\right)\hat{x}_i + \tilde{H}_i\tilde{D}_i y_{\text{ref}};$$

if y_{ref} is bounded, then $\hat{x}_i$ will be as well as long as the gain is unchanged. If we start $\hat{x}_i$ off at zero, an upper bound on $\|\tilde{x}(0)\|$ together with an upper bound on $\|w(t)\|$ can be used to generate an upper bound on $\phi_i(t)$. But $\hat{y}_i(t) - \tilde{y}(t) = \tilde{C}_i\phi_i(t) - \tilde{F}_i w(t)$ if the plant is $\mathbf{P}_i$, so the bound on $\phi_i(t)$ and the bound on $w(t)$ can be used to generate a bound on $\|\hat{y}_i(t) - \tilde{y}(t)\|$ which cannot be violated if the plant is $\mathbf{P}_i$. As in Controller 1 where we filter $\|\tilde{y}(t) - \tilde{D}y_{\text{ref}}(t)\|$, here we filter $\|\tilde{y}(t) - \hat{y}_i(t)\|$, since this signal, is, in general, discontinuous.

We define Controller 2 as follows. Fix $T > 0$, $\bar{b} > 0$, and $\varepsilon > 0$, and choose $\alpha_{i_1}, \alpha_{i_2}$, $i \in \mathbf{p}$, so that the inequality given in Lemma 1 is satisfied.

Also, for each $i \in \mathbf{p}$, choose γ_i and λ_i as above, and then choose $\tilde{\lambda}_i < \lambda_i$. The controller can be effected in two phases.

Phase 1: Finding a bound on the initial condition.
With $\tilde{u}(t) = 0$ and $z(0) = 0$, find

$$\theta = \int_0^T \|\tilde{y}(\tau)\|^2 \, d\tau,$$

and define

$$\mu_j := [\alpha_{j_1}\theta + \alpha_{j_2}\bar{b}^2]^{\frac{1}{2}}, j \in \mathbf{p}.$$

Assuming that $\|w(t)\| \leq \bar{b}$ for $t \in [0, T]$, it follows from Lemma 1 that if the plant is $\mathbf{P}_j$, then $\|\tilde{x}(0)\| \leq \mu_j$.

Phase 2: Searching the gains.
Now we apply the control action. Set

$$\tilde{u}(t) = \tilde{K}_i\tilde{y}(t), t \in (t_i, t_{i+1}],$$

and implement the p observers:

$$\begin{aligned} \dot{\hat{x}}_j(t) &= (\tilde{A}_j + \tilde{B}_j\tilde{K}_j\tilde{C}_j)\hat{x}_j(t) + \tilde{B}_j\tilde{u}(t) - \tilde{B}_j\tilde{K}_j\tilde{y}(t) + \tilde{B}_j\tilde{K}_j\tilde{D}_j y_{\text{ref}}(t), \\ &\quad \hat{x}_j(0) = 0, \\ \hat{y}_j(t) &= \tilde{C}_j\hat{x}_j(t) + \tilde{D}_j y_{\text{ref}}(t), \end{aligned}$$

the p signals which generate the bounds on the error between the observer state and the plant state:

$$\begin{aligned} \dot{\hat{\phi}}_j(t) &= \lambda_j\hat{\phi}_j(t) + \gamma_j\|\tilde{E}_j + \tilde{B}_j\tilde{K}_j\tilde{F}_j\|\bar{b}, \\ \hat{\phi}_j(T^+) &= \gamma_j e^{\lambda_j T}\mu_j + (1 - e^{\lambda_j T})\gamma_j\|\tilde{E}_j + \tilde{B}_j\tilde{K}_j\tilde{F}_j\|\bar{b}/|\lambda_j|, \end{aligned}$$

and the p filtered observer errors:

$$\dot{\tilde{\phi}}_j(t) = \tilde{\lambda}_j\tilde{\phi}_j(t) + (\lambda_j - \tilde{\lambda}_j)\|[\tilde{y}(t) - \hat{y}_j(t)]\|, \quad \tilde{\phi}_j(T^+) = 0.$$

We define the gain switching times recursively as follows: set $t_1 := T$, and for every $i \in \{2, \ldots, p+1\}$ for which $t_{i-1} \neq \infty$, let t_i be defined by

$$\begin{aligned} t_i := \min\{t \geq t_{i-1}: \quad & \text{there exists a time } \tilde{t} \in [T, t] \text{ for which} \\ & \tilde{\phi}_i(\tilde{t}) \geq \|\tilde{C}_{i-1}\|\hat{\phi}_i(\tilde{t}) + \|\tilde{F}_{i-1}\|\bar{b} + \varepsilon\} \end{aligned}$$

if the set is nonempty, and ∞ otherwise; if t_{p+1} is defined and finite, then set $t_{p+2} = \infty$ and $\tilde{K}_{p+1} = 0$.

Theorem 3. *Suppose that $y_{\text{ref}}, w \in PC_\infty$, and that $\|w(t)\| \leq \bar{b}$ for $t \geq 0$. For every initial condition $x(0)$, when Controller 2 is applied to the plant, the closed loop system has the following properties:*

(i) the gain eventually remains constant at an element of $\{\tilde{K}_i : i \in \mathbf{p}\}$, and

(ii) the state $\tilde{x}(t)$ is bounded.

PROOF. Let $y_{\text{ref}}, w \in PC_\infty$ be such that $\|w(t)\| \leq \bar{b}$ for $t \geq 0$, and let $x(0)$ be arbitrary. Choose $j \in \mathbf{p}$ so that the plant is $\mathbf{P}_j$.

It follows from Lemma 1 that

$$\|\tilde{x}(0)\|^2 \leq \alpha_{j_1}\theta + \alpha_{j_2}\bar{b}^2,$$

so $\|\tilde{x}(0)\| \leq \mu_j$. It follows from (11) that

$$\begin{aligned}\|\tilde{x}(t) - \hat{x}_j(t)\| &\leq \gamma_j e^{\lambda_j t}\mu_j + \int_0^t \gamma_j e^{\lambda_j(t-\tau)} \left\|\tilde{E}_j + \tilde{B}_j\tilde{K}_j\tilde{C}_j\right\| \cdot \|w(\tau)\|\, d\tau \\ &\leq \hat{\phi}_j(t), t \in [T, \infty).\end{aligned}$$

So

$$\begin{aligned}\|\tilde{y}(t) - \hat{y}_j(t)\| &\leq \|\tilde{C}_j[\tilde{x}(t) - \hat{x}_j(t)] + \tilde{F}_j w(t)\| \\ &\leq \|\tilde{C}_j\| \cdot \hat{\phi}_j(t) + \|\tilde{F}_j\|\bar{b}, t \in [T, \infty).\end{aligned}$$

It follows from Lemma 3 that

$$\|\tilde{\phi}_j(t)\| \leq \|\tilde{C}_j\|\hat{\phi}_j(t) + \|\tilde{F}_j\|\bar{b}, t \in [T, \infty)$$

as well. Hence, if t_{j+1} is defined, then it is necessarily ∞. We conclude the property (i) holds.

It remains to show that $\tilde{x}(t)$ is bounded. Suppose that the final gain is $\tilde{K}_{\hat{i}}$. Since $\hat{\phi}_{\hat{i}}$ is bounded, it follows from the definition of $t_{\hat{i}+1}$ that $\tilde{\phi}_{\hat{i}}$ is bounded. But $\hat{y}_{\hat{i}}$ is bounded since $y_{\text{ref}} \in PC_\infty$. So

$$\int_0^t e^{\lambda_{\hat{i}}(t-\tau)}\|\tilde{y}(\tau)\|\, d\tau$$

is bounded on $t \geq 0$, which means that

$$\int_0^t e^{\lambda_j(t-\tau)}\|\tilde{y}(\tau)\|\, d\tau$$

is as well. It follows from Lemma 2 that $\tilde{x}(t)$ is bounded on $t \geq 0$ as well.

Q.E.D.

Remark 4. This controller suffers from the same problems as Controller 1, namely that the final gain may not provide closed loop stability. However, if Assumptions 1 and 2 hold then this is highly unlikely. It is clear that a dual of Corollary 1 holds for Controller 2. Also, Controller 2 can be modified in a manner analogous to the modification made to Controller 1 in the last section with similar results, but in a practical situation this is probably unnecessary. □

Remark 5. In the event that the F_i 's are all zero, we can dispense with the filtering of $\tilde{y} - \tilde{D}y_{\text{ref}}$ in Controllers 1 and 1$'$, and with the filtering of $\tilde{y} - \hat{y}_j$ in Controller 2. □

Remark 6. Whereas the switching part of Controller 1 uses only $p+1$ integrators, the switching part of Controller 2 uses substantially more, namely $2p + \sum_{j=1}^{p} n_j$. Furthermore, since observers are being used, the latter controller is very sensitive to errors in the plant models. On the other hand, as will be seen from the simulations, Controller 2 has a much better transient response (at least for our example), since more of the plant information is incorporated into the controller. □

5 A Simulation Example

Consider the following unstable plant, which, depending on the operating point, is described by 3 linear time-invariant models of the form (1):

Model 1 ($\mathbf{P}_1$):

$$\dot{x} = \begin{bmatrix} -0.75 & 7.25 \\ -2.25 & -8.25 \end{bmatrix} x + \begin{bmatrix} 0.1 \\ 0.4 \end{bmatrix} u + \begin{bmatrix} 0.1 \\ 0.1 \end{bmatrix} w,$$

$$y = \begin{bmatrix} -1.2 & 4.1 \end{bmatrix} x,$$

with open loop eigenvalues of $-4.5 \pm j1.5$.

Model 2 ($\mathbf{P}_2$):

$$\dot{x} = \begin{bmatrix} 1.4 & 21.1 \\ -1.65 & -25.4 \end{bmatrix} x + \begin{bmatrix} -0.18 \\ 0.33 \end{bmatrix} u + \begin{bmatrix} 0.1 \\ 0.1 \end{bmatrix} w,$$

$$y = \begin{bmatrix} -0.45 & 7.0 \end{bmatrix} x,$$

with open loop eigenvalue of 0.031 and -24.031.

Model 3 ($\mathbf{P}_3$):

$$\dot{x} = \begin{bmatrix} -1.9 & 10.9 \\ -1.3 & -12.2 \end{bmatrix} x + \begin{bmatrix} 0 \\ 0.1 \end{bmatrix} u + \begin{bmatrix} 0.1 \\ 0.1 \end{bmatrix} w,$$

$$y = \begin{bmatrix} 2 & -5 \end{bmatrix} x,$$

with open loop eigenvalues of -3.535 and -10.565. The control objective is to provide error regulation for all constant reference and disturbance signals.

For each $i \in \mathbf{p}$, the parameter optimization method of [9] was used to design a robust controller $\mathbf{K}_i$ with a "fast speed of response" for the case of tracking/regulation of constant references/disturbances, subject to engineering constraints on the control input u. The controllers are each described by LTI models of form (2):

Controller 1 ($\mathbf{K}_1$): $\dot{z} = e,\ u = -2.75z.$

Controller 2 (K_2): $\dot{z} = e,\ u = -2z + 7y.$

Controller 3 ($\mathbf{K}_3$): $\dot{z} = e,\ u = 25z - 6y.$

It turns out that the controller $\mathbf{K}_i$ stabilizes the plant $\mathbf{P}_j$ iff $i = j$, and that no plant-controller combination has a pole on the imaginary axis. Hence, both Assumptions 1 and 2 hold.

Set $T = 1$, $\bar{b} = 2$, and $\varepsilon = 1$. The following constants satisfy the inequalities given in Lemmas 1 and 2:

$$\alpha_{1_1} := 2.1, \alpha_{1_2} := 0.015, \gamma_{1_1} := 21.1, \gamma_{1_2} := 21.1, \gamma_{1_3} := 3.0, \lambda_1 := -1.25,$$

$$\alpha_{2_1} := 3.6, \alpha_{2_2} := 0.036, \gamma_{2_1} := 33.1, \gamma_{2_2} := 33.1, \gamma_{2_3} := 4.7, \lambda_2 := -0.82,$$

$$\alpha_{3_1} := 1.95, \alpha_{3_2} := 0.015, \gamma_{3_1} := 19.9, \gamma_{3_2} := 19.9, \gamma_{3_3} := 2.81, \lambda_3 := -1.93.$$

Here we dispense with $\tilde{r}$ since the F_i 's are all zero. Suppose that the unknown plant is $\mathbf{P}_3$, and we apply Controller 1 with with $x(0) = \begin{bmatrix} 1 & 2 \end{bmatrix}'$, $z(0) = 0$, $w(t) = 2$, and with $y_{\text{ref}}(t)$ a square wave of magnitude 10 and period 20. The closed loop system response is given in Figure 1; $g(t)$ is used to indicate that $\tilde{K}_{g(t)}$ is in use at time t, and we define $\tilde{K}_0 := 0$. Notice that the transients are extremely large; this is due to the fact that the controller is changed *only if* the gain is incorrect. That is, we must detect instability before switching so the signals become quite large; observe that $\tilde{K}_1$ and $\tilde{K}_2$ have been ruled out. This differs from the controllers of [2] and [4] where the gain may change even if it is correct. The large transient response here implies that this controller has practical limitations.

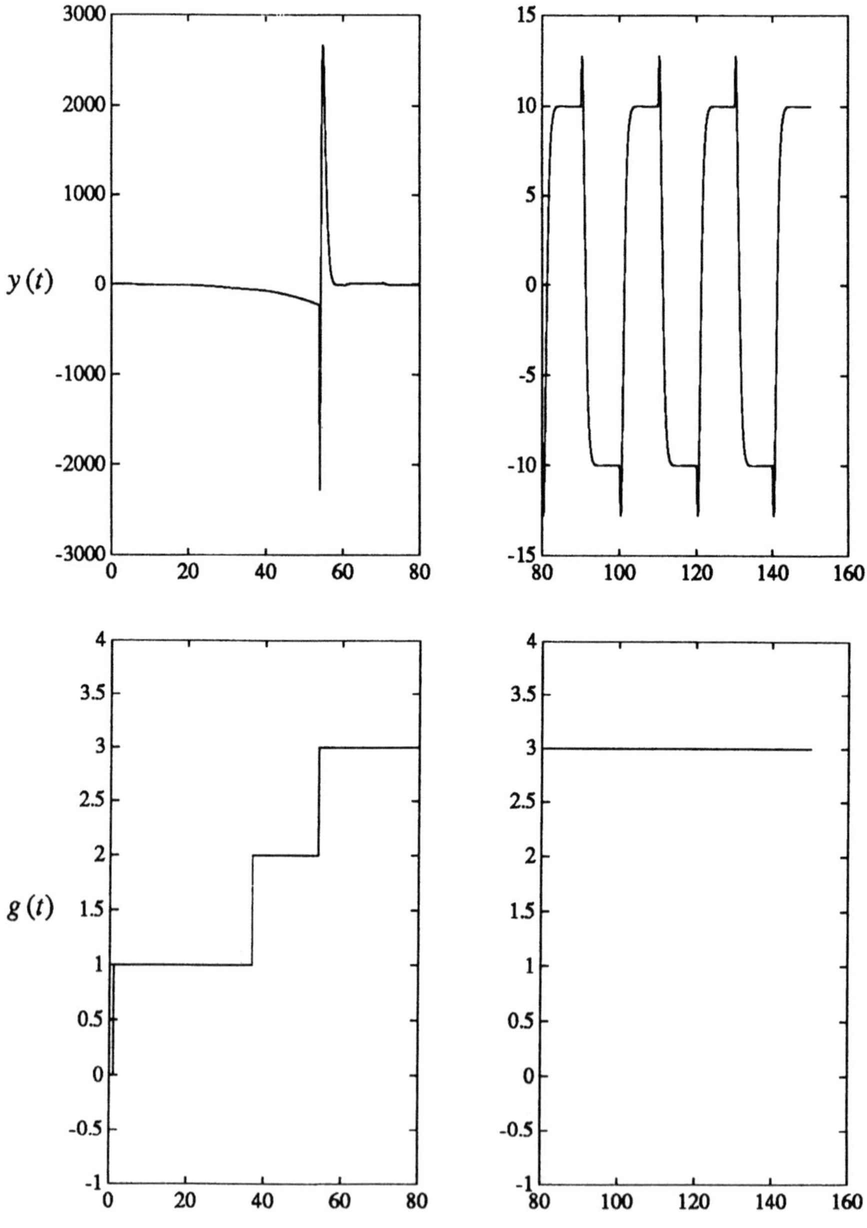

Figure 1. The response of $\mathbf{P}_3$ with Controller 1 applied.

Now set $T = 1$, $\bar{b} = 2$, and $\varepsilon = 1$. Choose

$$\gamma_1 := 21.1, \quad \gamma_2 = 33.1, \quad \gamma_3 = 19.9,$$

and

$$\lambda_1 = -1.25, \quad \lambda_2 = -0.82, \quad \lambda_3 = -1.93.$$

Here we dispense with the $\tilde{\phi}_i$'s since the F_i 's are all zero. Apply Controller 2 with the same choice of $x(0)$, $z(0)$, $w(t)$, and $y_{\text{ref}}(t)$ as above; the closed loop system response is given in Figure 2; $g(t)$ is used to indicate that $\tilde{K}_{g(t)}$ is in use at time t, and we define $\tilde{K}_0 := 0$. Here the closed loop system response is much more reasonable than before; this is due to the fact that we use more plant information in the design of the controller. Although it is not as tolerant to errors in the models of the possible plants $\{\mathbf{P}_i : i \in \mathbf{p}\}$, this switching controller looks much more practical than the one above.

6 Conclusions

In this paper we have considered the problem of controlling a partially known plant. It is assumed that the plant belongs to a finite set of known plants $\{\mathbf{P}_i : i \in \mathbf{p}\}$, and that a well-tuned LTI controller $\mathbf{K}_i$ has been designed off-line to carry out the desired control objective for each plant $\mathbf{P}_i$. Assuming that an upper bound on the unmeasurable disturbance is available in advance, several mechanisms are provided to switch between these controllers and finally stop at a particular one; if the reference is bounded, then the plant and controller state are bounded as well.

Unfortunately, we can ensure that the final controller is the correct one only if (i) $\mathbf{K}_j$ stabilizes $\mathbf{P}_i$ iff $i = j$, and (ii) no plant-controller combination has a pole on the imaginary axis; we do so by ensuring that if any unstable modes exist, then some are excited. It would be advantageous if we could use a more direct way of evaluating closed loop performance. This, of course, would require that we make our control objective more precise, and even then it would be possible only for those control objectives which have an easily computable performance measure.

One of the problems with the first switching controller is that the transient response may be extremely large, thereby making it somewhat impractical; further study may suggest modifications which result in an improvement in performance. The second controller seems to provide a much more reasonable transient response, at least for the example presented here.

We have not fully examined the tolerance of the switching controllers to plant changes. It would certainly be advantageous if such a plant change did not require a restart of the switching controller. This is a topic of further study.

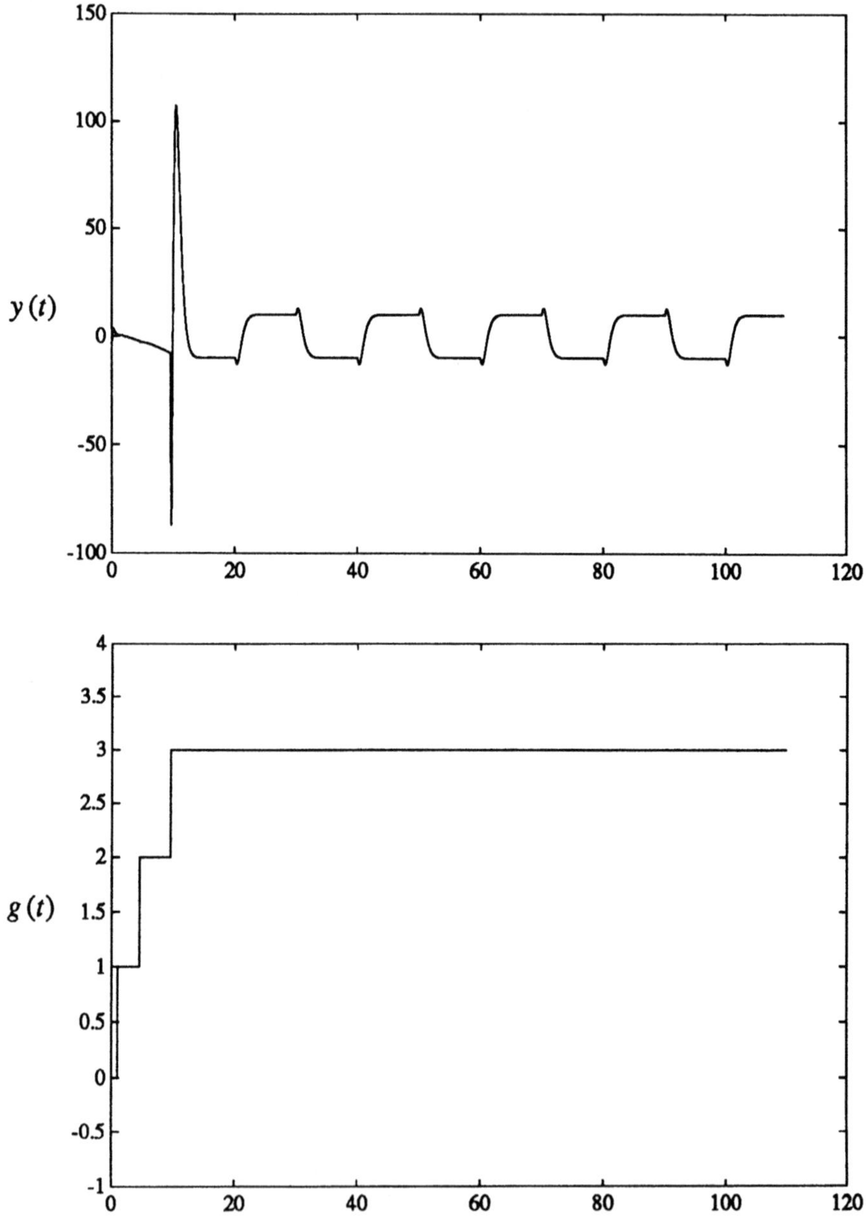

Figure 2. The response of $\mathbf{P}_3$ with Controller 2 applied.

REFERENCES

[1] B. Mårtensson, "The Order of any Stabilizing Regulator is Sufficient *a priori* Information for Adaptive Stabilization", *Syst. Contr. Lett.,* pp. 87–91, July 1985.

[2] D. E. Miller and E.J. Davison, "An Adaptive Controller which can Stabilize any Stabilizable and Detectable LTI System", *Analysis and Control of Nonlinear Systems,* (editors: C.I. Byrnes, C.F. Byrnes, R.E. Saeks), North-Holland, 1988, pp.51–58.

[3] M. Fu and B.R. Barmish, "Adaptive Stabilization of Linear Systems via Switching Control", *IEEE Trans. Automat. Contr.,* vol. AC-31, pp.1097–1103, Dec. 1986.

[4] D.E. Miller and E.J. Davison, "An Adaptive Controller Which Provides Lyapunov Stability", *IEEE Trans. Automat. Contr.,* vol. AC-34, pp. 599–609, June 1989.

[5] B. Mårtensson, "Adaptive Stabilization Without High-Gain", in *Proc. IIASA Conf.,* July 1986 (Lecture Notes in Control and Information Sciences, Vol. 105). New York: Springer-Verlag, 1988, pp. 226–238.

[6] D.E. Miller and E.J. Davison, "A New Self-Tuning Controller to Solve the Servomechanism Problem", *Proc. of the 26th CDC,* Dec. 1987, pp. 843–849.

[7] E.J. Davison and A. Goldenberg, "Robust Control of a General Problem: The Servocompensator", *Automatica,* vol. 11, pp. 461–471.

[8] E.J. Davison and S.H. Wang, "Properties and Calculation of Transmission Zeros of Linear Multivariable Systems", *Automatica,* vol. 10, pp. 643–658, Dec. 1974.

[9] E.J. Davison and I. Ferguson, "The Design of Controllers for the Multivariable Robust Servomechanism Problem Using Parameter Optimization Methods", *IEEE Trans. Autom. Contr.,* vol. 26, pp. 93–110, Feb. 1981.

D. E. Miller and E. J. Davison
Department of Electrical Engineering
University of Toronto
Toronto M5S 1A4
Canada

Predictive and LQG Optimal Control: Equivalences, Differences and Improvements

V. Peterka

The LQG feedback optimal control, extended for the case of preprogrammed setpoint sequence, is compared with open-loop optimal Predictive control (GPC) [4]. It is proved that, for any linear stochastic model (with possible time delay), the first control input within the receding control horizon is the same for both cases.

In Predictive control it is required that a certain number of last increments of the control input be zero. It is shown that this condition, stabilizing the resulting control, is equivalent to a special choice of the matrix by which the last state is weighted in LQG design.

It is shown that the GPC design can be improved if the predictions of the future outputs, required in one-shot optimization of GPC, are based on the state estimate supplied by the Kalman filter. This makes it possible to cover a more general class of possible cases and the predictions are optimal for arbitrarily short observation time (not only asymptotically).

The Kalman filter can be considerably simplified by exploiting a suitable canonical state representation of the employed incremental ARMA model of the controlled process.

Introduction of the state estimate also simplifies the feedforward from measurable external disturbance.

Finally, it is shown that the main parameters of the input-output model (a and b) can be estimated in real time jointly with the state with no approximation involved, under Gaussian assumptions.

1 Introduction

Capability to predict the effect of the applied input on the future outputs is the basic prerequisite for successful control of a given process. To guarantee a good performance of the control algorithm also for non-minimum phase processes and for processes with time delay it is necessary to design the control for a sufficiently long control horizon and to ensure, for stability reasons, that at the end of this control horizon the process be as close as possible to a steady state.

Two different design procedures have been suggested for applications in self-tuning control of processes which can be described by linear stochastic input-output models. The first one, here denoted as LQG, is feedback optimal, i.e. it designs the future process inputs as functions of the data which *will be* available at the particular stage of the control horizon. A method, based on factorization and decomposition of certain band matrices and suitable for regression type process models is described in [6]. Its extension for ARMA type models can be found in [7]. Probably the most compact and numerically reliable algorithms can be obtained if a certain canonical state representation of the input-output model is used and matrix factorization methods are employed [8].

The alternative design procedure is open-loop optimal which means that the future inputs, entering the control horizon considered, are assumed to be functions only of the data available at the beginning of the control horizon. This procedure has been generalized in [4] and analyzed in [5] where also further references can be found. Following [4] it is denoted as GPC (Generalized Predictive Control).

Both the above procedures consider the receding control horizon. They determine, at least implicitly, the control strategy for the entire control horizon, but only the first input is actually applied. In self-tuning control applications both the procedures rely on the certainty equivalence hypothesis and take the current point estimates of the model parameters as the true parameters, not considering their uncertainty.

The purpose of this paper is to compare these two approaches and to suggest some improvements. The paper is organized in the following manner.

In Section 2 the differences in the problem statement are discussed considering the process model, admissible strategies, and criterion for both cases. The main result can be found in Section 3, where the both problems are solved using the method of dynamic programming. It is proved that for arbitrary linear stochastic model the applied optimal input (the first one within the receding horizon) is the same for both cases. It is also shown that the settling of the process input, enforced by GPC, is equivalent to a certain penalization of the final state in the LQG design.

In Section 4 the one-shot optimization, intrinsic for GPC, is discussed. The procedure requires to express the predictions of the future outputs as linear functions of the inputs which are to be determined. It is shown that it is advantageous to base these predictions on the last observed output and on the current state estimate. Then the recursive relations, by which the predictions can be calculated, are independent of the c-parameters of the input-output model. The possible time delay of the process can be handled separately which reduces the computational load.

In Section 5 the general form of the Kalman filter is specialized and considerably simplified for a suitable canonical state representation of the input-output model considered. It turns out that the evolution of the covariance matrix of the estimated state (Riccati equation of the Kalman filter) depends only on the MA-part of the input-output model and produces certain modifications of its c-parameters . These time-varying auxiliary parameters determine the dynamics of the filter by which the state estimate is propagated, and guarantee its stability. This makes it possible to handle also cases when the corresponding c-polynomial has some of its roots on the unit circle. Such cases are practically important and cannot be handled by the asymptotic polynomial theory.

The introduction of the state estimate is advantageous also when a feedforward from measurable external disturbance has to be designed. In most practical cases of industrial process control it appears to be sufficient and adequate to model the measurable external disturbance as a generalized random walk (a stochastic process with independent increments having arbitrarily time-varying variances). Then, as shown in Section 6, the feedforward influences the optimal control only as a correction of the state estimate and does not enter the control law operating on this estimate.

In Section 7 the problem of joint estimation of the state and of the parameters a and b is considered. It turns out that, under Gaussian assumptions, the problem can be solved with no approximation involved.

The paper can be considered as a comparative study. The understanding and the insight are emphasized, not the numerical and algorithmic aspects. Reliable numerical algorithms for LQG can be found in [8].

2 Differences in problem formulation

It is appropriate to start the comparison by clearly stating the control problem in both cases. To formulate a control problem means to define 1) the process model considered, 2) the admissible control strategies among which the optimal one is to be chosen, and 3) the criterion of optimality.

Process model

In GPC the controlled single-input single-output process is represented by an incremental ARMA model (in [4, 5] called CARIMA to emphasize the presence of the manipulated input)

$$\Delta y(k) + \sum_{i=1}^{n} a_i \Delta y(k-i) = \sum_{i=0}^{n} b_i \Delta u(k-d-i) + e(t) + \sum_{i=1}^{n} c_i e(k-i) \quad (1)$$

where, as usually, $\{y(t)\}$, $\{u(t)\}$, and $\{e(t)\}$ mean the output, the input, and a discrete white noise with constant variance, respectively, and Δ denotes the backward difference, e.g. $\Delta y(k) = y(k) - y(k-1)$. Unlike [4, 5] the time indexing is chosen so that $u(t)$ *precedes* $y(t)$. (This convention appears to be more convenient in several places.) $d \geq 0$ is the possible time delay in the input-output relation ($d = 0$ means no dead time).

The standard LQG approach (see e.g. [2]) usually operates with state models. For the purpose of this paper the following incremental form of the state space model will be used.

$$y(k) = y(k-1) + \Psi s(k-1) + \Lambda\, \Delta u(k-d) + \eta(k) \quad (2)$$

$$s(k) = \Phi s(k-1) + \Gamma\, \Delta u(k-d) + \varsigma(k) \quad (3)$$

where $\eta(t)$ and $\varsigma(t)$ are white, but mutually correlated stochastic components of the model.

$$\mathrm{E}[\eta(t)\eta'(t)] = R_{\eta\eta} \qquad \mathrm{E}[\varsigma(t)\varsigma'(t)] = R_{\varsigma\varsigma} \qquad \mathrm{E}[\varsigma(t)\eta'(t)] = R_{\varsigma\eta} \quad (4)$$

Strictly taken, the vector $s(t)$ is only the significant component of the state of the model. To obtain the full state, $s(t)$ should be extended with $y(t-1)$ and, in case of time delay $d > 0$, also with $\{u(t-1), \cdots, u(t-d)\}$.

The representation of the input-output model (1) by a state model (2, 3) is not unique. It appears advantageous to use a canonical state representation which can be constructed in the following way. Rewrite (1) into the form

$$\Delta y(t) = b_0 \Delta u(t-d) + s_1(t-1) + e(t) \quad (5)$$

where

$$s_1(t-1) = \sum_{i=1}^{n} [-a_i \Delta y(t-i) + b_i \Delta u(t-d-i) + c_i e(t-i)]$$

is introduced as the first component of the state. By shifting the time index one step ahead and by extracting the components with the index i=1 it can be expressed through the second state component introduced in a similar way.

$$s_1(t) = -a_1\Delta y(t) + b_1\Delta u(t-d) + c_1 e(t) + s_2(t-1)$$

Continuing this procedure it is obtained generally for $j < n$

$$s_j(t) = -a_j\Delta y(t) + b_j\Delta u(t-d) + c_j e(t) + s_{j+1}(t-1) \tag{6}$$

and for $j = n$

$$s_n(t) = -a_n\Delta y(t) + b_n\Delta u(t-d) + c_n e(t) \tag{7}$$

The substitution for $\Delta y(t)$ in (6) from (5) finally gives for $j < n$

$$s_j(t) = -a_j s_1(t-1) + s_{j+1}(t-1) + [b_j - a_j b_0]\,\Delta u(t-d) + [c_j - a_j]\,e(t) \tag{8}$$

and similarly

$$s_n(t) = -a_n s_1(t-1) + [b_n - a_n b_0]\,\Delta u(t-d) + [c_n - a_n]\,e(t) \tag{9}$$

Hence the matrix coefficients in the state model (2) and (3) get the form

$$\Phi = H - \bar{a}h \qquad \Gamma = \bar{b} - \bar{a}b_0 \qquad \Psi = h \qquad \Lambda = b_0 \tag{10}$$

where H and h are nonparametric matrices

$$H = \begin{bmatrix} 0 & 1 & 0 & \cdots & 0 \\ 0 & 0 & 1 & \cdots & 0 \\ \vdots & \vdots & \vdots & & \vdots \\ 0 & 0 & 0 & \cdots & 1 \\ 0 & 0 & 0 & \cdots & 0 \end{bmatrix} \qquad h' = \begin{bmatrix} 1 \\ 0 \\ 0 \\ \vdots \\ 0 \\ 0 \end{bmatrix} \tag{11}$$

while the parameters, except b_0, are arranged into the vectors

$$\bar{a} = \begin{bmatrix} a_1 \\ a_2 \\ \vdots \\ a_n \end{bmatrix} \qquad \bar{b} = \begin{bmatrix} b_1 \\ b_2 \\ \vdots \\ b_n \end{bmatrix} \qquad \bar{c} = \begin{bmatrix} c_1 \\ c_2 \\ \vdots \\ c_n \end{bmatrix} \tag{12}$$

In this representation the white noise terms become

$$\eta(k) = e(k) \qquad \varsigma(k) = [\bar{c} - \bar{a}]e(k) \tag{13}$$

$$R_{\eta\eta} = \mathrm{E}[e^2(t)] = \rho, \qquad R_{\varsigma\varsigma} = [\bar{c}-\bar{a}][\bar{c}-\bar{a}]'\rho \qquad R_{\varsigma\eta} = [\bar{c}-\bar{a}]\rho \tag{14}$$

Admissible control strategies

It is assumed :

- The past history of the process is known up to and including the time index t.

$$D_t = \{y(t), u(t), y(t-1), u(t-1), \cdots, y(1), u(1)\} \tag{15}$$

In GPC according to [4, 5] the predictions are calculated using the asymptotical polynomial theory and therefore, strictly taken, infinite observation of the past input-output data is assumed. In this comparative study both the control horizon and the observation interval will be considered to be finite.

- Receding control horizon : The control has to be designed for a finite number N steps ahead, i.e. the generation of $u(t+1), u(t+2), \cdots, u(t+N)$ is to be determined. However, only $u(t+1)$ will be really applied and in the next control step the procedure is repeated.

- The expected future set-point sequence is given (designed) for the entire control horizon

$$W_t = \{w(t+d+1), w(t+d+2), \cdots, w(t+d+N)\}$$

and is available for the controller generating $u(t+1)$.

In GPC the future inputs $\{u(t+1), u(t+2), \cdots, u(t+N)\}$ have to be determined as functions of only the data D_t and W_t which are available at the time instant when $u(t+1)$ is to be generated.

$$u(t+k) = f(D_t, W_t) \tag{16}$$

This means that for given D_t and W_t the designed future inputs can be calculated explicitly as numbers.

In LQG design the set of admissible strategies is different. The optimal strategy is sought for among all sequences of control laws operating on the data which will be available at the moment when the particular input has to be generated. Hence, the future inputs have to be expressed as functions of data which are not yet known when the control for the entire control horizon is planned.

$$u(t+k) = F(D_{t+k-1}, W_t) \tag{17}$$

Criteria

Both GPC and LQG try to minimize the control errors

$$r(k) = y(k) - w(k) \qquad\qquad k > t \tag{18}$$

however, in somewhat different ways. Both of them also introduce penalization on the movements of the actuator in order to have the possibility to damp suitably the control process.

The GPC design relies on the predictions of the control errors made on the basis only of the data D_t

$$\hat{r}(t+j|t) = \hat{y}(t+j|t) - w(t+j) \tag{19}$$

$$\hat{y}(t+j|t) = \mathrm{E}\left[y(t+j) \,\middle|\, D_t\right] \tag{20}$$

and minimizes the criterion

$$J_{GPC} = \sum_{k=1}^{M} q_r(k)\ \hat{r}(t+d+k|t)^2 + \sum_{k=1}^{N} q_u(k)\ \Delta u(t+k)^2 \tag{21}$$

with

$$M > N \qquad\qquad q_r(k) \geq 0 \qquad\qquad q_u(k) \geq 0$$

under the condition

$$\Delta u(t+k) = 0 \qquad\qquad N < k \leq M \tag{22}$$

which is introduced to settle down the control process at the end of the control horizon. Later on it will be seen that $M - N$ must be sufficiently large in order to guarantee the required stabilizing effect of the condition (22). Note that the first control error which can be influenced by $\Delta u(t+1)$ is $r(t+d+1)$. Therefore the predictions of control errors in the criterion (21) are shifted by d ahead. Note also that all variables entering the criterion are deterministic functions of the data D_t which are assumed to be given. Hence, unlike LQG, no averaging is required.

In LQG design, as formulated in [8], the following criterion is minimized

$$J_{LQG} = \mathrm{E}\left\{\sum_{k=1}^{N} \left[q_r(k)\ r(t+d+k)^2 + q_u(k)\ \Delta u(t+k)^2\right] + \right.$$
$$\left. + x'(t+d+N)S(N)x(t+d+N) \,\middle|\, D_t\right\} \tag{23}$$

$$S(N) \gg 0$$

where $x(t+d+N)$ is the final state $s(t+d+N)$ extended with the final control error $r(t+d+N)$

$$x(t+d+N) = \begin{bmatrix} r(t+d+N) \\ s(t+d+N) \end{bmatrix}$$

It is important to note that for arbitrary but constant levels of the input and of the output the steady mean value of the state of the incremental state space model (3) is zero. To drive the state towards zero means to quiet down the transient process. This is the role of the last term in the criterion (23) and also the reason for the recommendation to penalize heavily the final state together with the final control error. (This holds only for incremental models. In case of positional models the final increment of the state has to be penalized [8].)

3 Solution by Dynamic Programming

The control problems formulated in the previous Section can be solved using various methods. However, not all known methods are suited for both cases we want to compare. For instance, the one-shot optimization, which is used in [4] to solve the GPC problem and which will be discussed in Section 4, is not directly applicable to the LQG formulation. In this Section the method of dynamic programming, well suited for both cases, will be used for the purpose of comparison. For this method it is advantageous to extend $s(k)$ with $r(k)$

$$x(k) = \begin{bmatrix} r(k) \\ s(k) \end{bmatrix} \tag{24}$$

Then, using the relation

$$r(k) = y(k) - w(k) = r(k-1) + \Psi s(k-1) + \Lambda\ \Delta u(k-d) - \Delta w(k) + e(k) \tag{25}$$

the overall state model becomes

$$x(k) = \Phi_x x(k-1) + \Gamma_x \Delta u(k-d) - h' \Delta w(k) + \xi(k) \tag{26}$$

where $h = [1, 0, 0, \cdots, 0]$ and

$$\Phi_x = \begin{bmatrix} \mathrm{I} & \Psi \\ 0 & \Phi \end{bmatrix} \qquad \Gamma_x = \begin{bmatrix} \Lambda \\ \Gamma \end{bmatrix} \tag{27}$$

$$\xi(k) = \begin{bmatrix} 1 \\ \bar{c} - \bar{a} \end{bmatrix} e(k) \tag{28}$$

To start with the simpler case, consider the GPC formulation first.

GPC :

Rewrite the criterion (21) in the following way

$$J_{GPC} = \sum_{k=1}^{j-1} \left[q_r(k)\hat{r}(t+d+k|t)^2 + q_u(k)\Delta u(t+k)^2\right] + \\ + q_r(j)\hat{r}(t+d+j|t)^2 + q_u(j)\Delta u(t+j)^2 + J_j(D_t) \tag{29}$$

where

$$J_j(D_t) = \sum_{k=j+1}^{M} q_r(k)\hat{r}(t+d+k|t)^2 + \sum_{k=j+1}^{N} q_u(k)\Delta u(t+k)^2 \tag{30}$$

To obtain the recursion called dynamic programming, assume that the part $J_j(D_t)$ of the criterion has been somehow minimized, under the condition (22), by the choice of the inputs $\{u(t+j+1), u(t+j+2), \cdots, u(t+N)\}$ determined as functions of D_t.

$$J_j^*(D_t) = \min_{u(t+j+1),\cdots,u(t+N)} J_j(D_t)$$

Since the terms for $k < j$ of the partially minimized criterion do not depend on $u(t+j)$, the optimal choice $u^*(t+j)$ of this input can be made by the following step of dynamic programming.

$$u^*(t+j) = \arg\min_{u(t+j)} \left[q_r(j)\hat{r}(t+d+j|t)^2 + q_u(j)\Delta u(t+j)^2 + J_j^*(D_t)\right] \tag{31}$$

$$J_{j-1}^*(D_t) = \min_{u(t+j)} \left[q_r(j)\hat{r}(t+d+j|t)^2 + q_u(j)\Delta u(t+j)^2 + J_j^*(D_t)\right] \tag{32}$$

From (26) it is easily seen that

$$\hat{x}(t+j+d|t) = \Phi_x\hat{x}(t+j-1+d|t) + \Gamma_x\Delta u(t+j) - h'\Delta w(t+j+d) \tag{33}$$

Performing just one step of this dynamic programming it can be proved that $J_j^*(D_t)$, sometimes called the Bellman function, can be propagated as the following quadratic form

$$J_j^*(D_t) = \hat{x}'(t+j+d|t)S(j)\hat{x}(t+j+d|t) + 2\hat{x}'(t+j+d|t)F(j) + \beta(j) \tag{34}$$

which is reproduced in each step and where $S(j)$ is a nonnegative definite matrix, $F(j)$ is a vector, and $\beta(j)$ is a scalar into which the components of the minimized criterion, which do not depend on control, are accumulated.

The recursion (31) and (32) for $j = N, N-1, \cdots, 1$ solves the optimization problem. The minimum of the criterion is $J_{GPS}^* = J_1^*(D_t)$.

LQG :

Following a similar route as in GPC, rewrite the criterion (23) into the form

$$J_{LQG} = \mathrm{E}\Bigg\{\sum_{k=1}^{j-1}\left[q_r(k)r(t+k+d)^2 + q_u(k)\Delta u(t+k)^2\right] + \\ + q_r(j)r(t+d+j)^2 + q_u(j)\Delta u(t+j)^2 + J_j(D_{t+j}) \Bigg| D_t\Bigg\} \tag{35}$$

where

$$J_j(D_{t+j}) = \mathrm{E}\Bigg\{\sum_{k=j+1}^{N}\left[q_r(k)r(t+k+d)^2 + q_u(k)\Delta u(t+k)^2\right] + \\ + x'(t+N+d)S(N)x(t+N+d) \Bigg| D_{t+j}\Bigg\} \tag{36}$$

Note that the equality $\mathrm{E}[\cdot \mid D_t] = \mathrm{E}\{\mathrm{E}[\cdot \mid D_{t+j}] \mid D_t\}$ has been applied and that $J_j(D_{t+j})$ is a function of the input-output data up to and including the time index $t+j$. Assume again that the functions (17) generating the inputs $\{u(t+j+1), u(t+j+2), \cdots, u(t+N)\}$ and minimizing $J_j(D_{t+j})$ have been already determined. With the Bellman function

$$J_j^*(D_{t+j}) = \min_{u(t+j+1),\cdots,u(t+N)} J_j(D_{t+j})$$

one step of dynamic programming now reads [1]

$$u^*(t+j) = \arg\min_{u(t+j)} \mathrm{E}\left\{q_r(j)r(t+d+j)^2 + q_u(j)\Delta u(t+j)^2 + J_j^* \mid D_{t+j-1}\right\} \tag{37}$$

$$J_{j-1}^*(D_{t+j-1}) = \min_{u(t+j)} \mathrm{E}\left\{q_r(j)r(t+d+j)^2 + q_u(j)\Delta u(t+j)^2 + J_j^* \mid D_{t+j-1}\right\} \tag{38}$$

Again, performing just one step of this dynamic programming it can be proved that for the stochastic model (26) the Bellman function can be propagated as the quadratic form

$$J_j^*(D_{t+j}) = \hat{x}'(t+j+d|t+j)S(j)\hat{x}(t+j+d|t+j) + \\ + 2\hat{x}'(t+j+d|t+j)F(j) + \tilde{\beta}(j) \tag{39}$$

where $S(j)$ and $F(j)$ evolve in the same way as in GPC case while into $\tilde{\beta}(j)$ also covariance terms are accumulated.

[1]For a more detailed discussion of stochastic dynamic programming see [8].

Start of Dynamic Programming

To complete the solution it is necessary to define the proper start of dynamic programming in both cases.

In the LQG case the dynamic programming starts in a straightforward way with

$$S(N) \gg 0 \qquad\qquad F(N) = 0 \tag{40}$$

The nonzero $\tilde{\beta}(N)$ is not significant for the control synthesis.

The start of dynamic programming in the GPC case is somewhat more complex. According to (30)

$$J_N(D_t) = \sum_{k=N+1}^{M} q_r(k)\hat{r}(t+k+d|t)^2 = \sum_{k=N+1}^{M} \hat{x}'(t+k+d|t)Q_x(k)\hat{x}(t+k+d|t) \tag{41}$$

where

$$Q_x(j) = \begin{bmatrix} q_r(j) & 0 \\ 0 & 0 \end{bmatrix}$$

However, in the final part of the control horizon for $k > N$ the input is prescribed to have zero increments (22). In order to achieve the stabilizing effect it is natural to require that also the set-point be fixed.

$$\Delta u(t+k) = 0 \qquad\qquad \Delta w(t+k+d) = 0 \qquad\qquad N < k \le M \tag{42}$$

Then the relation (33) gets the form

$$\hat{x}(t+j+d|t) = \Phi_x \hat{x}(t+j-1+d|t)$$

and (41) can be expressed as

$$J_N(D_t) = \hat{x}'(t+N+d|t)S(N)\hat{x}(t+N+d|t) \tag{43}$$

where $S(N)$ is determined by the following recursion.

$$j = M, \cdots, N{+}1 \quad : \quad S(j-1) = \Phi'_x S(j)\Phi_x + Q_x(j) \qquad\qquad S(M) = 0 \tag{44}$$

The following *Result 1* summarizes the main facts.

Result 1. (GPC and LQG optimum control synthesis) *The optimal sequence of inputs within the receding horizon is :*
for GPC

$$\Delta u(t+j) = L_x(j)\hat{x}(t+j-1+d \mid t) + L_w(j) \tag{45}$$

for LQG

$$\Delta u(t+j) = L_x(j)\hat{x}(t+j-1+d \mid t+j-1) + L_w(j) \tag{46}$$

In both cases the control law can be generated by the recursion

$$\bar{S}(j) = S(j) + Q_x(j) \qquad Q_x(j) = \begin{bmatrix} q_r(j) & 0 \\ 0 & 0 \end{bmatrix} \tag{47}$$

$$L_x(j) = -G(j)^{-1}\Gamma_x'\bar{S}(j)\Phi_x \tag{48}$$

$$G(j) = \Gamma_x'\bar{S}(j)\Gamma_x + q_u(j) \tag{49}$$

$$S(j-1) = \Phi_x'\bar{S}(j)\Phi_x - L_x'(j)G(j)L_x(j) \tag{50}$$

$$F_w(j) = \bar{S}(j)h'\Delta w(t{+}j{+}d) \tag{51}$$

$$L_w(j) = G(j)^{-1}\Gamma_x'\left[F(j) - F_w(j)\right] \tag{52}$$

$$F(j-1) = \left[\Phi_x - \Gamma_x L_x(j)\right]'\left[F(j) - F_w(j)\right] \tag{53}$$

The applied $\Delta u(t+1)$ is the same for GPC and LQG

$$\Delta u(t+1) = L_x(1)\hat{x}(t+d \mid t) + L_w(1) \tag{54}$$

if the recursion is started with $F(N) = 0$ and with $S(N) > 0$ determined by (44) to satisfy the conditions (42).

Remark a. The author does not see any reason why $S(N)$ determined according to (44) should be preferred to simply choosing it as a diagonal matrix with large positive diagonal entries. Note that in GPC the difference $M - N$ must be chosen sufficiently large in order to guarantee the positive definiteness of $S(N)$. The stability of GPC control has recently been studied in [3]. □

Remark b. The recursion (47 – 50) is equivalent to the well known Riccati equation. In case of time-varying sequence of the given (or expected) future setpoints it must be supplemented by the recursion (51 – 53) generating the additional term $L_w(1)$. □

Remark c. The formulae listed in the Result 1 give an insight and support analysis but cannot be recommended for numerical calculations. The relation (50) does not guarantee the positive definiteness of the propagated matrix $S(j)$ in the presence of round-off errors. This may be critical in some often met cases (e.g. in case of a minimum-phase process and no penalization of inputs when $S(j)$ converges to a zero matrix). A numerically save and efficient way is to propagate the Bellman function in a factorized form as reported in [8]. □

4 One-shot optimization of GPC

In the GPC case the future process inputs,entering the criterion (21), have to be determined only on the basis of the data up to and including the time index t. Then it is possible to express the predictions (19, 20) as linear functions of these future inputs and to reduce the minimization of the GPC criterion to a simple least square problem.

In [4] the required predictions are determined using the polynomial theory. Since this theory is asymptotical, it assumes the infinitely long observation of the past history of the process and excludes practically important cases when the roots of the c-polynomial, related to the moving average part of the model, lie on the unit circle. In this paper it will be shown that arbitrarily short observations and arbitrary c-polynomials can be considered and that some further advantages can be attained if the predictions are based on state estimation. The Kalman filter, which is required for this purpose, will be discussed in Section 5 where it will be shown how it can be significantly simplified by exploiting the canonical state representation (2,3,10) of the input-output model (1). In this Section it will be assumed that at the beginning of the control horizon, when the input $u(t+1)$ has to be decided, the current estimate of the state $\hat{s}(t|t) = \mathrm{E}[s(t)|D_t]$, the last measured process output $y(t)$ and, in case of process delay $d > 0$, also d previous inputs $\{u(t), u(t-1), \cdots, u(t-d+1)\}$ are available.

From the model (2,3) it directly follows for $j > 0$ and with $\hat{y}(t|t) = y(t)$

$$\hat{y}(t+j|t) = \hat{y}(t+j-1|t) + \Psi\hat{s}(t+j-1|t) + \Lambda\Delta u(t+j-d) \qquad (55)$$

$$\hat{s}(t+j|t) = \Phi\hat{s}(t+j-1|t) + \Gamma\Delta u(t+j-d) \qquad (56)$$

For $j = 1, \cdots, d$ the recursion (55) and (56) can be used to calculate explicitly $\hat{y}(t+d|t)$ and $\hat{s}(t+d|t)$. However, further predictions for $j > d$ are functions of the future inputs and can be determined according to the following

Result 2. (Predictions) *For the input-output model (1) in the canonical state representation (2,3,10) the output predictions* $\{\hat{y}(t+d+k|t) : k = 1, \cdots, M\}$, *required by GPC, are*

$$\hat{y}(t+d+k|t) = \sum_{i=0}^{k-1} g_i \Delta u(t+k-i) + \hat{y}(t+d|t) + f_k \hat{s}(t+d|t) \qquad (57)$$

where the coefficients g_i *are determined by the recursion*

$$g_0 = \Lambda \qquad m_0 = \Gamma$$
$$g_i = g_{i-1} + \Psi m_{i-1} \qquad m_i = \Phi m_{i-1} \qquad i > 0 \qquad (58)$$

while for the row vectors f_k the following recursion holds

$$f_1 = \Psi \qquad M_1 = \Phi$$
$$f_k = f_{k-1} + \Psi M_{k-1} \qquad M_k = \Phi M_{k-1} \qquad k > 1 \tag{59}$$

The above recursions determine also the predictions of the states

$$\hat{s}(t+d+k|t) = \sum_{i=0}^{k-1} m_i \Delta u(t+k-i) + M_k \hat{s}(t+d|t) \tag{60}$$

PROOF. By induction using (55) and (56) for $j = k + d$. Q.E.D.

Remark d. Note that, unlike the recursions which can be obtained from the asymptotical polynomial theory [4, Appendix A of Part II], neither the recursions (58) and (59) nor the recursions (55) and (56) depend on the c-parameters of the model. The c-parameters enter only the state estimation, which will be discussed in Section 5, and determine the dynamics of the Kalman filter (or state observer) producing the state estimate $\hat{s}(t|t)$. □

Introduce the vectors

$$\hat{y} = \begin{bmatrix} \hat{y}(t+d+1|t) \\ \hat{y}(t+d+2|t) \\ \vdots \\ \hat{y}(t+d+M|t) \end{bmatrix}, \qquad w = \begin{bmatrix} w(t+d+1) \\ w(t+d+2) \\ \vdots \\ w(t+d+M) \end{bmatrix} \tag{61}$$

$$\Delta u = \begin{bmatrix} \Delta u(t+1) \\ \Delta u(t+2) \\ \vdots \\ \Delta u(t+N) \end{bmatrix}, \qquad F = \begin{bmatrix} \hat{y}(t+d|t) + f_1 \hat{s}(t+d|t) \\ \hat{y}(t+d|t) + f_2 \hat{s}(t+d|t) \\ \vdots \\ \hat{y}(t+d|t) + f_M \hat{s}(t+d|t) \end{bmatrix} \tag{62}$$

and define the $M \times N$ matrix

$$G = \begin{bmatrix} g_0 & 0 & 0 & \cdots & 0 \\ g_1 & g_0 & 0 & \cdots & 0 \\ g_2 & g_1 & g_0 & \cdots & 0 \\ \vdots & \vdots & \vdots & \vdots & \vdots \\ g_{N-1} & g_{N-2} & g_{N-3} & \cdots & g_0 \\ \vdots & \vdots & \vdots & \vdots & \vdots \\ g_{M-1} & g_{M-2} & g_{M-3} & \cdots & g_{M-N} \end{bmatrix} \tag{63}$$

Using this notation, the predicted input-output relation can be written in the form

$$\hat{y} = G\,\Delta u + F \tag{64}$$

and the GPC criterion (21) can be expressed as follows

$$J = (\hat{y} - w)'Q_r(\hat{y} - w) + \Delta u' Q_u \Delta u \tag{65}$$

where Q_r and Q_u are diagonal matrices

$$Q_r = \mathrm{diag}(q_r(1), q_r(2), \cdots, q_r(M)), \qquad Q_u = \mathrm{diag}(q_u(1), q_u(2), \cdots, q_r(N)) \tag{66}$$

Then the sequence of the control actions minimizing the criterion is obtained simply as

$$\Delta u^* = (G_r' G + Q_u)^{-1} G' Q_r (w - F) \tag{67}$$

and the sought for control law, generating the input $u(t+1)$, is given by the first row in (67).

5 State estimation

To complete the above solutions of the given control problems it is necessary to provide $\hat{s}(t|t) = \mathrm{E}[s(t)|D_t]$, for both GPC and LQG. Recall that in case of time delay, $d > 0$, the predictions $\hat{y}(t+d|t)$ and $\hat{s}(t+d|t)$ required in (57) can be determined, for given $\hat{s}(t|t)$, by the recursion (55) and (56). Recall also that $\hat{x}(t+d|t)$ required in (54) is defined as $\hat{x}'(t+d|t) = [(\hat{y}(t+d|t) - w(t+d)), \hat{s}'(t+d|t)]$.

In general setting, the problem of real-time (recursive) state estimation can be formulated as follows : *Given* $\mathrm{p}(s(t-1)|D_{t-1})$ *and the newly observed input-output pair* $\{y(t), u(t)\}$ *determine* $\mathrm{p}(s(t)|D_t)$. To be able to solve this problem it is sufficient to have a model which makes it possible to determine the joint probability distribution

$$\mathrm{p}(y(t), s(t)|u(t), D_{t-1}) \tag{68}$$

Then the solution is given by the conditioning

$$\mathrm{p}(s(t)|D_t) = \frac{\mathrm{p}(y(t), s(t)|u(t), D_{t-1})}{\int \mathrm{p}(y(t), s(t)|u(t), D_{t-1})\mathrm{d}s(t)} \tag{69}$$

From the state model (2, 3) it follows directly

$$\hat{y}(t|t-1) = y(t-1) + \Psi\hat{s}(t-1|t-1) + \Lambda\Delta u(t-d) \tag{70}$$

$$\hat{s}(t|t-1) = \Phi\hat{s}(t-1|t-1) + \Gamma\Delta u(t-d) \tag{71}$$

and also

$$y(t) - \hat{y}(t|t-1) = \Psi(s(t-1) - \hat{s}(t-1|t-1)) + \eta(t)$$

$$s(t) - \hat{s}(t|t-1) = \Phi(s(t-1) - \hat{s}(t-1|t-1)) + \varsigma(t)$$

Hence the covariances of the joint probability distribution (68) are

$$R_{yy}(t|t-1) = \Psi R_{ss}(t-1|t-1)\Psi' + R_{\eta\eta} \quad (72)$$

$$R_{ss}(t|t-1) = \Phi R_{ss}(t-1|t-1)\Phi' + R_{\varsigma\varsigma} \quad (73)$$

$$R_{sy}(t|t-1) = \Phi R_{ss}(t-1|t-1)\Psi' + R_{\varsigma\eta} \quad (74)$$

If it is assumed that $\mathrm{p}(s(t-1)|D_{t-1})$ is normal (Gaussian) and that $\eta(t)$ and $\varsigma(t)$ are also normally distributed, then the normality is reproduced and the conditioning (69) yields [1, §2.5]

$$\hat{s}(t|t) = \hat{s}(t|t-1) + R_{sy}(t|t-1)R_{yy}^{-1}(t|t-1)(y(t) - \hat{y}(t|t-1)) \quad (75)$$

$$R_{ss}(t|t) = R_{ss}(t|t-1) - R_{sy}(t|t-1)R_{yy}^{-1}(t|t-1)R_{sy}'(t|t-1) \quad (76)$$

The substitution of (72 – 74) into (75,76) gives the Kalman filter for the case of correlated noises η and ς in the following form

$$\hat{s}(t|t) = \Phi\hat{s}(t-1|t-1) + \Gamma\,\Delta u(t-d) + K(t)\,[y(t) - \hat{y}(t|t-1)] \quad (77)$$

$$K(t) = [\Phi R_{ss}(t-1|t-1)\Psi' + R_{\varsigma\eta}]\,[\Psi R_{ss}(t-1|t-1)\Psi' + R_{\eta\eta}]^{-1} \quad (78)$$

$$R_{ss}(t|t) = \Phi R_{ss}(t-1|t-1)\Phi' + R_{\varsigma\varsigma} - K(t)\,[\Psi R_{ss}(t-1|t-1)\Psi' + R_{\eta\eta}]\,K'(t) \quad (79)$$

Now this Kalman filter will be specialized for the canonical state representation of the input-output model (1) introduced in Section 1. This means to substitute (10) and (14) into (77–79). At the same time it is advantages to express all covariances in proportion to ρ so that, for instance,

$$R_{ss}(t|t) = \rho\bar{R}_{ss}(t|t) \quad (80)$$

After this substitution and after some algebraic manipulation and rearrangements the following result is obtained.

Result 3. (Real-time state estimation) *For the state model (2, 3) with the parameters (10) and (14) the real-time estimation of the state $s(t)$ can be reduced to the following linear difference equation*

$$\hat{s}(t|t) = C(t)\hat{s}(t-1|t-1) - [\bar{a} - \tilde{c}(t)]\Delta y(t) + [\bar{b} - \tilde{c}(t)b_0]\Delta u(t-d) \quad (81)$$

$$C(t) = H - \tilde{c}(t)h = \begin{bmatrix} -\tilde{c}_1(t) & 1\ 0 \cdots 0 \\ -\tilde{c}_2(t) & 0\ 1 \cdots 0 \\ \vdots & \vdots \quad \vdots \\ -\tilde{c}_{n-1}(t) & 0\ 0 \cdots 1 \\ -\tilde{c}_n(t) & 0\ 0 \cdots 0 \end{bmatrix} \quad (82)$$

where the time varying parameters $\tilde{c}(t)$ are generated by the following algebraic recursion

$$\tilde{c}(t) = [H\bar{R}_{ss}(t-1|t-1)h' + \bar{c}]/(1 + h\bar{R}_{ss}(t-1|t-1)h') \qquad (83)$$

$$\bar{R}_{ss}(t|t) = C(t)\bar{R}_{ss}(t-1|t-1)C'(t) + [\tilde{c}(t) - \bar{c}][\tilde{c}(t) - \bar{c}]' \qquad (84)$$

Remark e. The Kalman gain gets the form

$$K(t) = \tilde{c}(t) - \bar{a} \qquad (85)$$

which has been used to rewrite (77) into (81). □

Remark f. Note that the variance ρ of the white noise $\{e(t)\}$ does not need to be known if the covariance matrix, by which the prior uncertainty of the initial state $s(0)$ is characterized, is chosen also in proportion to this, possibly unknown variance ρ , $R_{ss}(0) = \rho\bar{R}_{ss}(0)$. □

Remark g. The time-varying parameters $\tilde{c}(t)$ determine the dynamics of the filter (81). Note that the evolution of this parameters, governed by the recursion (83, 84), does not depend on the parameters a and b. It depends solely on the c-parameters of the moving-average part of the model and on the prior covariance of the initial state. □

Remark h. The recursion (83, 84) is a counterpart to the spectral factorization in the theory of stationary stochastic processes. The time-varying parameters $\tilde{c}(t)$, it generates for $\bar{R}_{ss}(0) > 0$, ensure the stability of the filter (81) for arbitrary c-parameters of the model. The following first order example gives some insight. For $n = 1$ the covariance matrix $\bar{R}_{ss}(t|t)$ is reduced to a nonnegative scalar $\rho_s(t)$ and the recursion gets the form

$$\tilde{c}_1(t) = \frac{c_1}{1 + \rho_s(t-1)} \quad , \qquad \rho_s(t) = \tilde{c}_1(t)\rho_s(t-1) + (\tilde{c}_1(t) - c_1)^2$$

and after substituting for $\tilde{c}_1(t)$

$$\rho_s(t) = \frac{c_1^2\rho_s(t-1)}{1 + \rho_s(t-1)}$$

Hence the inverse of $\rho_s(t)$ evolves according to the linear first order difference equation

$$\rho_s^{-1}(t) = c_1^{-2}\rho_s^{-1}(t-1) + c_1^{-2}$$

which clearly shows that

$$\begin{array}{lll} \text{for } |c_1| \le 1 & \rho_s(\infty) = 0 & \tilde{c}_1(\infty) = c_1 \\ \text{for } |c_1| > 1 & \rho_s(\infty) = c_1^2 - 1 & \tilde{c}_1(\infty) = \dfrac{1}{c_1} \end{array}$$

It may be also interesting to see that in the special, but practically important case $c_1 = -1$ the coefficient $\tilde{c}_1(t)$ is approaching the unit circle relatively slowly and from the stable side

$$\tilde{c}_1(t) = -1 + \frac{1}{\rho_s^{-1}(0) + t}$$

starting, for $\rho_s(0) \gg 0$, close from zero. □

Remark i. The formula (84) is not the best way how to compute numerically. If the roots of the c-polynomial lie inside the unit disk, which is a common case, then, for growing t, the covariance matrix $\bar{R}_{ss}(t|t)$ approaches a zero matrix (the state can be reconstructed exactly) and the round-off errors may cause the loss of positive definiteness of the propagated matrix. A numerically save algorithm, which performs the task and guarantees the positive definiteness, can be found in [8] and [9]. □

6 Feedforward from measurable external disturbance

In industrial applications cases are often met when the controlled process is corrupted by an external disturbance which can be measured. Then the quality of control can be improved if this information is introduced into the controller as a feedforward. In this Section it will be shown how this can be easily done if the state estimator is a part of the controller.

A measurable external disturbance, say v, can be introduced into the model of the controlled process as an additional input which cannot be manipulated, but can be measured. Hence, instead of (1) we now have

$$\Delta y(k) + \sum_{i=1}^{n} a_i \Delta y(k-i) = \sum_{i=0}^{n} b_i \Delta u(k-d-i) + \sum_{i=0}^{n} b_{vi} \Delta v(k-i) + \\ + e(t) + \sum_{i=1}^{n} c_i e(k-i) \qquad (86)$$

and instead of (2, 3) we have

$$y(k) = y(k-1) + \Psi s(k-1) + \Lambda\, \Delta u(k-d) + \Lambda_v\, \Delta v(k) + \eta(k) \qquad (87)$$

$$s(k) = \Phi s(k-1) + \Gamma\, \Delta u(k-d) + \Gamma_v\, \Delta v(k) + \varsigma(k) \qquad (88)$$

Let the time indexing of the sampled signal v be so that $u(k)$ *precedes* $v(k)$. Then it is sufficient to redefine the set of data D_t, introduced by (15), as follows

$$D_t = \{y(t), u(t), v(t), y(t-1), u(t-1), v(t-1), \cdots, y(1), u(1), v(1)\} \qquad (89)$$

and the definitions of admissible strategies from Section 2 remain the same, for both GPC and LQG.

First, let us consider the question how the observation of the external disturbance has to be incorporated into the real-time estimation of the process state $s(t)$. To be able to answer this question it is necessary to define what the adjective "external" means. A stochastic signal is called external, with respect to a given control loop, if there does not exist a hidden channel from the controlled process to the external signal, or, in other words, if the evolution of the external process depends only on its own past history. This can be formally introduced as the assumption

$$\begin{aligned} \mathrm{p}(v(t)|s(t-1),u(t),D_{t-1}) &= \mathrm{p}(v(t)|u(t),D_{t-1}) = \\ &= \mathrm{p}(v(t)|v(t-1),v(t-2),\cdots,v(1)) \end{aligned} \tag{90}$$

From the obvious identity

$$\begin{aligned} \mathrm{p}(v(t)|s(t-1),u(t),D_{t-1})\ \ \mathrm{p}(s(t-1)|u(t),D_{t-1}) &= \\ = \mathrm{p}(s(t-1)|v(t),u(t),D_{t-1})\ \ \mathrm{p}(v(t)|u(t),D_{t-1}) \end{aligned}$$

it follows that if (90) is true then it also holds

$$\mathrm{p}(s(t-1)|v(t),u(t),D_{t-1}) = \mathrm{p}(s(t-1)|u(t),D_{t-1})$$

and, since $u(t)$, in general, can be a function only of data D_{t-1}

$$\mathrm{p}(s(t-1)|v(t),u(t),D_{t-1}) = \mathrm{p}(s(t-1)|D_{t-1}) \tag{91}$$

This means that observations of $v(t)$ and $u(t)$ do not bring any additional information which could change the probability distribution by which the uncertainty of $s(t-1)$ is characterized. Under this condition the state model (87, 88) defines, for given $\mathrm{p}(s(t-1)|D_{t-1})$, the joint probability distribution $\mathrm{p}(y(t), s(t)|v(t), u(t), D_{t-1})$ which makes it possible to determine $\mathrm{p}(s(t)|D_t)$ similarly to (69). Following the same route as in Section 5 it is obtained that the Result 3 holds also for the case of measurable external disturbance if the difference equation (81), by which the conditional mean $\hat{s}(t|t)$ is propagated, is replaced by

$$\begin{aligned} \hat{s}(t|t) = C(t)\hat{s}(t-1|t-1) - \\ - [\bar{a} - \tilde{c}(t)]\Delta y(t) + [\bar{b} - \tilde{c}(t)b_0]\Delta u(t-d) + [\bar{b}_v - \tilde{c}(t)b_{v0}]\Delta v(t) \end{aligned} \tag{92}$$

It remains to revise the control synthesis. As the criterion, which is to be minimized by the choice of the control strategy, depends on the future course of the disturbance it is necessary to have a stochastic model available which makes it possible to predict the disturbance for the control horizon considered. For many industrial applications it appears suitable to model the disturbance as a generalized random walk, i.e. as a stochastic process with independent increments the variances of which can vary in an arbitrary and unknown way.

$$v(k) = v(k-1) + e_v(k) \tag{93}$$

$$\mathrm{E}[e_v(k)|v(k-1), v(k-2), \cdots] = 0, \quad \mathrm{E}[e_v^2(k)|v(k-1), v(k-2), \cdots] = \rho_v(k)$$

The advantage of such a simple model is that it does not need to be identified and at the same time covers a rather broad class of practical cases, including nonstationary disturbances.

It is easily seen that for a generalized random walk and $k > 0$ it holds

$$\hat{v}(t+k|t) = \mathrm{E}[v(t+k)|D_t] = v(t), \quad \mathrm{E}[\Delta v(t+k)|D_t] = 0 \tag{94}$$

$$\mathrm{E}[\Delta v(t+k)|D_{t+k-1}] = 0, \quad \mathrm{E}\left[(\Delta v(t+k))^2 \,|D_{t+k-1}\right] = \rho_v(t+k) \tag{95}$$

Hence, the recursive relations (55, 56), which determine the predictions of the future process outputs, as well as the Result 1 and the Result 2 still hold. Summing up we come to the following

Result 4. (Feedforward from measurable external disturbance) *If an external disturbance v is accessible for measurement then it can be used to improve the estimation of the state $s(t)$ of the process model (87, 88) by replacing the difference equation (81), propagating the state estimate, by the equation (92).*

If the future development of the measurable external disturbance is modelled as a generalized random walk, then the optimal feedforward enters only the estimation of the process state leaving the control law, operating on this improved state estimate, unchanged.

Remark j. The generalized random walk is a rather rough model for the future development of the measurable external disturbance. However, the simulation studies indicate that usually, in realistic cases, not much can be gained if a more sophisticated model is used for the control law synthesis. It should be also emphasized that a more detailed model, which could be used for this purpose, has to be identified relying on some kind of stationarity. □

7 Joint estimation of the state and of the parameters a and b

In the previous sections it has been assumed that the model parameters are given. In this Section the problem of real-time estimation of the parameters a and b jointly with the state $s(t)$ will be considered. It is assumed that the parameters c, as well as the model structure determined by n and d, are given. The case of more then one prior choices of $\{n, d, c\}$ is considered in [10]. As the parameters b_v can be estimated in the same way as the parameters b, they will be omitted here for brevity. Hence, the set of unknown parameters is

$$\Theta = \{a_1, a_2, \cdots, a_n, b_0, b_1, \cdots, b_n\}$$

and when ordered into a vector

$$\Theta = \begin{bmatrix} \bar{a} \\ b \end{bmatrix} \qquad b = \begin{bmatrix} b_0 \\ \bar{b} \end{bmatrix} \tag{96}$$

To estimate the parameters Θ simultaneously with the state $s(t)$ means, in Bayesian view, to propagate the joint probability distribution

$$\mathrm{p}(s(t), \Theta | D_t) = \mathrm{p}(s(t)|\Theta, D_t) \ \mathrm{p}(\Theta|D_t) \tag{97}$$

In Section 5 we were dealing with the first factor on the right-hand side of (97). However, in the Result 3, which determines this normal probability distribution, the parameters Θ are incorporated into the conditional mean $\hat{s}(t|t)$. For the present purpose it is necessary to express $\hat{s}(t|t)$ explicitly as a function of these, now unknown, parameters. This can be done in the following way.

As the difference equation (81) is linear, its explicit solution can be given. Denoting

$$\hat{s}(t|t, \Theta) = \mathrm{E}\left[s(t)|D_t, \Theta\right] \tag{98}$$

we obtain in this way

$$\hat{s}(t|t, \Theta) = G(t, 0)\hat{s}(0) +$$

$$+ \sum_{k=1}^{t} G(t, k) \left[-\left(\bar{a} - \tilde{c}(k)\right) \Delta y(k) + \left(\bar{b} - \tilde{c}(k) b_0\right) \Delta u(k-d) \right] \tag{99}$$

where

$$\text{for} \quad k < t \quad G(t, k) = \prod_{j=k+1}^{t} C(j); \qquad G(t, t) = I \tag{100}$$

First, let us consider the term depending on the output.

$$\sum_{k=1}^{t} G(t, k) \left(\bar{a} - \tilde{c}(k)\right) \Delta y(k) = Y^{(1)}(t)\bar{a} - Y^{(0)}(t) \tag{101}$$

where the data are compressed into the $n \times n$ matrix $Y^{(1)}(t)$ and into the vector $Y^{(0)}(t)$. Exploiting the obvious property of the transition matrix $G(t,k)$

$$G(t,k) = C(t)G(t-1,k) \tag{102}$$

it is possible to express these statistics as follows.

$$Y^{(1)}(t) = \sum_{k=1}^{t} G(t,k)\Delta y(k) = C(t)Y^{(1)}(t-1) + I\Delta y(t) \tag{103}$$

$$Y^{(0)}(t) = \sum_{k=1}^{t} G(t,k)\tilde{c}(k)\Delta y(k) = C(t)Y^{(0)}(t-1) + \tilde{c}(t)\Delta y(t) \tag{104}$$

For compactness, it is advantageous to introduce the statistic

$$Y(t) = \left[-Y^{(0)}(t), Y^{(1)}(t)\right]$$

into which all output data are compressed and which can be updated in the following way.

$$Y(t) = C(t)Y(t-1) + [-\tilde{c}(t), I]\Delta y(t) \tag{105}$$

In a very similar way the terms in (99) depending on the input can be expressed.

$$U(t) = C(t)U(t-1) + [-\tilde{c}(t), I]\Delta u(t-d) \tag{106}$$

and the conditional mean (99) gets the required form

$$\hat{s}(t|t,\Theta) = G(t,0)\hat{s}(0) - Y(t)\begin{bmatrix} 1 \\ \bar{a} \end{bmatrix} + U(t)b \tag{107}$$

In the sequel only the standard choice $\hat{s}(0) = 0$ will be considered. Then the conditional mean of the state can be expressed in the following compact form

$$\hat{s}(t|t,\Theta) = X(t)\begin{bmatrix} 1 \\ \Theta \end{bmatrix} = Y^{(0)}(t) + Z(t)\Theta \tag{108}$$

where

$$X(t) = [-Y(t), U(t)]; \qquad Z(t) = \left[-Y^{(1)}(t), U(t)\right] \tag{109}$$

Now everything is prepared for parameter estimation. The problem is : *Given the normal probability density* $\mathrm{p}(\Theta|D_{t-t})$ *with the mean* $\hat{\Theta}(t-1)$ *and with the covariance matrix* $\rho\bar{R}_{\Theta\Theta}(t-1)$ *determine, for a newly observed output* $y(t)$, $\mathrm{p}(\Theta|D_t)$. It can be solved using the conditioning

$$\mathrm{p}(\Theta|D_t) = \frac{\mathrm{p}(y(t),\Theta|u(t),D_{t-1})}{\int \mathrm{p}(y(t),\Theta|u(t),D_{t-1})\mathrm{d}\Theta} \tag{110}$$

The required joint probability distribution can be determined in the following way. Using (108) we have

$$\hat{y}(t|t-1,\Theta) = y(t-1) + \hat{s}_1(t-1|t-1) + b_0\Delta u(t-d) =$$
$$= y(t-1) + Y_1^{(0)}(t-1) + z(t)\Theta \tag{111}$$

where $Y_1^{(0)}(t-1)$ is the first entry of the column vector $Y^{(0)(t-1)}$, and the row vector $z(t)$ is the first row of $Z(t-1)$ with $\Delta u(t)$ added to the entry $Z_{1,n+1}(t-1)$ in order to incorporate the term $b_0\Delta u(t-d)$.

$$z_i(t) = Z_{1,i}(t-1) \qquad i \neq n+1$$
$$z_{n+1}(t) = Z_{1,n+1}(t-1) + \Delta u(t-d) \tag{112}$$

Applying the small trick

$$\mathrm{E}\left[\cdot|u(t), D_{t-1}\right] = \mathrm{E}\left[\mathrm{E}\left[\cdot|\Theta\right]|u(t), D_{t-1}\right]$$

it is obtained (Star indicates the lack of knowledge of Θ.)

$$\hat{y}^*(t|t-1) = y(t-1) + Y_1^{(0)}(t-1) + z(t)\hat{\Theta}(t-1) \tag{113}$$
$$\bar{R}^*_{yy}(t|t-1) = 1 + hR_{ss}(t-1|t-1)h' + z(t)\bar{R}_{\Theta\Theta}(t-1)z'(t) \tag{114}$$
$$\bar{R}_{\Theta y}(t|t-1) = \bar{R}_{\Theta\Theta}(t-1)z'(t) \tag{115}$$

Application of the standard result [1, §2.5] now gives the following

Result 5. (Joint parameter and state estimation) *Let Θ and $s(t)$ be the parameters (96) and the state of an input-output model in the canonical state representation (2, 3, 10). Suppose that $\mathrm{p}(\Theta|D_{t-1})$ is normal with the mean $\hat{\Theta}(t-1)$ and with the covariance matrix $\rho\bar{R}_{\Theta\Theta}(t-1)$. If the stochastic term $e(t)$ of the model is normally distributed with the variance ρ, then $\mathrm{p}(\Theta|D_t)$ is also normal and given by*

$$\hat{\Theta}(t) = \hat{\Theta}(t-1) + K_\Theta(t)\left[\Delta y(t) - Y_1^{(0)}(t-1) - z(t)\hat{\Theta}(t-1)\right] \tag{116}$$

$$K(t) = \bar{R}_{\Theta\Theta}(t-1)z'(t)\left(\bar{R}^*_{yy}(t|t-1)\right)^{-1} \tag{117}$$
$$\bar{R}_{\Theta\Theta}(t) = \bar{R}_{\Theta\Theta}(t-1) - K_\Theta(t)\bar{R}^*_{yy}(t|t-1)K'_\Theta(t) \tag{118}$$

where $\bar{R}^*_{yy}(t|t-1)$ is given by (114). $Y_1^{(0)}(t-1)$ and $z(t)$ (112) come from the first row of the statistic $X(t-1) = [Y(t-1), U(t-1)]$ which is to be updated according to (105) and (106). $R_{ss}(t|t)$ is provided by the recursion (83, 84) which also generates $\tilde{c}(t)$ used in updating $Y(t)$ and $U(t)$.

The state estimate, under lack of knowledge of the parameters Θ, then is

$$\hat{s}(t|t) = X(t)\begin{bmatrix} 1 \\ \hat{\Theta}(t) \end{bmatrix} \tag{119}$$

Remark k. Parameter estimation according to the above Result is nothing else than weighted recursive least squares regression with the regressant $\Delta y(t) - Y_1^{(0)(t-1)}$ and with the regressor $z(t)$. The LD filter, described in [8] and [9], is a reliable algorithm for the given purpose. □

REFERENCES

[1] Anderson T.W. : *An Introduction to multivariate statistical analysis*, John Wiley & Sons, 1958

[2] Åström K.J. : *Introduction to Stochastic Control Theory*. Academic Press, 1970.

[3] Bitmead R.R., M.Gevers and V.Wertz : Optimal control redesign of generalized predictive control. Preprints of the IFAC symposium *Adaptive Systems in Control and Signal Processing*, Glasgow, UK, April 1989, Vol.I, 129–134.

[4] Clarke D.W., C. Mohtadi and P.S. Tuffs: Generalized predictive control. *Automatica 23* (1987), No.2, 137–160.

[5] Clarke D.W. and C. Mohtadi : Properties of generalized predictive control. IFAC Congress 1987, Preprints Vol.10, 63–74.

[6] Peterka V.: Predictor based self-tuning control. *Automatica* **20** (1984), 39–50.

[7] Peterka V.: On LQ optimum control of ARMAX processes. Proc. 9th IFAC Congress, Budapest, 1986, Vol.IX

[8] Peterka V.: Control of uncertain processes : Applied theory and algorithms. Supplement to the journal *Kybernetika 22* (1986), No 3,4,5,6.

[9] Peterka V.: Algorithms for LQG self-tuning control based on input-output Delta models. 2nd IFAC Workshop on Adaptive Systems and Signal processing, Lund, Sweden, 1986, Preprints 13–18

[10] Peterka V.: Self-tuning control with alternative sets of uncertain process models. IFAC Symposium Adaptive Systems in Control and Signal Processing, Glasgow, April 1989, Preprints, 409–414

V. Peterka
Institute of Information Theory and Automation
Pod vodárenskou věží 4
CS 182 08 Prague
Czechoslovakia

Discontinuous Feedback and Universal Adaptive Stabilization

E.P. Ryan*

An adaptive stabilizer, universal for a class of nonlinear systems, is described. The stabilizer is of discontinuous feedback form and incorporates gains of Nussbaum type. The framework is that of differential inclusions and the stability analysis draws on an extension, to that framework, of LaSalle's invariance principle for ordinary differential equations.

1 Introduction

As a motivating class $\mathcal{L}$, consider planar controlled linear systems of the form

$$\dot{x}(t) = \alpha x(t) + a y(t), \quad \dot{y}(t) = \beta_1 x(t) + \beta_2 y(t) + b u(t), \quad ab \neq 0,$$

where α, a, β_1, β_2, $b \in \mathbb{R}$ are unknown parameters. Evidently, $ab \neq 0$ is a necessary condition for existence of an *$\mathcal{L}$-universal stabilizer* (since, if $ab = 0$, then there are systems of the above type which are not stabilizable: for example, systems with $\alpha > 0$ and $\beta_1 = 0$). This condition is equivalent to controllability of every member of $\mathcal{L}$ and so, in accordance with [1], is also sufficient for universal stabilization.

*The author is indebted to J.F. Toland of the University of Bath for many helpful discussions. The work was supported, in part, by the UK Science & Engineering Research Council.

In our framework, $\mathcal{L}$ is a subclass of a larger class $\mathcal{N}$ of nonlinear systems of the form

$$\dot{x}(t) = f(x(t)) + ay(t), \quad \dot{y}(t) = g(x(t), y(t)) + bu(t), \quad ab \neq 0,$$
$$(x(0), y(0)) = (x^0, y^0) \in \mathbb{R}^2,$$

on which only the following structure is imposed: f and g are continuous with f linearly bounded, that is, $|f(x)| \le \alpha|x|$ (α unknown), and g bounded, modulo an unknown scalar multiplier $\beta \ge 0$, by a known continuous function γ, that is, $|g(x,y)| \le \beta\gamma(x,y)$. It is assumed that the state $(x(t), y(t))$ is available for feedback purposes. We will construct an adaptive discontinuous feedback strategy which ensures that, for each (x^0, y^0), every solution of the above initial value problem approaches zero.

This prototype class $\mathcal{N}$ and a higher dimensional generalization are the main concerns of the paper. The proposed universal stabilizers incorporate adaptive gains of Nussbaum type [2–4]. The underlying feedbacks are *discontinuous*; the attendant technical difficulties are handled by embedding the strategies in set-valued maps and formulating the feedback controlled system as a differential inclusion [5–7]. The stability analysis draws on an extension, to differential inclusions, of LaSalle's invariance principle for ordinary differential equations [8]; we begin by describing this extension.

2 An invariance principle for differential inclusions

Digressing from the main subject, we first investigate qualitative properties of solutions of the general initial-value problem

$$\dot{x}(t) \in F(x(t)), \quad x(t) \in G \subset \mathbb{R}^N, \tag{1a}$$

$$x(0) = x^0, \tag{1b}$$

where $G \neq \emptyset$ is open. The *set-valued map* $x \mapsto F(x) \subset \mathbb{R}^N$ is assumed to be *upper semicontinuous* on G with *non-empty, convex* and *compact* values. This is sufficient to ensure that, for each $x^0 \in G$, (1) admits a solution, that is, an absolutely continuous function $x : [0, \omega) \to G$ satisfying (1a) almost everywhere and (1b).

Recall that F is upper semicontinuous on G if it is upper semicontinuous at every $\bar{x} \in G$ in the sense of compact-set-valued maps, that is, for each $\varepsilon > 0$ there exists $\delta > 0$ such that $F(x) \subset F(\bar{x}) + \varepsilon B$ for all $x \in \bar{x} + \delta B$, where B denotes the open unit ball (centred at 0) in $\mathbb{R}^N$. In the context of real-valued functions, the standard definition of upper semicontinuity applies.

Definition 1. Suppose $0 < \omega \leq \infty$ and that $x: [0, \omega) \mapsto G$ is a solution of (1). The interval $[0, \omega)$ is *a maximal interval of existence*, and the solution x is said to be *maximal*, if x does not have a proper extension which is also a solution of (1).

The following proposition is implicit in the literature (see, for example, [7: Chapter 2] and [9]); here we give a simple proof using Zorn's lemma.

Proposition 1. *Every solution of (1) can be extended into a maximal solution.*

PROOF. Let $x: [0, \omega) \mapsto G$ be a solution of (1) for some $0 < \omega \leq \infty$ (recall that the existence of at least one solution is assured by [5: p.98, Theorem 3]). Let

$$\mathcal{A} = \{(\rho, y) : \omega \leq \rho \leq \infty;\ y : [0, \rho) \to G \text{ solves (1)};\ y(t) = x(t)\ \forall t \in [0, \omega)\}$$

and, on this non-empty set $\mathcal{A}$, define a partial ordering $\preceq$ by

$$(\rho_1, y_1) \preceq (\rho_2, y_2) \iff \rho_1 \leq \rho_2 \text{ and } y_2(t) = y_1(t)\ \forall t \in [0, \rho_1).$$

The proposition is proved if we can show that $\mathcal{A}$ has a maximal element. This we do by Zorn's lemma as follows. Let $\mathcal{T}$ be any totally ordered subset of $\mathcal{A}$. Let $P = \sup\{\rho : (\rho, y) \in \mathcal{T}\}$ and let $Y : [0, P) \to G$ be defined by the property that, for every $(\rho, y) \in \mathcal{T}$, $Y(t) = y(t)$ for all $t \in [0, \rho)$. Then (P, Y) is in $\mathcal{A}$ and is an upper bound for $\mathcal{T}$. By Zorn's lemma, it follows that $\mathcal{A}$ contains at least one maximal element. Q.E.D.

Definition 2. A solution $x: [0, \omega) \mapsto G$ of (1) is said to be *precompact* if it is maximal and the closure $cl(x([0, \omega))) \subset \mathbb{R}^N$ of its trajectory $x([0, \omega))$ is a compact subset of G.

Proposition 2. *If x is a precompact solution of (1), then $\omega = \infty$.*

PROOF. Let $x : [0, \omega) \to G$ be a precompact solution of (1). Then, by [5: p.42, Proposition 3], $F(cl(x([0, \omega)))) \subset \mathbb{R}^N$ is compact; this implies uniform continuity of x. Seeking a contradiction, suppose $\omega < \infty$. Then x extends into a continuous function $[0, \omega] \to G$. By the existence theorem [5: p.98, Theorem 3], this function extends into a solution of (1) on some interval $[0, \omega^e)$, $\omega^e > \omega$, contradicting the maximality of x. Q.E.D.

Definition 3. Let $x : [0, \omega) \to G$ be a maximal solution of (1). A point $\bar{x} \in \mathbb{R}^N$ is a *ω-limit point of x* if there exists an increasing sequence $\{t_n\} \subset [0, \omega)$ such that $t_n \to \omega$ and $x(t_n) \to \bar{x}$ as $n \to \infty$. The set $\Omega(x)$ of all ω-limit points of x is the *ω-limit set of x*.

Definition 4. Let $C \subset \mathbb{R}^N$ be non-empty. A maximal solution $x : [0,\omega) \mapsto G$ of (1) is said to *approach* C if $d_C(x(t)) \to 0$ as $t \to \omega$, where d_C is the (Euclidean) distance function for C defined (on $\mathbb{R}^N$) by $d_C(v) := \inf\{|v - c| : c \in C\}$.

Definition 5. Relative to (1), $S \subset \mathbb{R}^N$ is said to be a *quasi-invariant set* if, for each $x^0 \in S \cap G$, there exists at least one maximal solution of (1) with $\omega = \infty$ and with trajectory in S.

The final proposition summarizes properties of ω-limit sets (see, for example, [7: pp.129-131]).

Proposition 3. *If x is a precompact solution of (1), then $\Omega(x)$ is a non-empty, compact, connected subset of G and is a quasi-invariant set.*

Theorem 1. *Let $V : G \to \mathbb{R}$ be continuously differentiable. Define*

$$q : G \to \mathbb{R}, \quad z \mapsto q(z) := \max\{\langle \nabla V(z), \phi\rangle : \phi \in F(z)\}.$$

Suppose that $U \subset G$ is non-empty and that

$$q(z) \le 0 \quad \forall z \in U. \tag{2}$$

If x is a precompact solution of (1) with trajectory in U, then, for some $c \in V(cl(U) \cap G)$, x approaches the maximal quasi-invariant set M_c in the set $\Sigma \cap V^{-1}(c)$, where $\Sigma = \{z \in cl(U) \cap G : q(z) \ge 0\}$.

PROOF. For later convenience, we first show that q is upper semicontinuous (in the usual sense). Let $\{z_n\} \subset G$ be such that $z_n \to z \in G$ as $n \to \infty$. From $\{q(z_n)\}$ we extract a subsequence $\{q(z_{n_k})\}$ with $q(z_{n_k}) \to \limsup_{n\to\infty} q(z_n)$ as $k \to \infty$. For each k, let ϕ_k be a maximizer of continuous $\langle \nabla V(z_{n_k}), \cdot\,\rangle$ over compact $F(z_{n_k})$, and so $q(z_{n_k}) = \langle \nabla V(z_{n_k}), \phi_k\rangle$. By upper semicontinuity of F, for each $\epsilon > 0$ we have $F(z_{n_k}) \subset F(z) + \epsilon B$ for all k sufficiently large. Since $\phi_k \in F(z_{n_k})$ and $F(z)$ is compact, $\{\phi_k\}$ contains a subsequence (which we do not relabel) converging to $\phi^* \in cl(F(z) + \epsilon B)$. Since $\epsilon > 0$ is arbitrary and $F(z)$ is compact, $\phi^* \in F(z)$. Thus, we may conclude that

$$\limsup_{n\to\infty} q(z_n) = \langle \nabla V(z), \phi^*\rangle \le q(z).$$

Therefore q is upper semicontinuous.

By Proposition 2, x has interval of existence $\mathbb{R}^+$. By definition of q and (2), we have

$$V(x(t)) - V(x(\tau)) \le \int_\tau^t q(x(s))ds \le 0 \quad \forall \tau, t \in \mathbb{R}^+,\ \tau \le t \tag{3}$$

and so $V(x(\cdot))$ is monotone decreasing. We can now show that Σ is non-empty and is approached by x. Since, by Proposition 3, precompact x approaches its non-empty ω-limit set $\Omega(x)$ and since the trajectory of x lies in U (and so $\Omega(x) \subset cl(U) \cap G$), it suffices to show that $q(\bar{x}) \geq 0$ for each $\bar{x} \in \Omega(x)$. Seeking a contradiction, suppose that this is false. Then there exists $\bar{x} \in \Omega(x)$ and $\epsilon > 0$ such that $q(\bar{x}) \leq -2\epsilon$. By upper semicontinuity of q, there exists $\delta_1 > 0$ such that $q(z) < -\epsilon$ for all $z \in \bar{x} + \delta_1 B$. By upper semicontinuity of F and compactness of its values, there exist $\delta > 0$ and $r > 0$ such that $F(z) \subset F(\bar{x}) + \epsilon B \subset rB$ for all $z \in \bar{x} + \delta B$. We may assume that $\delta \leq \delta_1$. Since $\bar{x} \in \Omega(x)$, there exists an increasing sequence $\{t_n\} \subset \mathbb{R}^+$ with $t_n \to \infty$ and $x(t_n) \to \bar{x}$ as $n \to \infty$. By continuity of V we have

$$V(x(t_n)) - V(\bar{x}) < \frac{\delta\epsilon}{4r} \quad \text{for all } n \text{ sufficiently large.} \tag{4}$$

Let n^* be such that $x(t_n) \in \bar{x} + \frac{1}{2}\delta B$ for all $n > n^*$. Since $F(z) \subset rB$ for all $z \in \bar{x} + \delta B$, it follows from (1) that, for all $n > n^*$, $x(t) \in \bar{x} + \delta B$ for all $t \in [t_n, t_n + (\delta/3r)]$. Hence, using the monotonicity of $V(x(\cdot))$ and (3), we conclude

$$V(x(t_n)) - V(\bar{x}) \geq -\int_{t_n}^{t_n+(\delta/3r)} q(x(s))ds \;\geq\; \frac{\delta\epsilon}{3r}$$

for all $n > n^*$. This contradicts (4). Thus, $q(\bar{x}) \geq 0$ for all $\bar{x} \in \Omega(x)$ and so $\Omega(x) \subset \Sigma$. By continuity of V and precompactness of x, we conclude that the monotone function $V(x(\cdot))$ is bounded. Therefore, $V(x(t)) \to c$ as $t \to \infty$. By definition of $\Omega(x)$ and continuity of V, it follows that $V(\bar{x}) = c$ for all $\bar{x} \in \Omega(x)$. Therefore, $\Omega(x) \subset \Sigma \cap V^{-1}(c)$. Finally, to conclude that x approaches the largest quasi-invariant set M_c in $\Sigma \cap V^{-1}(c)$, we recall that, by Proposition 3, $\Omega(x)$ is a quasi-invariant set. Q.E.D.

2.1 Remarks.

The C^1 assumption on V in Theorem 4 can, in fact, be weakened to requiring only that V be locally Lipschitz. The theorem then remains valid if, in the definition of the function q, $\langle \nabla V(z), \phi \rangle$ is replaced by the Clarke generalized directional derivative [6] of V at z in direction ϕ, that is

$$V^o(z;\phi) := \limsup_{\substack{y \to z \\ h \downarrow 0}} \frac{V(y + h\phi) - V(y)}{h} .$$

However, for our immediate purpose, the class of C^1 functions V suffices. Finally, we remark that Theorem 1 subsumes the invariance principle of [10] wherein, in effect, *continuity* of the map F is assumed.

3 $\mathcal{N}$-universal adaptive stabilizer

We now return to our prototype class $\mathcal{N}$ of uncertain controlled systems:

$$\dot{x}(t) = f(x(t)) + ay(t), \quad \dot{y}(t) = g(x(t), y(t)) + bu(t), \tag{5a}$$

$$(x(0), y(0)) = (x^0, y^0) \in \mathbb{R}^2. \tag{5b}$$

Recall that only the following *a priori* information is assumed.

Assumption 1. $ab \neq 0$.

Assumption 2. f is continuous and linearly bounded, that is, $|f(x)| \leq \alpha|x| \ \forall x$ (α unknown);

Assumption 3. g is continuous and bounded, modulo an unknown multiplier β, by a known continuous function $\gamma : \mathbb{R}^2 \to \mathbb{R}^+$, that is, $|g(x,y)| \leq \beta\,\gamma(x,y) \ \forall \ (x,y)$.

Our goal is to construct a control strategy (independent of the unknown quantities a, b, f and g) which ensures that, for each $(x^0, y^0) \in \mathbb{R}^2$, every solution of (5) approaches $\{(0,0)\}$.

We will show that one such strategy is (formally) given by:

$$\begin{aligned}
\dot{\kappa}_1(t) &= x^2(t), \qquad \kappa_1(0) = \kappa_1^0, \\
\dot{\kappa}_2(t) &= [y(t) - \nu(\kappa_1(t))x(t)]^2 + \xi(x(t), y(t), \kappa_1(t))|y(t) - \nu(\kappa_1(t))x(t)|, \\
\kappa_2(0) &= \kappa_2^0, \\
u(t) &= \nu(\kappa_2(t))[y(t) - \nu(\kappa_1(t))x(t) \\
&\qquad + \xi(x(t), y(t), \kappa_1(t))\,\mathrm{sgn}[y(t) - \nu(\kappa_1(t))x(t)]],
\end{aligned}$$

where $\xi(x, y, \kappa_1) := \gamma(x,y) + |\nu'(\kappa_1)x^3| + |\nu(\kappa_1)|[|x| + |y|]$ and ν is any C^1 function of Nussbaum type, that is, any C^1 function $\mathbb{R} \to \mathbb{R}$ with the properties

$$\limsup_{\eta\to\infty} \frac{1}{\eta}\int_0^\eta \nu \ = \infty\,, \qquad \liminf_{\eta\to\infty} \frac{1}{\eta}\int_0^\eta \nu \ = -\infty\,. \tag{6}$$

For example, $\nu : \theta \mapsto \theta^2 \cos\theta$ suffices.

3.1 Remarks

The rationale underlying the proposed strategy is as follows. First consider the subsystem $\dot{x}(t) = f(x(t)) + ay(t)$ in isolation, that is, as an independent system with $y(t)$ regarded as control input. It is known that this system is stabilized by the adaptive strategy: $y(t) = \nu(\kappa_1(t))x(t)$, $\dot{\kappa}_1(t) = x^2(t)$. Now, in the context of the full system (5), if a control $u(\cdot)$ can be found which ensures that the "error" $e(t) = y(t) - \nu(\kappa_1(t))x(t) \to 0$ as $t \to \infty$, then intuition would suggest that $x(t) \to 0$ as $t \to \infty$; moreover, if boundedness of $\kappa_1(\cdot)$ can also be achieved, then $y(t) \to 0$ as $t \to \infty$. The proposed strategy essentially generates one such control $u(\cdot)$. Additional insight into the control structure may be gained by supposing that the values of parameters a, b, α, β are known. In this hypothetical case, adaptation is obviated: the time-varying gains $\nu(\kappa_1(t))$ and $\nu(\kappa_2(t))$ may be replaced by suitably defined *constant* gains κ_1^* and κ_2^*, respectively (for example, as defined in the proof of Theorem 2 below). The resulting non-adaptive feedback control then has the effect of rendering the subspace $\mathcal{S} = \{(x,y) : y = \kappa_1^* x\}$ a globally *finite-time attractive* (by virtue of the discontinuous feedback component) *stable manifold* for the system: in other words, every solution of the (hypothetical) feedback system is such that the state ultimately attains $\mathcal{S}$, subsequently remains therein, and approaches the origin asymptotically.

Finally, some remarks on the discontinuous feedback nature of the proposed strategy are warranted. Although attractivity of $\{(0,0)\}$ is our goal, this point is not necessarily an equilibrium of the uncontrolled system. Therefore, in general, an *unknown* control bias $u^0 = -b^{-1}g(0,0)$ is required to render $\{(0,0)\}$ an equilibrium. The discontinuous feedback implicitly realizes this control component (in effect, as the "average" of a "chattering" mode). In the special case for which $\gamma(0,0) = 0$ (and so $g(0,0) = 0$), the origin is an equilibrium point of the uncontrolled system: in this case, we see that $\xi(0,0,\kappa_1) = 0$ which quenches "chatter" at the origin, as might be expected.

3.2 Stability analysis

Writing $e(t) = y(t) - \nu(\kappa_1(t))x(t)$ and $z(t) = (x(t), e(t), \kappa_1(t), \kappa_2(t))$, we interpret the above discontinuous adaptive control in the following generalized sense:

$$u(t) \in \Psi(z(t)), \tag{7a}$$

where

$$\Psi(z) := \nu(\kappa_2)[e + \xi(x, e + \nu(\kappa_1)x, \kappa_1)\psi(e)] \tag{7b}$$

with

$$\psi(e) := \begin{cases} \{+1\}, & e > 0 \\ [-1,+1], & e = 0 \\ \{-1\}, & e < 0. \end{cases} \tag{7c}$$

The overall adaptively controlled system (5,7) is now embedded in the following initial-value problem

$$\dot{z}(t) \in F(z(t)), \tag{8a}$$

$$z(0) = z^0 = (x^0, y^0 - \nu(\kappa_1^0)x^0, \kappa_1^0, \kappa_2^0) \in \mathbb{R}^4, \tag{8b}$$

which is of type (1) with $N = 4$, $G = \mathbb{R}^4$, and with the set-valued map $(x, e, \kappa_1, \kappa_2) = z \mapsto F(z) \subset \mathbb{R}^4$ given by

$$F(z) := \{h_1(z)\} \times E(z) \times \{x^2\} \times \{h_2(z)\}$$

where $h_1(z) := f(x) + a[e + \nu(\kappa_1)x]$, $h_2(z) := e^2 + \xi(x, e + \nu(\kappa_1)x, \kappa_1)|e|$ and $E(z) := \{g(x, e + \nu(\kappa_1)x) - \nu(\kappa_1)h_1(z) - \nu'(\kappa_1)x^3 + bu : \ u \in \Psi(z)\}$.

By upper semicontinuity of ψ and continuity of ν, f, g and ξ, the set-valued map F is upper semicontinuous on $\mathbb{R}^4$. Clearly, F takes convex and compact values. Thus, for each $z^0 \in \mathbb{R}^4$, (8) admits a solution and, by Proposition 1, every such solution has a maximal extension.

Theorem 2. *Let $z(\cdot) = (x(\cdot), e(\cdot), \kappa_1(\cdot), \kappa_2(\cdot)) : [0, \omega) \to \mathbb{R}^4$ be a maximal solution of (8). Then*

(i) $\omega = \infty$,

(ii) $\lim_{t\to\infty} \kappa_1(t)$ *exists and is finite,*

(iii) $\lim_{t\to\infty} \kappa_2(t)$ *exists and is finite,*

(iv) $\lim_{t\to\infty} ||(x(t), e(t))|| = 0$.

PROOF. We first show that $\kappa_2(\cdot)$ is bounded. Define

$$\kappa_2^* := -b^{-1}\max\{1, \alpha, \beta, |a|\}.$$

By direct calculation,

$$\max\{ep : \ p \in E(z)\} \le -e^2 + b(\nu(\kappa_2) - \kappa_2^*)h_2(z) \tag{9}$$

for all $z = (x, e, \kappa_1, \kappa_2) \in \mathbb{R}^4$. Therefore, for the solution $z(\cdot)$, we have $e(t)\dot{e}(t) \le b(\nu(\kappa_2(t)) - \kappa_2^*)\dot{\kappa}_2(t)$ a.e., which, on integration, yields

$$e^2(t) \le e^2(\tau) - 2b\kappa_2^*[\kappa_2(t) - \kappa_2(\tau)] + 2b\int_{\kappa_2(\tau)}^{\kappa_2(t)} \nu \qquad \forall\, t \ge \tau.$$

Suppose that the monotone function $\kappa_2(\cdot)$ is unbounded. Let $\tau \in [0, \omega)$ be such that $\kappa_2(\tau) > 0$. Then

$$0 \le \liminf_{t\uparrow\omega} \frac{e^2(t)}{2\kappa_2(t)} \le \frac{e^2(\tau)}{2\kappa_2(\tau)} - b\kappa_2^* + \liminf_{\eta\to\infty} \frac{b}{\eta}\int_{\kappa_2(\tau)}^{\eta} \nu$$

which, depending on the sign of b (non-zero), contradicts one or the other of properties (6). Therefore, $\kappa_2(\cdot)$ is bounded.

Secondly, we show that $\kappa_1(\cdot)$ is bounded. Let $W : (x,e) \mapsto \frac{1}{2}(x^2 + e^2)$ and define

$$\kappa_1^* := -(2a)^{-1}(1 + 2\alpha + a^2).$$

By direct calculation, $xh_1(z) \leq -\frac{1}{2}(x^2 - e^2) + a(\nu(\kappa_1) - \kappa_1^*)x^2$ which, when combined with (9), yields

$$\langle \nabla W(x,e), p \rangle \leq -W(x,e) + a(\nu(\kappa_1) - \kappa_1^*)x^2 \\ + b(\nu(\kappa_2) - \kappa_2^*)h_2(z) \quad \forall\, p \in F(z) \tag{10}$$

for all $z = (x, e, \kappa_1, \kappa_2) \in \mathbb{R}^4$. Therefore, for the solution $z(\cdot)$, we have

$$\frac{d}{dt}W(x(t), e(t)) \leq a(\nu(\kappa_1(t)) - \kappa_1^*)\dot{\kappa}_1(t) + b(\nu(\kappa_2(t)) - \kappa_2^*)\dot{\kappa}_2(t) \quad \text{a.e.}$$

Integrating and using boundedness of $\kappa_2(\cdot)$, it follows that, for some $M \in \mathbb{R}$ and for all $t \geq \tau$,

$$W(x(t), e(t)) \leq W(x(\tau), e(\tau)) + M - a\kappa_1^*[\kappa_1(t) - \kappa_1(\tau)] + a\int_{\kappa_1(\tau)}^{\kappa_1(t)} \nu.$$

By the same contradiction argument as that used in establishing boundedness of $\kappa_2(\cdot)$, we may conclude that the monotone function $\kappa_1(\cdot)$ is also bounded. It follows that $W(x(\cdot), e(\cdot))$ is bounded, whence boundedness of $z(\cdot)$, and so, by Proposition 2, we have $\omega = \infty$. This proves assertions (i)–(iii) of the theorem.

As the final step, define

$$V : z \mapsto W(x,e) - \int_0^{\kappa_1} a[\nu(\theta) - \kappa_1^*]d\theta - \int_0^{\kappa_2} b[\nu(\theta) - \kappa_2^*]d\theta\ .$$

From (10) it follows that, for all $z = (x, e, \kappa_1, \kappa_2) \in \mathbb{R}^4$,

$$\langle \nabla V(z), p \rangle \leq -W(x,e) = -\tfrac{1}{2}(x^2 + e^2) \quad \forall\, p \in F(z).$$

By the invariance principle (Theorem 1), we conclude that $z(\cdot)$ approaches the set

$$\Sigma = \{z = (x, e, \kappa_1, \kappa_2) : x = 0 = e\}$$

whence assertion (iv) of the theorem. Q.E.D.

3.3 Example

For purposes of illustration, consider the case for which $\gamma(\cdot,\cdot) \equiv 1$ and $\nu : \theta \mapsto \theta^2 \cos\theta$. Suppose that $f : x \mapsto \sin x$, $g : (x,y) \mapsto \cos x + \cos y$ and $a = 1 = b$. Observe that $\{(0,0)\}$ is not an equilibrium of the uncontrolled system. Figures 1–3 depict the behaviour of the adaptively controlled system, with initial data $(x^0, y^0, \kappa_1^0, \kappa_2^0) = (1, -2, 0, 0)$.

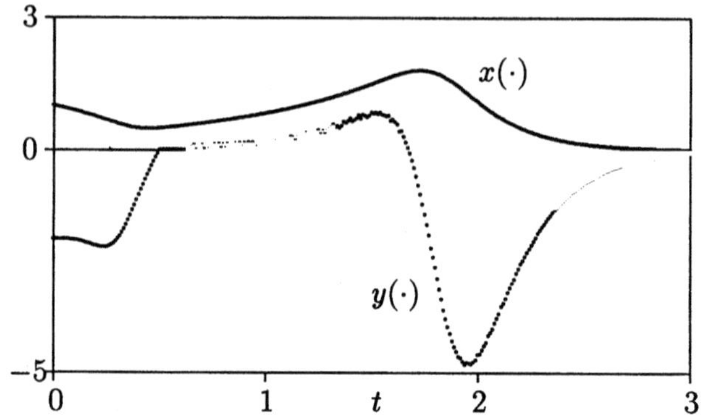

Figure 1. State evolution

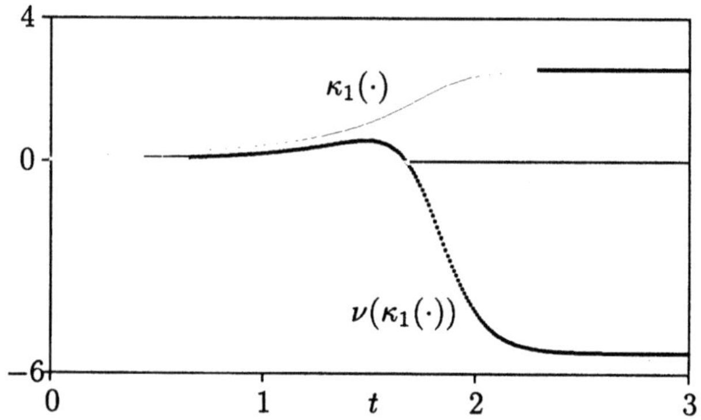

Figure 2. Evolution of parameter κ_1 and Nussbaum gain

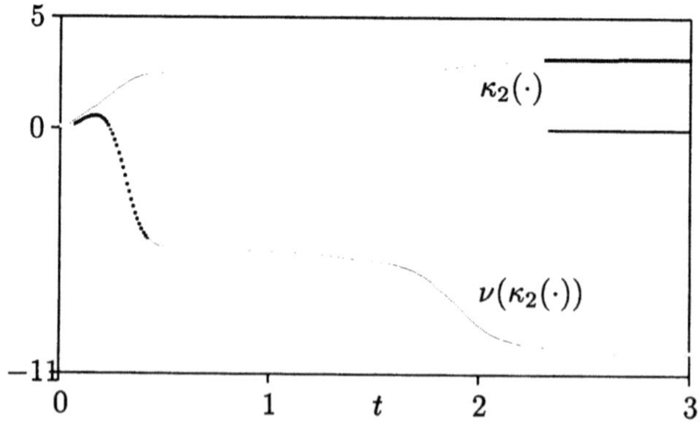

Figure 3. Evolution of parameter κ_2 and Nussbaum gain

4 Generalization

In this final section we describe an extension of the results on planar systems to higher-dimensional systems. Specifically, the analysis of §3 generalizes to systems on $\mathbb{R}^{n+1}$ of the form

$$\dot{x}(t) = Ax(t) + f(x(t)) + ay(t), \quad \dot{y}(t) = g(x(t), y(t)) + bu(t), \tag{11a}$$

$$(x(0), y(0)) = (x^0, y^0) \in \mathbb{R}^n \times \mathbb{R}, \tag{11b}$$

for which $f : \mathbb{R}^n \to \operatorname{im} a$ and $g : \mathbb{R}^n \times \mathbb{R} \to \mathbb{R}$ have the same structure as before, that is, they satisfy the higher-dimensional analogues of Assumptions 2 and 3. Note that f takes its values in $\operatorname{im} a$. Only the following additional *a priori* information is required (equivalent to Assumption 1 in the planar case).

Assumption 1a. For some known $c \in \mathbb{R}^n$, (i) $\langle c, a\rangle b \neq 0$, and (ii) the single-input, single-output linear system (with output map $\langle c, \cdot\rangle$) defined by the triple (c, A, a) has no zeros in the closed right half complex plane. Without loss of generality, we assume that c has unit norm $||c|| = 1$.

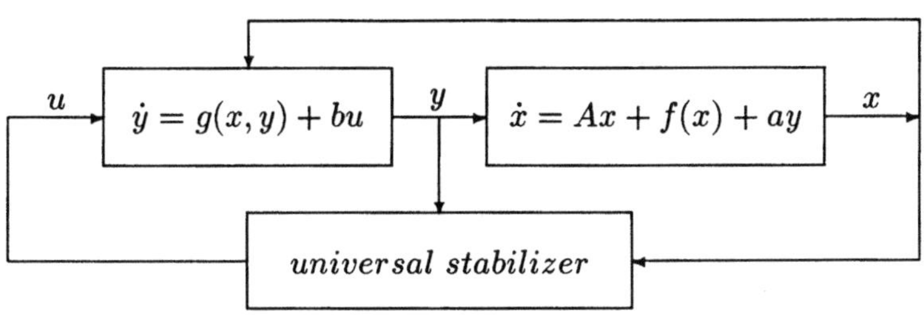

Figure 4. General system structure

We will show that a universal stabilizer for this class of systems is given (formally) by the adaptive strategy:

$$\begin{aligned}
&\dot{\kappa}_1(t) = \langle c, x(t)\rangle^2, \quad \kappa_1(0) = \kappa_1^0, \\
&\dot{\kappa}_2(t) = [y(t) - \nu(\kappa_1(t))\langle c, x(t)\rangle]^2 \\
&\qquad\qquad + \xi(x(t), y(t), \kappa_1(t))|y(t) - \nu(\kappa_1(t))\langle c, x(t)\rangle|, \\
&\kappa_2(0) = \kappa_2^0, \\
&u(t) = \nu(\kappa_2(t))[y(t) - \nu(\kappa_1(t))\langle c, x(t)\rangle \\
&\qquad\qquad + \xi(x(t), y(t), \kappa_1(t)) \operatorname{sgn}[y(t) - \nu(\kappa_1(t)\langle c, x(t)\rangle]],
\end{aligned}$$

where $\xi(x,y,\kappa_1) := \gamma(x,y) + |\nu'(\kappa_1)\langle c, x(t)\rangle^3| + |\nu(\kappa_1)|[\|x\| + |y|]$ and, as before, ν is any C^1 function of Nussbaum type.

4.1 Stability analysis

Let $T_1 : \mathbb{R}^n \to \mathbb{R}^{n-1}$ be such that $\ker T_1 = \operatorname{im} a$. Then the coordinate transformation $T : x \mapsto (w, \tilde{x}) := (T_1 x, \langle c, x\rangle)$, with inverse

$$T^{-1} : (w, \tilde{x}) \mapsto x = S_1 w + \langle c, a\rangle^{-1} a\tilde{x},$$

takes (11) into the form

$$\begin{aligned}
\dot{w}(t) &= A_{11}w(t) + A_{12}\tilde{x}(t), \quad w(t) \in \mathbb{R}^{n-1},\\
\dot{\tilde{x}}(t) &= A_{21}w(t) + A_{22}\tilde{x}(t) + \tilde{f}(w(t), \tilde{x}(t)) + \tilde{a}y(t), \quad \tilde{x}(t) \in \mathbb{R},\\
\dot{y}(t) &= \tilde{g}(w(t), \tilde{x}(t), y(t)) + bu(t), \quad y(t) \in \mathbb{R},\\
(w(0), \tilde{x}(0), y(0)) &= (w^0, \tilde{x}^0, y^0) := (Tx^0, y^0),
\end{aligned}$$

where $\tilde{a} := \langle c, a\rangle$, $\tilde{f}(\cdot,\cdot) := \langle c, f(T^{-1}(\cdot,\cdot))\rangle$ and $\tilde{g}(\cdot,\cdot,\cdot) := g(T^{-1}(\cdot,\cdot),\cdot)$.

Writing $e(t) = y(t) - \nu(\kappa_1(t))\tilde{x}(t)$ and $z(t) = (w(t), \tilde{x}(t), e(t), \kappa_1(t), \kappa_2(t))$, we interpret the above discontinuous adaptive control in the generalized sense of (7), where now

$$\Psi(z) := \nu(\kappa_2)[e + \tilde{\xi}(w, \tilde{x}, e + \nu(\kappa_1)\tilde{x}, \kappa_1)\psi(e)]$$

with ψ defined as before, and $\tilde{\xi}(\cdot,\cdot,\cdot,\cdot) := \xi(T^{-1}(\cdot,\cdot),\cdot,\cdot)$. We embed the overall controlled system in the following initial-value problem

$$\dot{z}(t) \in F(z(t)), \tag{12a}$$

$$z(0) = z^0 = (w^0, \tilde{x}^0 - \nu(\kappa_1^0)w^0, \kappa_1^0, \kappa_2^0), \tag{12b}$$

where now the set-valued map $(w, \tilde{x}, e, \kappa_1, \kappa_2) = z \mapsto F(z) \subset \mathbb{R}^{n+3}$ is given by

$$F(z) := \{A_{11}w + A_{12}\tilde{x}\} \times \{h_1(z)\} \times E(z) \times \{\tilde{x}^2\} \times \{h_2(z)\}$$

with

$$\begin{aligned}
h_1(z) &:= A_{21}w + A_{22}\tilde{x} + \tilde{f}(w, \tilde{x}) + \tilde{a}[e + \nu(\kappa_1)\tilde{x}]\\
h_2(z) &:= e^2 + \tilde{\xi}(w, \tilde{x}, e + \nu(\kappa_1)\tilde{x}, \kappa_1)|e|
\end{aligned}$$

$$\begin{aligned}
E(z) := \{\tilde{g}(w, \tilde{x}, e + \nu(\kappa_1)\tilde{x}) &- \nu(\kappa_1)h_1(z)\\
&- \nu'(\kappa_1)\tilde{x}^3 + bu : \ u \in \Psi(z)\}.
\end{aligned}$$

Noting that the spectrum of A_{11} coincides with the set of zeros of the linear system (c, A, a), it follows, by Assumption 1a(ii), that the Lyapunov equation

$$PA_{11} + A_{11}^* P + 2I = 0$$

has unique positive solution $P > 0$.

Now, writing $k = 2|A_{22}| + [||PA_{12}|| + ||A_{21}|| + \alpha||S_1||]^2$, we define κ_1^*, κ_2^*, W and V as follows

$$\begin{aligned}
&\kappa_1^* := -(2\tilde{a})^{-1}(1 + 2\alpha|\tilde{a}|^{-1}||a|| + \tilde{a}^2 + k), \\
&\kappa_2^* := -b^{-1}\max\{1, \tilde{\alpha}, \beta, |\tilde{a}|\}, \quad \text{where } \tilde{\alpha} := \alpha + ||A_{21}T_1|| + |A_{22}|, \\
&W : (w, \tilde{x}, e) \mapsto \tfrac{1}{2}\langle w, Pw\rangle + \tfrac{1}{2}(\tilde{x}^2 + e^2), \\
&V : z \mapsto W(w, \tilde{x}, e) - \textstyle\int_0^{\kappa_1} \tilde{a}[\nu(\theta) - \kappa_1^*]d\theta - \int_0^{\kappa_2} b[\nu(\theta) - \kappa_2^*]d\theta.
\end{aligned}$$

By direct calculation (analogous to that outlined in the proof of Theorem 2) we find that, for all $z = (w, \tilde{x}, e, \kappa_1, \kappa_2) \in \mathbb{R}^{n+3}$,

$$\langle \nabla V(z), p\rangle \le -\tfrac{1}{2}(||w||^2 + \tilde{x}^2 + e^2) \quad \forall\, p \in F(z),$$

and the proof of Theorem 2 generalizes in a straightforward manner to give:

Theorem 3. *Let $z(\cdot) = (w(\cdot), \tilde{x}(\cdot), e(\cdot), \kappa_1(\cdot), \kappa_2(\cdot)) : [0, \omega) \to \mathbb{R}^{n+3}$ be a maximal solution of (12). Then*

(i) $\omega = \infty$,

(ii) $\lim_{t\to\infty} \kappa_1(t)$ *exists and is finite,*

(iii) $\lim_{t\to\infty} \kappa_2(t)$ *exists and is finite,*

(iv) $\lim_{t\to\infty} ||(w(t), \tilde{x}(t), e(t))|| = 0.$

REFERENCES

[1] B. Mårtensson, “The order of any stabilizing regulator is sufficient a priori information for adaptive stabilization”, Systems & Control Letters, **6** (1985), 87–91.

[2] R.D. Nussbaum, “Some remarks on a conjecture in parameter adaptive control”, Systems & Control Letters, **3** (1983), 243–246.

[3] C.I. Byrnes & J.C.Willems, “Adaptive stabilization of multivariable linear systems”, Proc 23rd IEEE Conf. Decision & Control (1984), 1574–1577.

[4] A.S. Morse, "A three-dimensional universal controller for the adaptive stabilization of any strictly proper minimum-phase system with relative degree not exceeding two", IEEE Trans Autom Control, **AC-30** (1985), 1188–1191.

[5] J.P. Aubin & A. Cellina, *Differential Inclusions*, Springer-Verlag, 1984.

[6] F.H. Clarke, *Optimization and Nonsmooth Analysis*, Wiley-Interscience, 1983.

[7] A.F. Filippov, *Differential Equations with Discontinuous Righthand Sides*, Kluwer, 1988.

[8] J.P. LaSalle, *The Stability of Dynamical Systems*, Regional Conf. Series in Applied Math., SIAM, 1976.

[9] J.A. Yorke, "Invariance for contingent equations", in *Mathematical Systems Theory and Economics II* (W.H. Kuhn & G.P. Szegö, eds), 379–381, Springer-Verlag, 1969.

[10] T.K.C. Peng, "Invariance and stability for bounded uncertain systems", SIAM J Control, **10** (1972), 679–690.

E.P. Ryan
School of Mathematical Sciences
University of Bath
Claverton Down
Bath BA2 7AY
UK

Robust Stability of Complex Families of Matrices and Polynomials

L. Saydy A.L. Tits E.H. Abed

Recently, the authors introduced the "guardian map" approach as a unifying tool in the study of robust generalized stability questions for parametrized families of matrices and polynomials. Real matrices and polynomials have been emphasized in previous reports on this approach. In the present note, the approach is discussed in the context of complex matrices and polynomials. In the case of polynomials, some algebraic connections with other recent work are uncovered.

1 Introduction

In a recent paper [1], the authors developed the "guardian map" approach to the study of generalized stability of families of matrices and polynomials. The case of real matrices and polynomials was emphasized in [1]. For some applications however, such as network realizability and filter design, the stability tests involve polynomials and matrices with complex coefficients (see [2–4] and the references therein). The aim of this note is to explicate the results of [1] for problems involving complex matrices and polynomials. A systematic procedure for finding guardian and semiguardian maps for the complex case is given. We specialize the results to the Hurwitz and Schur stability of families of polynomials, and proceed to uncover algebraic connections between constructions of Bose [2] and the guardian maps employed here.

2 Guardian Maps and Robust Stability

The guardian map approach was introduced in [1,5] as a unifying tool for the study of generalized stability of parametrized families of matrices and polynomials. In this section we summarize the elements of this approach.

2.1 Notation

s^*: Conjugate of the complex number s
$\mathring{\mathbb{C}}_-, \overline{\mathbb{C}}_-$: Open, closed left-half complex plane
$\overline{\mathcal{D}}, \partial\mathcal{D}$: Closure, boundary of set $\mathcal{D}$
A^H: Conjugate transpose of matrix A
$\sigma(A)$: Eigenvalues of matrix A (counting multiplicities)
$\mathcal{Z}(p)$: Zeros of polynomial p (counting multiplicities)
$\mathcal{P}_n$: Set of all monic complex polynomials of degree n.
$\mathcal{C}(p)$: Companion matrix associated with polynomial p
$\otimes, \oplus$: Kronecker product, Kronecker sum ($A \oplus B = A \otimes I + I \otimes B$)
$\mathcal{S}(\Omega)$: Set of all $n \times n$ complex matrices with spectrum inside Ω. Also denotes the set of all polynomials in $\mathcal{P}_n$ with zeros inside Ω.

2.2 Guardian and Semiguardian Maps

The notions of guardian, semiguardian and polynomic map are recalled in the next definition [1].

Definition 1. Let $\mathcal{X}$ be either the set of $n \times n$ complex matrices, or the set of all monic polynomials of degree n with complex coefficients , and let $\mathcal{S}$ be an open subset of $\mathcal{X}$. Let ν map $\mathcal{X}$ into $\mathbb{C}$. We say that ν is *semiguarding* for $\mathcal{S}$ if for all $x \in \overline{\mathcal{S}}$, the implication

$$x \in \partial\mathcal{S} \quad \Longrightarrow \quad \nu(x) = 0 \tag{1}$$

holds. Moreover, ν *guards* $\mathcal{S}$ if the converse implication also holds. The map ν is said to be *polynomic* if it is a polynomial function of the entries (matrix case) or coefficients (polynomial case) of its argument and of their complex conjugates.[1]

In [1], guardian and semiguardian maps are exhibited for a large variety of sets of interest, the focus of the representation being the case of real matrices and polynomials. Analogous maps for corresponding sets of *complex* matrices and polynomials are constructed in a straightforward manner. For

[1]Complex conjugates are not explicit in the definition given in [1], as that paper focuses on the case of *real* matrices and polynomials.

example, the map $\nu : A \mapsto \det(A \oplus A^H)$ guards the set of $n \times n$ Hurwitz stable complex matrices. This map is obtained by noting that the spectrum of the Kronecker sum of two square matrices A and B consists of all pairwise sums of eigenvalues of A and B (see, e.g., [6]). Note that this same result also yields a guardian map for the Hurwitz stable complex polynomials, namely $\nu(p) = \det(\mathcal{C}(p) \oplus \mathcal{C}^H(p))$.

Let Ω be an open subset of the complex plane. Of particular importance are sets $\mathcal{S}$ of the form $\mathcal{S}(\Omega)$, where $\mathcal{S}(\Omega)$ is given, for matrix stability problems, by

$$\mathcal{S}(\Omega) = \{A \in \mathbb{C}^{n\times n} : \sigma(A) \subset \Omega\}, \tag{2}$$

and, for polynomial stability problems, by

$$\mathcal{S}(\Omega) = \{p \in \mathcal{P}_n : \mathcal{Z}(p) \subset \Omega\}. \tag{3}$$

Such sets $\mathcal{S}(\Omega)$ are referred to as *(generalized) stability sets.*

The example above dealing with Hurwitz stable families is a particular case of a more general result given next. This result gives a systematic procedure for constructing semiguardian and guardian maps for a large class of stability sets, corresponding to domains of the complex plane with polynomial boundaries.

Let $p(x, y)$ be a real bivariate polynomial of the form

$$p(x,y) = \sum_{k,\ell} p_{k\ell} x^k y^\ell. \tag{4}$$

Denote by Ω the subset of the complex plane

$$\Omega = \{s = x + jy : \ p(x,y) < 0\}, \tag{5}$$

and associate with p the complex polynomial

$$q(\lambda, \mu) := p\left(\frac{\lambda+\mu}{2}, \frac{\lambda-\mu}{2j}\right) \tag{6}$$

$$= \sum_{k,\ell} c_{k\ell} (\lambda+\mu)^k (\lambda-\mu)^\ell. \tag{7}$$

Here

$$c_{k\ell} := (-j)^\ell \left(\frac{1}{2}\right)^{k+\ell} p_{k\ell}. \tag{8}$$

Rewrite (6) as

$$q(\lambda, \mu) = \sum_{k,\ell} q_{k\ell} \lambda^k \mu^\ell. \tag{9}$$

With this notation, Ω and $\partial\Omega$ have the alternative expressions

$$\Omega = \{\lambda \in \mathbb{C} : \quad q(\lambda, \lambda^*) < 0\}, \tag{10}$$
$$\partial\Omega = \{\lambda \in \mathbb{C} : \quad q(\lambda, \lambda^*) = 0\}. \tag{11}$$

Throughout the remainder of this section we focus on the matrix case with the understanding that similar considerations apply in the polynomial case. For example, we may specialize any general matrix result to companion matrices, as in the example above.

Consider the mapping $\mathcal{F} : \mathbb{C}^{n\times n} \to \mathbb{C}^{n^2\times n^2}$ given by

$$\mathcal{F}(A) := \sum_{k,\ell} q_{k\ell} A^k \otimes (A^H)^\ell. \tag{12}$$

Let $\sigma(A) = \{\lambda_1, \ldots, \lambda_n\}$. Then Stéphanos' Theorem [1,6,7] implies that

$$\sigma(\mathcal{F}(A)) = \{q(\lambda_i, \lambda_j^*) : \quad i, j = 1, \ldots, n\}. \tag{13}$$

As direct consequences of this fact, we have the following two results.

Theorem 1. ([1]) *The map ν given by*

$$\nu : \quad A \mapsto \det\mathcal{F}(A) \tag{14}$$

is polynomic semiguarding for $\mathcal{S}(\Omega)$.

Proposition 2. ([1]) *The map ν of Eq.* (14) *guards $\mathcal{S}(\Omega)$ if and only if q satisfies the condition*

$$q(\lambda, \lambda^*) < 0 \;\; \textit{and} \;\; q(\mu, \mu^*) < 0 \;\; \Rightarrow \;\; q(\lambda, \mu^*) \neq 0. \qquad (\textit{Condition } C)$$

Condition C is in general difficult to check. A sufficient condition for it to hold is given by the next proposition.

Proposition 3. ([1]) *Assume that*

$$q_{kk} \geq 0, \qquad \text{for all } k \geq 1, \tag{15}$$

$$q_{k\ell} = -q_{\ell k}, \qquad \text{for all } k \neq \ell, \;\; k\ell \neq 0. \tag{16}$$

Then q satisfies Condition C.

2.3 Robust Stability

The robust stability problem for parametrized families of matrices or polynomials may be stated as follows. Let $r = (r_1, \dots, r_k) \in U$, where U is a pathwise connected subset of $\mathbb{R}^k$, and let $x(r)$ be an element of $\mathcal{X}$ which depends continuously on the parameter vector r. Given an open subset $\mathcal{S}$ of $\mathcal{X}$, we seek basic conditions for $x(r)$ to lie within $\mathcal{S}$ for all values of r in U. The next theorem gives a basic necessary and sufficient condition for this problem both for guarded and semiguarded sets $\mathcal{S}$. Typically, $\mathcal{S}$ is a stability set of the form $\mathcal{S}(\Omega)$ where Ω is a given subset of the complex plane.

Given a continuous map $\nu : \mathcal{X} \longrightarrow \mathbb{C}$, define the *critical set*

$$U_{\rm cr} := \{r \in U : \nu(x(r)) = 0\}. \tag{17}$$

Theorem 4. ([1]) *Assume that the family $\Phi := \{x(r) : r \in U\}$ is nominally stable relative to Ω, i.e., assume that $x(r^0) \in \mathcal{S}(\Omega)$ for some $r^0 \in U$. Then: (i) if $\mathcal{S}(\Omega)$ is guarded by ν, then Φ is stable relative to Ω if and only if $U_{\rm cr} = \emptyset$, (ii) if $\mathcal{S}(\Omega)$ is semiguarded by ν, then Φ is stable relative to Ω if and only if $x(r) \in \mathcal{S}(\Omega)$ for all $r \in U_{\rm cr}$.*

In the case of polynomic guardian or semiguardian maps, unless $U_{\rm cr}$ is all of U, it is finite and the theorem above yields computable conditions for robust stability of parametrized families of matrices or polynomials [1]. The guardian or semiguardian maps obtained in Section 2.2 are all of the polynomic type. It is not surprising that results analogous to those obtained for one- and two-parameter families of *real* matrices or polynomials in [1] hold for *complex* families . We give special attention to the cases of Hurwitz and Schur stability of families of complex polynomials, uncovering algebraic connections between recent results of Bose [2] and those which follow from the present approach.

3 Families of Polynomials

Let $\{p(r) : r \in U\}$ be a polynomic family of polynomials in $\mathcal{P}_n$ and let Ω be a given open subset of the complex plane. We seek conditions under which this family is stable relative to Ω.

Theorem 4 states that if ν is a polynomic guardian map for $\mathcal{S}(\Omega)$ and at least one polynomial $p(\hat{r})$, $\hat{r} \in U$, is stable relative to Ω, then a necessary and sufficient condition for the entire family to be stable relative to Ω is that the polynomial $\nu(p(r))$ have no zeros in U. In the cases of Hurwitz and Schur stability, this translates into the polynomials

$$\det\left(\mathcal{C}(p(r)) \oplus \mathcal{C}^H(p(r))\right) \tag{18}$$

$$\det\left(\mathcal{C}(p(r)) \otimes \mathcal{C}^H(p(r)) - I\right) \tag{19}$$

having no zeros in U.

We now proceed to show that (i) the results of [2] are particular cases of Theorem 4, and (ii) the corresponding "resultant-based" guardian maps are in fact identical to (18) and (19). The Hurwitz case is treated first.

Following [2], we associate with any complex polynomial

$$p(s) := p_0 + p_1 s + p_2 s^2 + \cdots + p_{n-1}s^{n-1} + s^n \tag{20}$$

the pair of polynomials

$$\alpha_p(s) := \frac{1}{2}\left(p(s) + (p(-s^*))^*\right), \tag{21}$$

$$\beta_p(s) := \frac{1}{2}\left(p(s) - (p(-s^*))^*\right). \tag{22}$$

Also, given two polynomials x and y with degrees n and m respectively, the *resultant* of x and y is the determinant of the $(m+n)\times(m+n)$ matrix $R(x,y)$ given by[2,3]

$$\begin{bmatrix}
x_n & x_{n-1} & \cdot & \cdot & \cdot & x_0 & 0 & \cdot & \cdot & \cdot & 0 \\
0 & x_n & x_{n-1} & \cdot & \cdot & \cdot & x_0 & 0 & \cdot & \cdot & 0 \\
\cdot & \cdot & \cdot & \cdot & \cdot & \cdot & \cdot & \cdot & \cdot & \cdot & \cdot \\
\cdot & \cdot & \cdot & \cdot & \cdot & \cdot & \cdot & \cdot & \cdot & x_0 & 0 \\
\cdot & \cdot & \cdot & \cdot & 0 & x_n & x_{n-1} & \cdot & \cdot & \cdot & x_0 \\
\cdot & \cdot & \cdot & \cdot & \cdot & 0 & y_m & y_{m-1} & \cdot & \cdot & y_0 \\
\cdot & \cdot & \cdot & \cdot & 0 & y_m & y_{m-1} & \cdot & \cdot & y_0 & 0 \\
\cdot & \cdot & \cdot & \cdot & \cdot & \cdot & \cdot & \cdot & \cdot & \cdot & \cdot \\
\cdot & \cdot & \cdot & \cdot & \cdot & \cdot & \cdot & \cdot & \cdot & \cdot & \cdot \\
0 & y_m & y_{m-1} & \cdot & \cdot & y_0 & 0 & \cdot & \cdot & \cdot & 0 \\
y_m & y_{m-1} & \cdot & \cdot & y_0 & 0 & \cdot & \cdot & \cdot & \cdot & 0
\end{bmatrix}. \tag{23}$$

Proposition 5. *The set of Hurwitz stable complex monic polynomials is guarded by the (polynomic) map δ given by*

$$\delta(p) = \det R(\alpha_p, \beta_p). \tag{24}$$

PROOF. Let $p \in \overline{\mathcal{S}(\mathring{\mathbb{C}}_-)}$, i.e., assume p has all its zeros in $\overline{\mathbb{C}}_-$. Then we have

$$\begin{aligned}
\delta(p) = 0 &\Leftrightarrow \alpha_p \text{ and } \beta_p \text{ have a common zero} \\
&\Leftrightarrow p(s) = (p(-s^*))^* = 0.
\end{aligned}$$

[2] A vanishing resultant signals that x and y have a common zero.

[3] Eq. (23) corresponds to the case $n = m + 1$.

Since p has all its zeros in $\overline{\mathbb{C}}_-$, it follows that

$$\begin{aligned}\delta(p) = 0 &\Leftrightarrow p(s) = 0 \text{ for some } s \in \partial\mathring{\mathbb{C}}_- \\ &\Leftrightarrow p \in \partial\mathcal{S}(\mathring{\mathbb{C}}_-).\end{aligned}$$

Q.E.D.

In the light of this proposition, an alternate necessary and sufficient condition for the convex hull of two polynomials p^0 and p^1 to be Hurwitz stable, given that p^0 is Hurwitz stable, is that the polynomial $\delta((1-r)p^0 + rp^1)$ in the indeterminate r be nonzero for all $r \in (0, 1]$ This appears as Theorem 1 of [2].[4]

The question now arises as to the relationship between guardian maps (18) and (24). It turns out that, essentially, these maps are identical, as is shown next.

Proposition 6. *Let $p \in \mathcal{P}_n$. Then the following identity holds:*

$$\det\left(\mathcal{C}(p) \oplus \mathcal{C}^H(p)\right) = (-1)^{\frac{n(n-1)}{2}} 2^n R\left(\alpha_p, \beta_p\right). \tag{25}$$

PROOF. Recall [2] that if x and y are polynomials with zeros given by $v_i,\ i = 1, \ldots, n$ and $w_k,\ k = 1, \ldots, m$, respectively, then

$$\det R(x, y) = (-1)^{\frac{n(n-1)}{2}} x_n^m y_m^n \prod_{i=1}^{n} \prod_{k=1}^{m} (v_i - w_k). \tag{26}$$

Here, the x_i's and y_i's denote the coefficients of x and y respectively. Note that this can also be written as

$$\det R(x, y) = (-1)^{\frac{n(n-1)}{2}} x_n^m \prod_{i=1}^{n} y(v_i). \tag{27}$$

Since $\alpha_p(s) + \beta_p(s) = p(s)$, it follows from (21)–(22) that at least one of $\alpha_p(s)$ and $\beta_p(s)$ has degree n. Without loss of generality, assume $\alpha_p(s)$ has degree n and denote by a_n the coefficient of the highest power. Let $\mathcal{Z}(p) =: \{z_1, \ldots, z_n\}$, $\mathcal{Z}(\alpha_p) =: \{\alpha_1, \ldots, \alpha_n\}$ and $\mathcal{Z}(\beta_p) =: \{\beta_1, \ldots, \beta_{n_2}\}$, where n_2 denotes the degree of p. Now let $\nu(p) = \det(C(p) \oplus C^H(p))$. Since

$$\sigma\left(\mathcal{C}(p) \oplus \mathcal{C}^H(p)\right) = \{z_i + z_k^* :\ i, k = 1, \ldots, n\}, \tag{28}$$

[4] The additional condition given in [2] on the polynomial coefficients is apparently superfluous.

we can write

$$\nu(p) = \prod_{i=1}^{n} \prod_{k=1}^{n} (z_i + z_k^*). \tag{29}$$

It also follows that

$$\nu(p) = (-1)^{n^2} \prod_{i=1}^{n} \widehat{p}(z_i), \tag{30}$$

where

$$\widehat{p}(s) := (p(-s^*))^*. \tag{31}$$

With this notation, Eqs. (21), (22) may be rewritten as

$$\alpha_p(s) = \frac{1}{2}\left(p(s) + \widehat{p}(s)\right), \tag{32}$$

$$\beta_p(s) = \frac{1}{2}\left(p(s) - \widehat{p}(s)\right). \tag{33}$$

Eq. (33) implies that $\widehat{p}(z_i) = -2\beta_p(z_i),\ i = 1, \ldots, n$, whence

$$\nu(p) = (-1)^{n^2} \prod_{n=1}^{n} \widehat{p}(z_i) = 2^n \prod_{i=1}^{n} \beta_p(z_i). \tag{34}$$

From the fact that $\alpha_p(s) + \beta_p(s) = p(s)$, we have $\alpha_p(z_i) = -\beta_p(z_i),\ i = 1, \ldots, n$. Hence

$$\nu(p) = (-1)^n 2^n \prod_{i=1}^{n} \alpha_p(z_i). \tag{35}$$

We now use the following easily proved fact: Given monic polynomials $A(s)$ and $B(s)$ with zeros $x_1, \ldots, x_n$ and $y_1, \ldots, y_m$, respectively,

$$\prod_{i=1}^{m} A(y_i) = (-1)^{mn} \prod_{i=1}^{n} B(x_i). \tag{36}$$

It follows from (35) and (36) that

$$\nu(p) = 2^n a_n^n \prod_{i=1}^{n} p(\alpha_i) \tag{37}$$

$$= 2^n a_n^n \prod_{i=1}^{n} \beta_p(\alpha_i). \tag{38}$$

On the other hand, Eq. (27) implies

$$\delta(p) = \det R(\alpha_p, \beta_p) = (-1)^{\frac{n(n-1)}{2}} a_n^{n_2} \prod_{i=1}^{n} \beta_p(\alpha_i). \tag{39}$$

Comparison of Eqs. (38) and (39) yields

$$\nu(p) = (-1)^{\frac{n(n-1)}{2}} 2^n a_n^{n-n_2} \delta(p). \tag{40}$$

Note that, if $n_2 < n$, then $a_n = p_n = 1$. Thus, $a_n^{n-n_2} = 1$ both for $n_2 = n$ and for $n_2 < n$. Q.E.D.

Remark 1. As a reviewer has pointed out to the authors, the guardian map $\delta(p)$ may alternatively be written as the determinant of the *Bezoutian* $B(p, \widehat{p})$, modulo a constant factor. The Bezoutian of two n^{th} degree polynomials is an $n \times n$ matrix [8]. The reduction in dimensionality as compared to the resultant formulation used here may lead to a savings in computation. □

Finally, it can be shown that similar results hold for the case of Schur stability where ν and δ now take the form

$$\nu(p) = \det\left(\mathcal{C}(p) \otimes \mathcal{C}^H(p) - I\right) \tag{41}$$

and

$$\delta(p) = \det R(\alpha_p, \beta_p). \tag{42}$$

Here, polynomials α_p and β_p are defined by

$$\alpha_p(s) := \frac{1}{2}\left(p(s) + s^n (p(1/s^*))^*\right), \tag{43}$$

$$\beta_p(s) := \frac{1}{2}\left(p(s) - s^n (p(1/s^*))^*\right). \tag{44}$$

The case of polynomials with real coefficients is considered in the following remark.

Remark 2. Bose [2] shows that if p^0 and p^1 have real coefficients and are both Hurwitz stable, then the convex hull $\{p(r) = (1-r)p^0 + rp^1, r \in [0,1]\}$ is Hurwitz stable if and only if $\widehat{\delta}(p(r))$ has no zeros in $[0,1]$, where $\widehat{\delta}(p(r))$ is the determinant of a resultant matrix of size $(n-1) \times (n-1)$, linear in the coefficients of p. It can be easily verified that $p_0\widehat{\delta}(p)$, with p_0 as in (20), is in fact a guardian map for the set of Hurwitz stable real monic polynomials. Białas' test [9] on the other hand, corresponds to the guardian map $\det H(p)$, where

$$H(p) = \begin{bmatrix} p_{n-1} & p_{n-3} & p_{n-5} & \cdots & \cdot & 0 \\ 1 & p_{n-2} & p_{n-4} & \cdots & \cdot & 0 \\ 0 & p_{n-1} & p_{n-3} & \cdots & \cdot & 0 \\ 0 & 1 & p_{n-2} & \cdots & \cdot & 0 \\ \cdot & \cdot & \cdot & \cdots & \cdot & \cdot \\ 0 & \cdot & \cdot & \cdots p_3 & p_1 & 0 \\ 0 & \cdot & \cdot & \cdots & p_2 & p_0 \end{bmatrix}$$

is the $n \times n$ Hurwitz test matrix associated with p, which too is linear in the coefficients of p. Since clearly $\det H(p) = p_0 \cdot \det \tilde{H}(p)$, with $\tilde{H}(p)$ an $(n-1) \times (n-1)$ matrix, it is clear that Bose's test and Białas' test are identical from a computational point of view. A similar comment holds in the Schur case regarding Bose's test and that proposed by Ackermann and Barmish [10]. □

We conclude with an example illustrating the use of Theorem 4 and Proposition 5.

Example 3. We test the one-parameter family of polynomials

$$p(r)(s) = s^3 + 3(1-r^2)s^2 + 3s + 1 - (1-j)r$$

for Hurwitz stability. Clearly $p(0)$ is stable. Using, e.g., the symbolic manipulation system MACSYMA,[5] one obtains

$$(p(r)(-s^*))^* = -\frac{1}{8}(s^3 + 3(r^2-1)s^2 + 3s + (1+j)r - 1),$$

$$\alpha_p(r)(s) = \frac{1}{16}(7s^3 + 27(1-r^2)s^2 + 21s + 9 - (9-7j)r),$$

and

$$\beta_p(r)(s) = \frac{1}{16}(9s^3 + 21(1-r^2)s^2 + 27s + 7 - (7-9j)r)$$

yielding

$$\begin{aligned}\delta(p(r)) = &-884736r^8 + 2654208r^6 + 2654208r^5 - 5898240r^4 \\ &-4096000r^3 + 6094848r^2 + 1572864r - 2097152.\end{aligned}$$

The real roots of $\delta(p(r))$ are given approximately by

$$-0.904, -0.864, 1.0, 1.126$$

Since $p(0)$ is stable, we now have that $p(r)$ is Hurwitz stable for any $r \in [-0.864, 1.0]$. It can be checked that $p(-0.864)$ and $p(1.0)$ do indeed have zeros on the imaginary axis. □

Acknowledgement: The authors are grateful to an anonymous reviewer for his detailed comments regarding Proposition 6. This work was supported in part by the National Science Foundation's Engineering Research Centers Program: NSFD CDR 8803012, as well as by the NSF under Grants ECS-86-57561 and DMC-84-51515.

[5]MACSYMA is a registered trademark of Symbolics, Inc., Cambridge, MA.

REFERENCES

[1] L. Saydy, A.L. Tits & E.H. Abed, "Guardian Maps and the Generalized Stability of Parametrized Families of Matrices and Polynomials," *Mathematics of Control, Signals and Systems* 3 (1990, in press).

[2] N.K. Bose, "Tests for Hurwitz and Schur Properties of Convex Combination of Complex Polynomials," *IEEE Trans. Circuits and Syst.* CAS-36 (1989), 1245–1247.

[3] N. K. Bose & Y. Q. Shi, "Network Realizability Theory Approach to Stability of Complex Polynomials," *IEEE Trans. Circuits and Systems* CAS-34 (1987), 216–218.

[4] A. Rantzer, "An Efficient Test for the Stability of Families of Polynomials," in *Realization and Modeling in System Theory. Proceedings of the International Symposium MTNS-89 – Volume 2*, M.A. Kaashoek, J.H. van Schuppen & A.C.M. Ran, eds., Birkhäuser, New York, 1990.

[5] L. Saydy, A.L. Tits & E.H. Abed, "Robust Stability of Linear Systems Relative to Guarded Domains," *Proc. 27th IEEE Conf. on Decision and Control*, Austin, Texas (December 1988).

[6] P. Lancaster & M. Tismenetsky, *The Theory of Matrices*, Academic Press, New York, 1985.

[7] C. Stéphanos, "Sur une Extension du Calcul des Substitutions Linéaires," *J. Math. Pures Appl.* 6 (1900), 73–128.

[8] U. Helmke & P.A. Fuhrmann, "Bezoutians," *Linear Algebra Appl.* 122/123/124 (1989), 1039–1097.

[9] S. Białas, "A Necessary and Sufficient Condition for the Stability of Convex Combinations of Polynomials or Matrices," *Bulletin of the Polish Academy of Sciences* 33 (1985), 473–480.

[10] J.E. Ackermann & B.R. Barmish, *IEEE Trans. Automat. Control* AC-33 (1988), 984–986.

Lahcen Saydy
Faculté des Sciences
Université Cadi Ayyad
Marrakech
Morocco

André L. Tits and Eyad H. Abed
Department of Electrical Engineering and Systems Research Center
University of Maryland
College Park, MD 20742
U.S.A.
andre@cacse.src.umd.edu -- abed@cacse.src.umd.edu

Application of Analytic Centers to Feedback Design for Systems with Uncertainties

G. Sonnevend*

Generalizing the classical notion of the "central", or "maximum entropy" solution of moment problems of the Nevanlinna-Pick type to the notion of a central solution for general, convex systems of inequalities, we present new algorithms for solving several classes of problems of robust feedback design, providing also new methods for ellipsoidal approximations of the state vector in the case of fixed saturation bounds for the controls and disturbances.

1 Introduction

In this paper we present a number of applications of the notion of the *"analytic centre"* (of a system of convex, analytic inequalities) for the numerical solution of several classes of problems of feedback design (i.e. for identification, observation and control, these are specified below in this section). In the last years the introduction and deeper study of these centers, see [20],[3],[6] led to new classes of numerical methods for the solution of linear (and more general convex) programming problems. From the already available theoretical, see [24],[3],[27] and practical (i.e. implementation) results (see e.g. [7]) it is obvious that the new algorithms represent a serious alternative to the classical algorithms arising from the simplex method (as well as to nonlinear programming methods, like SQP) especially for *large* systems. Since the basic engineering feedback design (e.g. robust stabilization) problems are — mathematically — feasibility problems, we shall

*Supported in part by the German Research Society, DFG

be basically concerned with the question: how to find a nice (stable, low complexity, easily updated,...) feasible solution of a system of inequalities. Having found the "right" solution we shall (in fact, can immediately) use it for the solution of optimization problems. Of course, all our concrete applications exploit the special features of the class of problems studied there, however the main aim of this paper is to demonstrate the strength of the proposed new algorithmic tools in dealing with several, difficult classes of problems of feedback control.

In section 2 we present the generalization of the notion of the analytic centre (of a finite set of linear inequalities) to the semi-infinite case, i.e. define this centre for a rather arbitrary (infinite) system of convex, algebraic-analytic inequalities assuming only that its feasible set in R^n

$$P^A := \{x \mid f(\alpha, x) \geq 0, \alpha \in A\} \tag{1}$$

is bounded.

We assume that the functions $f(\alpha, x)$ are

(1) easily computable for each $(\alpha, x) \in A \bigoplus R^n$,

(2) concave in x for each $\alpha \in A$ on $P_\alpha := \{x \mid f(\alpha, x) > 0, x \in R^n\}$

(3) there is a canonical measure on $A, d\alpha$ such that the integrals of the form — formed from derivatives of $f(\alpha, x)$ with respect to x —

$$\int_A D(f(\alpha, x))\, d\alpha \tag{2}$$

can be computed easily — by suitable quadrature rules — (the precise form of the functionals $D(\cdot)$ will be given below).

The method (of picking up the "right" notion of a central solution) we propose here originates from the theory of the Nevanlinna-Pick type moment problems and the theory of "central" extensions and dilatations of operators see [15],[6]; we shall show that it leads — in important cases — to solutions which posses a "rich", i.e. "simple" (algebraic) structure and nice geometric properties, mainly because they inherit the invariance properties shared by the given class of problems. We shall not discuss the connections to "maximum entropy" (reconstruction) methods, see e.g. [13] and [14].

While real life feasibility problems (for the unknown object x) are often "nonconvex" in the above sense, we shall apply the methods of dealing with (homotopies of) simple convex feasibility problems (with a larger set A) to solve some *nonconvex* ones (with a smaller set A), see the results concerning "structured singular values" (studied by Doyle and others — we shall not encounter the difficulty to be trapped in a local optimum see

[9] —) in section 3, where we also present a new method for computing the H^∞-norm of a transfer function (a basic problem in the theory of robust stabilization.

We shall show that in many problems one can (in fact should whenever possible) *avoid nonsmooth in x functions* $f(\alpha,\cdot)$ by building up equivalent problems, again with a larger set A, where each of the elementary functions are analytic, i e. rather smooth. This will be illustrated in a novel approach for the solution of H^1 (and H^∞ optimal feedback design problems, where we do not need to compute realizations neither the corresponding "zero structure", but work directly with the values of the underlying transfer functions (of possibly infinite degree).

In section 4 we apply the proposed method to construct observers in linear, time invariant systems with unknown, but *bounded* disturbance inputs and measurement noise, providing a new method for constructing dynamic observers and ellipsoidal (inner-outer) approximations for the current set of localization of the unknown state.

2 Basic properties of analytic centers

The (analytic) center $x(f^A)$ of the convex inequality system (1) — with a bounded feasible set P^A having a nonempty interior in R^n — is defined as the (in general) unique solution of the supremum problem

$$\sup\left\{\int_A \log f(\alpha,x)\,d\alpha \mid x\in P^A\right\}. \tag{3}$$

In fact if A is a finite set and if all $f(\alpha,\cdot)$ are linear, or if all $f(\alpha,\cdot)$ are (at most) quadratic at least one being negative definite, the function

$$\Phi(x)=\int \log f(\alpha,x)\,d\alpha \tag{4}$$

is strongly concave over P^A. To assure the existence of the integrals in (3) and those appearing later below it would be enough to assume that $f(\cdot,x)$ and its derivatives (up to order two) are continuous and uniformly (in $x\in P^A$) bounded over A. More important is that in the proposed methodology (using analytic homotopies along analytic centers) we need (at least we should like to get) a much higher degree of smoothness and algebraic simplicity, therefore a nonsmooth constraint of the type, say

$$\max_i g_i(y)\le 1,\;\; g_i(y):=\sum_{1\le j\le k}|f_{ij}(y)|,\;\; i=1,\ldots,n, \tag{5}$$

will be replaced by a set of smooth constraints

$$\sum_j \eta_{ij}<1,\;\; -\eta_{ij}\le f_{ij}(y)\le\eta_j,\;\; 1\le j\le k,\;\; i=1,\ldots,n$$

i.e. we set $x = (\eta_1, \ldots, \eta_k, y)$. The following invariance properties of this "solution" concept are important

(1) *affine invariance*: if we replace $f(\alpha, \cdot)$ by $\hat{f}(\alpha, \cdot), \hat{f}(\alpha, y) := f(\alpha, Ty + t)$, where $t \in R^n, T : R^n \to R^n, \det T \neq 0$ then

$$t + Tx(\hat{f}^A) = x(f^A)$$

(2) invariance under *scaling*: if we replace $f(\alpha, \cdot)$ by $\check{f}(\alpha, \cdot) := k(\alpha) f(\alpha, \cdot)$, for an arbitrary function $k(\alpha)$ then

$$x(\check{f}^A) = x(f^A)$$

The proof of (1) is easy obtained from the characterization of the optimal solutions for (3)

$$\int_A \frac{\nabla_x f(\alpha, x)}{f(\alpha, x)} \, d\alpha = 0 \tag{6}$$

The proof of (2) is immediate from the "additivity" of the log function.

A further, rather useful property of this solution concept is that it often allows to find good *ellipsoidal inner-other approximations* for the set P^A. The idea is simple: consider the function

$$\Psi^A(x) := \exp(\phi(x)) \tag{7}$$

and the set — where $H = D_2 \Psi^A(x(f^A))$ is the Hessian of Ψ^A at $x(f^A)$ —

$$E(f^A) := \left\{ z \mid -\frac{1}{2} z^* H z \leq \Psi^A \left(x \left(f^A \right) \right) \right\}. \tag{8}$$

It is natural to expect that $E = E(f^A)$ is a good ellipsoidal approximation of P^A (which of course, shares the invariance properties (1) and (2) of $x(f^A)$).

Theorem 1. *Suppose that A is a finite set of cardinality m and that the functions $f(\alpha, \cdot)$, $\alpha \in A$ are concave and quadratic, then*

$$x(f^A) + \frac{1}{2\sqrt{m}} E(f^A) \leq P^A \leq x(f^A) + \sqrt{2m} E(f^A). \tag{9}$$

This is proved in [24]. The surprising point is here the independence of the quality of this approximation on the specific data, i.e. the form of the functions f_α, $\alpha \in A$.

An important property of the function Ψ is that it is *concave* on P^A, if

$$\int_A d\alpha = 1,$$

a condition (normalization) which we shall assume below (obviously without loss of generality).This can be proved by computing the Hessian of Ψ

$$H = \int_A \left(\frac{D_2 f(\alpha, x)}{f(\alpha, x)} - a(\alpha)a^*(\alpha) \right) d\alpha + \int_A a(\alpha)\, d\alpha \left(\int_A a(\alpha)\, d\alpha \right)^*$$

— where $a(\alpha) = \frac{\nabla_x f(\alpha,x)}{f(\alpha,x)}$ — using the inequality

$$\left(\int_A d\alpha \right)^2 \int_A a(\alpha)a(\alpha)^* \, d\alpha \geq \int_A a(\alpha)\, d\alpha \left(\int_A a(\alpha)\, d\alpha \right)^*,$$

which is obtained from the Cauchy-Schwartz inequality (multiplying — from both sides — with a vector).

Many convex feasibility problems are written in "dual" form, known as a finite, or restricted moment problem: given $c^T = \{c(t) \mid t \in T\}$, the "set of feasible solutions" is formed by the nonnegative mass distributions (densities over a set S

$$P_K(c^T) = \left\{ \mu \mid c(t) = \int_S K(t,s)\, d\mu(s), \;\; t \in T \;\; d\mu \geq 0 \right\}. \qquad (10)$$

The central feasible solution is defined as that element (if it exists and is unique) which solves the supremum problem

$$\sup \left\{ \int_S \log \mu'(s)\, ds \mid \mu \in \operatorname{int} P^K(c^T) \right\} \qquad (11)$$

here "int" stands (intuitively) for "interior of" and means (precisely) that only those mass distributions are regarded which have a log integrable density.

If A is a finite set, $A = \{1, \ldots, m\}$ and $f(\alpha, \cdot)$ are linear in x, say,

$$f(\alpha, x) = b_\alpha - a_\alpha^T x, \;\; \alpha = 1, \ldots, m,$$

then introducing

$$\mu_i := b_i - a_i^T x, \;\; i = 1, \ldots, m$$

we can find vectors $k_1, \ldots, k_{m-n} \in R^m$ and scalars $c_1, \ldots, c_{m-n}$ such that

$$P^K(c^N) = \{\mu \mid \langle k_j, \mu \rangle = c_j, \;\; \mu \in R^m_+\} \qquad (12)$$

is identical with the set of vectors $\{f^m(x) \mid x \in P(f^A)\}$ (note that $|T| = N = m - n$). Obviously $f^m(x(f^A))$ yields then the solution of problem (11).

A further strong motivation for the importance of the solution concept (11) — thus for (3) — is that for some rather interesting special cases (of the kernel function $K(\cdot,\cdot)$), e.g. in the class of the Nevanlinna Pick type moment problems, the solution (11) — known as the maximum entropy solution — can be *exactly*(!) computed in a very simple way in $O(N^2)$ (in fact even in $O(N \log N)$ operations and recursively in N), see In fact this classical example is at the root of recent more general results about the role (existence, applications) of central or "maximum entropy" solutions to the basic H^∞ optimization problems, see below. There is a geometrically appealing formulation of these class of problems in terms of optimal extensions (and dilatation) of operators (with a given "symmetry" structure: self-adjoint, Toeplitz, Hankel,...) given on some subspace of a Hilbert space to whole space, which we shall touch upon in the next and the last section.

Returning to Theorem 1, note that such a good ellipsoidal approximation indicates a good, global approximability of the function Ψ by the quadratic function Q. It is thus not a big surprise that for the computation of the center $x(f^A)$ Newton's method (with a suitable homotopy, see below) is very efficient.

Theorem 2. *Under the same conditions as in Theorem 1 there exist an universal (explicitly computed) constant γ, such that Newton's method — applied to find $x(f^A)$, i.e. to solve (6) — converges superlinearly (with order at least 1.5) for every starting position within the domain obtained from the interior ellipsoid by contraction with γ around $x(f^A)$.*

The proof is given in [11].

An important application of centers is for solving optimization problems of the type

$$\inf \{f_0(x) \mid x \in P^A\} =: f_0^* =? \tag{13}$$

Using the observation that $f_0(x(\lambda)) \searrow f_0^*$ for $\lambda \searrow f_0^*$, where $x(\lambda)$ is the center of the extended system

$$\sup \left\{ \log(\lambda - f_0(x)) + \int_A \log f(\alpha, x)\, d\alpha \mid x \in P^A, \lambda \geq f_0(x) \right\}, \tag{14}$$

the idea is to follow the homotopy path $x(\cdot)$ from λ_0 to f_0^* by a predictor-corrector method, where in the corrector steps Newton's method is applied to the system

$$\frac{\nabla f_0(x)}{r} + \int_A \frac{\nabla f(\alpha, x)}{f(\alpha, x)}\, d\alpha = 0, \quad r > 0. \tag{15}$$

Here we introduced a new parametrization of the central path by $r = \lambda - f_0(x(\lambda))$ such that $r = 0$ corresponds to the optimum. Under the assumptions of (and using) Theorem 1 it can be easily proved, that for a linear (in x) function $f_0(x)$

$$f_0(x(r)) - f_0^* \leq (m+1)r.$$

This can be generalized immediately to semiinfinite, linear systems, where $f(\alpha, x) = b(\alpha) - a^*(\alpha)x, \alpha \in A$ to

$$c^*x(r) - f_0^* \leq r \int_A d\alpha.$$

For finite sets A with cardinality m and linear (in x) functions $f(\alpha, \cdot)$ the complexity of a continuation algorithm using *first order*, i.e. linear *extrapolation*, which adaptively selects the step size, based on the above explicit description of the domain of fast convergence for Newton's method can be estimated as follows. In order to reach $x(R)$ from $x(\delta)$ we need no more than

$$N = N(R, \delta) = \text{const} \cdot m^\alpha \log \frac{R}{\delta} + \text{const} \tag{16}$$

— where $\alpha \leq \frac{1}{2}$ — iterations (each requiring no more than const $\cdot\, m^3$ arithmetical operations). In fact, for interesting classes K of such problems one can show that $\alpha \leq \frac{1}{4}$. One obtains these estimations by computing and estimating the integral between δ and R of the square root of a suitably defined weighted curvature, $\kappa(r)$, of the path $x(\cdot)$, here $\kappa(r) = \sqrt{\langle \ddot{x}(r), H(x(r))\ddot{x}(r)\rangle}$ see [25], where the surprisingly rich structure of these paths is analyzed, it turns out that for all $r \geq 0$ and k

$$g^k(r) := \|x^{(k)}(r)\|_H(r)^{\frac{1}{k}} \leq \text{const}((g^2(r) + \text{const}).$$

These rigorous results, (see also [24] on increasingly higher order Pade type, i.e. rational extrapolation) indicate that suitable implementations of these, center-based homotopies are quite effective. It should be however emphasized that the selection of the right measure $d\alpha$, extrapolation algorithm and (matrix inverse) update techniques is a nontrivial problem. In order to briefly address the first of these problems note first that the analytic centre $x(f^A)$ depends on f^A, i.e. not only on P^A. An interesting definition of the analytic center of a convex, compact set K in R^n described by its support function is given in [23], it leads to the Santalo point of K, for this point the homothecity ratio of the corresponding inner-outer ellipsoids — constructed with the help of the method (7)-(8) — is, (i.e. can be chosen as) a constant multiple of n.

The positive definiteness of a matrix M can be expressed by requiring that

$$\langle Mx, x\rangle > O, \quad \text{for all } x \text{ such that} \quad \|x\| = 1.$$

the "barrier" function over the class $M_+^{n\times n}$ of positive definite symmetric matrices

$$f(M) := \int_{\|x\|=1} \log \langle Mx, x\rangle \, dx$$

is however not "(strictly) affine invariant" in the sense that

$$f(T^*MT) \not\equiv f(M) \quad \forall M \text{ and } T \in R^{n\times n}, \text{such that } \det T = 1$$

The "right" barrier function for expressing positive definiteness, i.e. the analogon of the function Φ over $M_+^{n\times n}$ turns out to be

$$\Phi(M) := \log\det M, \tag{17}$$

it will be one of the principal actor of the next section, note that the analogon of the function Φ is $\det^{\frac{1}{n}}(M)$, both of these functions are concave on $M_+^{n\times n}$.

3 Methods to solve structured robust stabilization problems

In order to motivate the use of det function, we begin with the following "minimum norm" matrix extension problem: find X such that

$$\min_X \left\| \begin{matrix} A & B \\ C & X \end{matrix} \right\| = \lambda^* \tag{18}$$

This problem can also be successfully approached by the above method. Let $\lambda > \lambda^*$, then

$$\sup_X \det \left(\lambda^2 I - \begin{pmatrix} A & B \\ C & X \end{pmatrix} \begin{pmatrix} A & B \\ C & X \end{pmatrix}^* \right) \tag{19}$$

has a unique exactly computable solution $X = C(\lambda^2 I - AA^*)^{-\frac{1}{2}} A^*(\lambda^2 I - AA^*)^{-\frac{1}{2}} B$. One of the encouraging motivation for choosing the barrier functions (17) is the fact that the optimal solution of the central extension problem for a partially given symmetric, pos. def. matrix A^0

$$\sup_x \left\{ \det(A(x)) \mid A(x) = A^*(x) \geq 0, A_{ij}(x) = A^0_{ij} = \text{fixed}, ij \in I \right\}$$

— where $I \subset \{n \times n\}$ is a fixed index set — has a simple characterization

$$A^{-1}(x^{\text{opt}})_{ij} = 0, \quad ij \in \{n \times n\} \setminus I.$$

Moreover if I is block diagonal x^{opt} can be computed recursively in I. For further generalizations see the references in [6] and [23]. We consider now — as an application — the problem of constructing a "maximally stable, partial realization; it is known that important classes of multivariate ("non-robust") feedback control problems can be reduced to a similar problem, requiring just (i.e. not "maximal") stability, see e.g. [1]. Suppose, that $\Sigma = (A_1, \ldots, A_N \in R^{s \times m})$ is a fixed, known sequence of matrices, "Markov parameters" of an unknown transfer function

$$\frac{P(z)}{q(z)} = C(zI - A)^{-1}B = \sum_{i=1}^{N} A_i z^{-i+1} + O(z^{N+1}),$$

the problem is to find a partial realization $A_i = CA^{i-1}B, I = 1, \ldots, N$, for which $\|A\|$, i.e. the spectral norm of A, becomes minimal. Denoting the corresponding, minimal value by r^*, let fix $r > r^*$; by scaling the data, we can assume that $r = 1$. From the theory of dilatations of contractions, see [15] we know the existence of a unitary operator U on a larger Hilbert space H^0 such, that $CA^kB = CU^kB$, for all k. Consider now the subspace H_0 formed by the vectors

$$\left\{U^j b_n, U^k c_r; 1 \leq n \leq m, 1 \leq r \leq s, 0 \leq j, k \leq \frac{N}{2}\right\}$$

assuming — for simplicity — that N is even. The projection of U onto H^0, provides now a desired contraction.

Proposition 3. *For the solvability of the above realization problem (for fixed* $r = 1$*) it is necessary and sufficient that the unspecified elements of the block Toeplitz matrix* $T = T_{k,l} = \langle U^{k-l}(B,C), (B,C)\rangle, 0 \leq k, l \leq N$ *can be selected so that* T *becomes nonnegative definite.*

The proof follows the (by now) well-known arguments, see e.g. [22], the Hilbert space H_0 is obtained by symmetric factorization of T. A method (based again on the use of center-based homotopies) to incorporate degree requirements on the realization — which could be used here — is proposed in [21].

An application of this homotopy technique to feedback design consider the problems of computation of the *structural singular value*

$$\varrho^* = \max_{z \in \Omega} \|a + bz(I + dz)^{-1}c\|, \tag{20}$$

where a, b, c, d, e, z are matrices of appropriate size and Ω is a region of uncertainty, say a box around some nominal value $\dot{z}$. It is easy to check that — if c is invertible — for any fixed $\varrho > \varrho^*$

$$\|M(z)\| = \|a + bz(I + dz)^{-1}c\| < \varrho \quad \forall z \in \Omega,$$

is equivalent to the positive definiteness of the matrix $\varrho^2 I - M(z)M^*(z)$, i.e. to

$$F_\varrho(z, a, b, c, d) > 0 \tag{21}$$

where F_ϱ is a simple polynomial function of its variables, which depends only on ϱ^2 and is linear in the latter. The function $\log\det F_\varrho(z)$ is therefore a concave function of ϱ, for all z satisfying (17). Let us consider the scalar function $\varrho(\lambda)$, defined, for $\lambda > \varrho^*$ as the solution of

$$\sup_{\varrho}\left(\log(\lambda-\varrho)+\int_{\Omega}\log\det F_\varrho(z,a,b,c,d)dz\right). \qquad (22)$$

Obviously ϱ^* is the unique solution of the equation $\lambda = \varrho(\lambda)$ in $\lambda \in [\varrho^*, \infty)$. Thus we propose to move along the homotopy path $\varrho(\cdot)$, which is an analytic, (rather smooth) function of ϱ. Comparing this method with other methods, see e.g. [9] our advantage is that everything is convex and smooth, we do not encounter the serious difficulty caused by the existence of isolated local optima, occurring when one is minimizing $x^T F(z,a,b,c,d)x$ over $\|x\| = 1$ and $z \in \Omega$ (to see whether it is larger than zero). The arising integrals, being rather smooth and algebraically simple, seem to be calculable — at least within sufficient approximations — easily for moderate dimensions of Ω. It would be interesting to look at whether "symbolic integration methods" like those in [5] could be used here for the integrals (22),(15).

A similar approach can be applied to the ("nonconvex") problem of how to compute the H^∞-norm of a transfer function G. This problem is a central one in the theory of robust stabilization of linear, time invariant systems: it is known that

$$\|G\|_\infty < \gamma, \qquad G = H(zI - A)^{-1}D, \qquad z = Gv$$

specifies a level in "energy gain" $\|z\|_2/\|v\|_2$, for $\gamma = 1$ it is the main condition in the "small gain theorem" see e.g. [17] and $\|G\|_\infty^{-1}$ appears as the "complex stability radius", r for structured perturbations, (i.e. $\tilde{A} = A + HKD$ is stable — for fixed (A, H, D), iff $\|K\| \leq r$, see [10].

Let us — for simplicity — consider the discrete time problem

$$\|G\|_\infty = \sup\{\|G(z)\| \mid |z| \leq 1\} = \lambda^* =?$$

Let $\lambda > \lambda^*$, and consider the central function $\gamma(\lambda)$ — T being the unit circle — and

$$g(\gamma) := \sup_{\gamma}\left\{\log(\lambda-\gamma)+\int_{T}\log\det(\gamma^2 I - G(z)G^*(z))dz\right\}. \qquad (23)$$

Note, that the values of the integrals, i.e. the "entropies" $I(\gamma, G)$ here can be computed *explicitly*, for each, fixed value of γ by solving two Riccati equations, see [14], this is a special case of the model matching problem $H_1 = 0, H_2 = 0$, see below. Since $I(\gamma, G)$ is an analytic, rather smooth function of γ, (23) is an easy, one-dimensional problem and we can use high order extrapolation techniques to find and follow the solution path $\gamma(\lambda)$ till $\lambda \cong \gamma(\lambda)$, i.e. $\lambda \cong \lambda^*$. We think that "γ iteration" techniques, see [17] using fixed step sizes, dyadic search or even low (e.g. first) order extrapolations could be improved dramatically by constructing high order extrapolants (results derived in [24] indicate that suitable rational, Pade type intrapolations should be quite effective). Of course, within this technology of center based homotopies, other methods could be conceived, e.g. we could try to fulfill the "continuation" (in γ) directly of the function values $I(\gamma, G)$, "detecting" that value $\gamma = \lambda^*$, where the (standard) methods used to solve the (corresponding) Riccati equations first become unapplicable.

The problem of "quadratic stability" (via linear control) of an uncertain system

$$\frac{dx}{dt} = [A + \Delta A(\omega(t))]x + Bv \quad t \geq 0.$$

where $\omega(t) \in \Omega$, a fixed compact set for all t, is known to be solvable, i.e. there exist a matrix K and a pos. def. symm. matrix P, such that — for some constant $\alpha > 0$, and the (Ljapunov) function $V(x) := x^* P^{-1} x$ —

$$\frac{dV(x(t))}{dt} = x^*(P^{-1}(A+\Delta A(\omega(t))+(A+\Delta A(\omega(t))^* P^{-1})x \leq -\alpha \|x\|^2 t \geq 0,$$

iff the following holds: there exist a positive definite, symmetric matrix S, $(S := P)$ such that

$$F(\omega, S) := \Theta^*(A + \Delta A(\omega)S + S(A + \Delta A(\omega))^T)\Theta < 0,$$

where Θ is a left unitary matrix, whose columns form a basis in the kernel space of B^*, i.e. satisfying

$$B^*\Theta = 0, \quad \Theta^T\Theta = I,$$

see [2]. The existence of such an S can be checked via "homotopy through central solutions" supposing that on Ω we have a measure $d\omega$ and defining a homotopy path $S(\lambda)$, $0 < \lambda < 1$ by

$$\sup_S \left\{ \int_\Omega \log\det(-F_\lambda(\omega, S))d\omega + \log\det S \mid \mathrm{Tr}(S) = 1 \right\}, \tag{24}$$

where the positivity of the arguments is assumed, and

$$F_\lambda(\omega, S) := \Theta^*((A + \Delta A_\lambda(\omega))S + S(A + \Delta A_\lambda(\omega))^*)\Theta$$

$$\Delta A_\lambda(\omega) := \lambda A(\omega) + (1 - \lambda)A(\omega_0),$$

where $\omega_0 \in \Omega$ is such that an S_0 is known (i.e. is computed to begin the homotopy) which solves the problem

$$\max_S \{\log\det SF_0(\omega_0, S) + \log\det S \;\; | \operatorname{Tr}(S) = 1\}$$

Since the Trace is a linear function, these supremum problems are "convex" (but unfortunately this formulation is not invariant under the affine group of T with $detT = 1$) since in general $\operatorname{Tr}(S) \neq \operatorname{Tr}(T^*ST)$.

4 Constructing feedback controls in linear systems with L^p bounded controls $p = 2, \infty$

In this section we shall study some problems of constructing (stabilizing) feedback control $u(\cdot) = U(y(\cdot))$ — to "minimize" an output $z(\cdot)$ — in the partially observed linear time invariant system, with state vector x, subjected to unknown disturbances $v(\cdot)$ and measurement errors $w(\cdot)$

$$\begin{aligned} \dot{x}(t) &= Ax(t) + Bu(t) + Dv(t) && (25)\\ y(t) &= Cx(t) + Ew(t) \\ z(t) &= Hx(t) && (26) \end{aligned}$$

where A, B, C, D, E, H are known, constant matrices. In the H^∞ resp. H^1 problems we suppose that

$$\|(w(t), v(t)\|_{L^p(0,\infty)} \leq 1 \tag{27}$$

and try to minimize

$$\sup \|z(t)\|_{L^p(0,\infty)} =: \gamma^*, \tag{28}$$

where $p = 2$ resp. $p = \infty$, and the feedback controller — which has to be optimized — is assumed to be the output of a linear time invariant system driven by $y(\cdot)$; we have used the above, slightly special form just for reasons of economy: the same system (of equations) will be used below to formulate other problems. Necessary and sufficient conditions for asymptotically exact regulation, i.e. one achieving $z(t) \to 0$, for $t \to \infty$ have been obtained already in [18]. Using the standard Youla-Kucera parametrization of internally stabilizing feedback controllers the problem is reduced — under standard assumptions, see e.g. [17],[14] — to a convex problem

$$\min_{Q \in H^p} \|H_0(z) - H_1(z)Q(z)H_2(z)\|_{H^p} = \gamma^* \tag{29}$$

where $p = \infty$, resp. $p = 1$, see [4] for a more precise statement of the H^1 problem. Here the functions H^0, H^1, H^2 are explicitly computed rational functions, depending on (A, B, C, D, E, H). Suppose now that we have to solve the H^∞ problem under the assumption that only the values of $H^i(z), i = 0, 1, 2$ are available for arbitrary $z = i\omega, \omega \in R^1$, thus we assume that these functions not only belong to H^∞ but are analytic there. Since computing realizations from (empirically given) transfer function data is a rather ill conditioned problem (especially if one wants to compute the zero structure of H^1 and H^2 our approach seems to provide a robust approximation for the solution of high (infinite) dimensional H^∞ and H^1 problems, see e.g. [8] for additional motivation. In the H^∞ problem (28) we are led — similarly as in (19) to consider, for $\gamma > \gamma^*$

$$\max_{Q \in H^\infty} \int_{-\infty}^{+\infty} \log\det(\gamma^2 I - \Phi(Q)\Phi^*(Q))[i\omega]d\omega \tag{30}$$

where $\Phi(Q)(z) = H_0(z) - H_1(z)Q(z)H_2(z)$. It turns out again, that (28) is almost *explicitly solvable*, its solution can be obtained in terms of the solutions of two (finite dimensional) algebraic Riccati equations, whenever the (finite dimensional) realizations for H_0, H_1, H_2 are known. The solution of (28) via (30) in our case — similarly to that of (14) or (22) — requires numerical integration and restricting Q to a suitable finite dimensional linear subspace in H^∞:

$$Q(z) = \sum_{i=1}^{n} x_i q_i(z),$$

then the problem (28) provides a convex, analytic homotopy. In the case of the H^1 optimization problem (27) we can — proceeding similarly as in (28) — obtain smooth integrand by dealing with the nonsmooth norm of H^1 as indicated in (5). In the discrete time case — assuming $\Phi(Q)(z) = \sum_{k=0}^{\infty} \Phi_k z^k \in M^{n\times m}$ — it is well known that

$$\|\Phi(\cdot)\|_{H^1} = \max_{i=1,\dots,n} \sum_{j=1}^{m} \|\Phi_{ij}\|_1,$$

where $\|h\|_1 = \sum_{k=0}^{\infty} |h_k|$ is the H_1-norm of a scalar transfer function h.

We note that in the more realistic "variant" of the H^1 problem, when we impose the saturation constraint $\|u\| \le \varrho =$ const the restriction to linear feedback controllers does not lead to an optimal solution, even simple nonlinear feedback controllers are, in general, superior to linear ones. Consider, for example, the following system

$$\begin{aligned} \dot{x}_1 &= x_2 - v\sigma, & z &= x_1, & \|v\| &\le 1, \\ \dot{x}_2 &= u\varrho, & y &= x_1, & \|u\| &\le 1, \end{aligned} \tag{31}$$

where the aim is to select the feedback control $u(\cdot)$ in such a way as to guarantee — say for $x(0) = 0$ —

$$\|x_1(s)\| \leq \gamma, \qquad 0 \leq s \leq \infty$$

for a possible small value of γ. Methods to solve this problem are proposed in [19]. Consider first the observation problem ($u = 0$) assuming in addition to (27), for example, that for given concave, e.g. linear or quadratic in x functions $h_1, \ldots, h_m$

$$h_i^* x(t) \geq 0, \quad i = 1, \ldots, m \quad 0 \leq t \leq T. \tag{32}$$

We propose to compute a "central" solution — i.e. to get a reconstructed path $x(T, \cdot)$, specially the present state $x(T, T)$ — by solving for a given observed output $y(\cdot)$

$$\max_{v(\cdot), w(\cdot)} \int_0^T (\log(1 - \|v(t)\|^2) + W_0 \log(1 - \|w(t)\|^2)) \, dt + \sum \log h_i(x(t))$$

under the side constraints (26),(27),(32). Note that an other interpretation of the above problem (of interest e.g. in robotics) is the construction of an admissible path in the controlled (with $w(t)$) system (26), where now $u(\cdot) = 0$, such that the true output stays within a given distance from a specified function and that the phase constraints (32) are fulfilled.

The above is a *classical*, i.e. analytic (smooth), convex variational problem, its extremals satisfy the canonical differential equations with the transversality conditions at the endpoints

$$\dot{x} = Ax + \frac{1}{2} BB^* \Psi \left(1 - \frac{z}{2+z}\right), \quad \text{where} \quad z = \sqrt{1 + | B^* \Psi(t) |^2} - 1 \tag{33}$$

$$\dot{\Psi} = -A^* \Psi + \frac{C^*(y - Cx)}{1 - | - Cx|^2} - \sum_i \frac{1}{h_i(x)} \frac{\partial h_i(x)}{\partial x}$$

$$\Psi(T, 0) = 0, \quad \Psi(T, T) = 0.$$

At this moment our solution is not dynamic. We can however derive an interesting dynamic observer by simply propagating the Euler-Lagrange equations: indeed replacing $y(s)$ by

$$\tilde{y}(s) := (1 - e^{-\lambda_0(s)s}) y(s),$$

where $\lambda_0(s)$ is appropriately chosen such that

$$\tilde{y}_{(0)}^{(k)} \equiv 0, \quad k = 0, \ldots, n \quad \text{and} \quad \lambda_(s)s \to \infty, s \to \infty,$$

we can (should) start with $\tilde{x}(0) = 0$ (and $\Psi(0) = 0$).

Whenever $\lambda_0(\cdot)$ is chosen so that $\tilde{y}(\cdot)$ is an allowed trajectory (i.e. compatible with some motion $\hat{x}(\cdot)$) for the system (26) the equations in (33) can be propagated for all $s > 0$. The corresponding path will be optimal on each interval $[0, s]$ under the "end point" condition $\tilde{x}(s) = \hat{x}(s)$. If we consider the discrete time analogon of the system (26), we can immediately apply Theorem 1 (and Theorem 2), since the unknown state will belong to a finite intersection of ellipsoids, as well as the reachable set for the control u (for $v \equiv 0, w \equiv 0$) is a finite intersection of ellipsoids, since we have quadratic constraints for the values of u and everything else is linear.

Now we use the previous method — assuming for simplicity that $m = 0$ in (32) — for constructing ellipsoidal approximations for the set $L(T)$ of all $x(T)$ compatible with the given measurements. Consider for fixed $z \in L(T)$

$$\Phi(z) = \sup \left\{ \frac{1}{T} \int_0^T \left(\log(1 - \|v\|^2) + w_0 \log(1 - \|w(t)\|^2) \right) dt \mid x(T) = z \right\}. \tag{34}$$

This is the classical Hamilton-Jacobi function, the analogon of the barrier function Φ for the domain $L(T)$, and pursuing the analogy we are led to look to the ellipsoid

$$E(T) := \left\{ z \mid \frac{1}{2}(z - x(T,T))^* D^2 F(x(T,T))(z - x(T,T)) \leq F(x(T,T)) \right\}, \tag{35}$$

where $F = \exp(\Phi)$. From the classical theory of second order variations it is known, that

$$D^2\Phi(x(T,T))\xi\xi = - \min_{v(\cdot),w(\cdot)} \left\{ \int_0^T \left(\|v(t)\|^2 + W_0 \|w(t)\|^2 \right) dt \mid x(t) = \xi \right\} \tag{36}$$

since

$$\frac{\partial^2}{\partial v^2} \log\left(1 - \|v\|^2\right) = 2I, \quad \frac{\partial^2}{\partial w^2} \log\left(1 - \|w\|^2\right) = 2I.$$

Moreover the ellipsoid, i.e. the matrix $B(T) := D^2\Phi(x(T,T)$ can be propagated, from the classical theory of variations we know that $B(T)$ satisfies — along each extremal $\tilde{x}(\cdot)$, i.e. for fixed $\tilde{y}(\cdot)$ and say $\tilde{x}(0) = 0$ — a Riccati type equation. In order to estimate how well the ellipsoid $E(T)$ approximates the set $L(T)$, suppose that $y(t) \equiv 0,\ 0 \leq t \leq T$ and $D = I$, i.e. $w(t) = Cx(t)$. In this case (36) is a classical quadratic-linear regulator problem and $\Phi(T,t), 0 \leq t \leq T$ satisfies a well known Riccati equation.

Proposition 4. *There exist a universal constant* $k_0 \leq \frac{1}{\sqrt{2}}(1 - e^{-\frac{1}{4}})^{-1}$. *such that*

$$k_0(E(T) - \bar{x}(T)) + \bar{x}(T) \supseteq L(T) \tag{37}$$

holds for all $T > 0$.

The proof is based on the concavity of Ψ and that of the function

$$\exp\left(-\frac{1}{2}\|x\|^2\right) \text{ over the ball } \|x\| \leq \frac{1}{\sqrt{2}}$$

see [23] for further details. In order to compare this approach of constructing ellipsoidal approximations with the approach of [26] (see also [16] and [12]), let suppose that $E = 0$, $C = 0$, and $B = 0$, i.e. that we are just approximating reachable sets; in the case, when

$$A = A^* \leq 0, x(0) = 0, T = \infty, \text{spect}(A) \subseteq [\lambda^0, \lambda^1],$$

the "locally optimal", (with respect to the volume of $E(t)$) method proposed there, see ch.5, provides internal ellipsoids, which vanish asymptotically for $T \to \infty$, whenever $\frac{\lambda^0}{\lambda_1} > 3$. Our ellipsoidal approximation $E(\infty)$ for $R(\infty)$ is such that — besides of (37) —

$$E(\infty) \subseteq R(\infty)\sqrt{\frac{\lambda_0}{2\lambda_1}}$$

holds, for all A, (i.e. λ_0, λ_1), see [23].

Acknowledgement: The author is happy to acknowledge the friendly support of colleagues at the Institute of Applied Mathematics of the University of Würzburg.

REFERENCES

[1] A. Antoulas. A new approach to synthesis problems in linear system theory. *IEEE Trans. Aut. Contr.*, 30:465–473, 1985.

[2] B. Barmish. Stabilization of uncertain systems via linear control. *IEEE Trans. Automat. Contr.*, 28:848–850, 1983.

[3] D. Bayer and J. Lagarias. The nonlinear geometry of linear programming I–III. Technical report, ATT Bell Lab., Murray Hill New Yersey, 1987.

[4] M. Dahleh and J. Pearson. Optimal rejection of bounded disturbances. *IEEE Trans. Aut. Contr.*, 33:722–731, 1988.

[5] H. Davenport. *On the integration of algebraic functions.* Springer Verlag, Berlin Heidelberg New York, 1981.

[6] H. Dym. *J-contractive matrix functions, reproducing kernel Hilbert spaces and interpolation.* CBMS Lecture Series. Amer. Math. Soc., 1988.

[7] P. Gill, W. Murray, and M. Saunders. A single phase dual barrier method for linear programming. Technical report, System Optim. Lab. Stanford Univ., 1988.

[8] G. Gu, P. Khargonekar, and E. Lee. Approximation of infinite dimensional systems. *IEEE Trans. Aut. Contr.*, 34:610–619, 1989.

[9] J. Helton. A numerical method for computing the structured singular value. *System and Control Letters*, pages 21–26, 1988.

[10] D. Hinrichsen and A. Pritchard. Stability radii of Hurwitz and Schur polynomials. Technical Report 185, Inst. f. Dynamische Systeme Univ. Bremen, 1988.

[11] F. Jarre. On the method of analytic centers when applied to convex quadratic programs. *to appear in Math. Progr.*, 1989.

[12] A. Kurshanskii and I. Valyi. Set valued solutions to control problems and their approximation. *Lect. N. in Contr. Inf. Sci.*, 111:775–785, 1988.

[13] H. Landau. Maximum entropy and the moment problem. *Bull. Amer. Math. Soc.*, pages 47–77, 1987.

[14] D. Mustafa. *Minimum Entropy H^∞ Control.* PhD thesis, St. John's College, Cambridge, 1989.

[15] B. Nagy and C. Foias. *Harmonic Analysis of Operators on Hilbert Space.* North Holland Amsterdam, 1970.

[16] J. Norton. Identification and application of bounded parameter models. *IEEE Trans. Aut. Contr.*, 32:497–507, 1987.

[17] I. Postlethwaite, S. Young, and D. Gu. Stable *h* user's guide. Technical Report 1687, Univ. of Oxford Dept. Eng. Sci., 1987.

[18] G. Sonnevend. Output regulation in partially observed linear systems under disturbances. *Lect. N. in Control and Inf.Sci.*, 6, 1977.

[19] G. Sonnevend. Existence and numerical computation of extremal invariant sets in linear differential games. *Lect. N. in Control and Inf.Sci.*, 22, 1980.

[20] G. Sonnevend. An analytic center for polyhedrons and its application for solving convex programming problems. *Lect. N. in Contr.Inf. Sci.*, 84:866–876, 1985.

[21] G. Sonnevend. Application of analytic centers. In P. von Dooren, editor, *Numerical Linear Algebra and Digital Signal Processing.* NATO ASI, REIDEL, 1988.

[22] G. Sonnevend. Existence and computation of solutions for the twodimensional moment problem. *Lect. N. in Contr. Inf. Sci.*, 121:469–478, 1988.

[23] G. Sonnevend. An analytic center for convex bodies and its application for ellipsoidal approximation in systems with bounded control. *to appear in Lect. N. in Contr. Inf. Sci.*, 1989.

[24] G. Sonnevend and J. Stoer. Global ellipsoidal approximations and homotopy methods for solving convex, analytic programs. *Applied Mathematics and Optimization*, 1989.

[25] G. Sonnevend, J. Stoer, and G. Zhao. On the complexity of linear extrapolation algorithms for solving linear programs. *to appear in Oper. Res. Verfahren (Methods of Oper. Research)*, 1989.

[26] F. Tchernoussko. *State estimation for dynamical systems.* Nauka (Moscow), 1988.

[27] M. Todd. Recent developments and new directions in linear programming. Technical Report 827, College of Engineering Cornell Univ., 1988.

György Sonnevend
(on leave from Inst. Math. Univ.Budapest, HUNGARY)
(Institut für Angewandte Mathematik
Universität Würzburg,
Am Hubland
D-8700 Würzburg
F.R.G.
ANGM048@DWUUNI21.BITNET

Constant McMillan Degree and the Continuous Stabilization of Families of Transfer Matrices

Eduardo D. Sontag*

This note deals with stabilization of parametric families using transfer matrices which are themselves parameterized in the same manner, a problem which is of some interest in the context of indirect adaptive control. It is shown that constancy of the McMillan degree is a necessary condition for stabilizability with arbitrary convergence rates.

1 Introduction

In indirect parameter adaptive control, one updates controller coefficients as new estimates of the plant are obtained. It is often of interest in that context to know if it is possible to design controllers that depend explicitly on plant parameters. For instance, if this dependence is polynomial or rational, the update itself consists simply of an evaluation, with no further computation. In addition to computational considerations, it is also of interest to know if one can design at least continuously on the parameters.

From the work of Delchamps ([Del]), we know that indeed continuous, and even analytic, dependence of controllers on parameters is possible. In [So1], combining ideas of Delchamps together with a result given in that reference, we showed that it is even possible to obtain rational or polynomial dependency. The basic assumption in this kind of result is always that the McMillan degree of the plant does not vary over the parameter space. As

*Research supported in part by US Air Force Grant AFOSR-88-0235

far as we know, the question of whether this hypothesis is really necessary was not studied before.

In this note we show that the constant degree condition is necessary if one wishes to obtain stabilization with arbitrary rates of convergence. We also make some preliminary remarks regarding just stabilization (without the arbitrary rate requirement).

2 Definitions and Statement of Main Result

Let Λ be any connected topological space. A *family of (p by m) transfer matrices over* Λ is a parameterized p by m matrix $\{\mathcal{W}^\lambda(s)\}$ all whose entries are rational functions of s whose coefficients depend continuously on $\lambda \in \Lambda$, so that for each λ every entry is strictly proper; more precisely, there is a representation

$$\{\mathcal{W}^\lambda(s)\} = \left\{\left(\frac{a_\lambda^{ij}(s)}{b_\lambda^{ij}(s)}\right)\right\}$$

where for each i, j, $a_\lambda^{ij}(s)$ and $b_\lambda^{ij}(s)$ are polynomials in s whose coefficients are continuous functions of λ,

$$\deg a_\lambda^{ij}(s) < \deg b_\lambda^{ij}(s) \quad \forall i, j$$

and the leading coefficient of $b_\lambda^{ij}(s)$ is independent of λ.

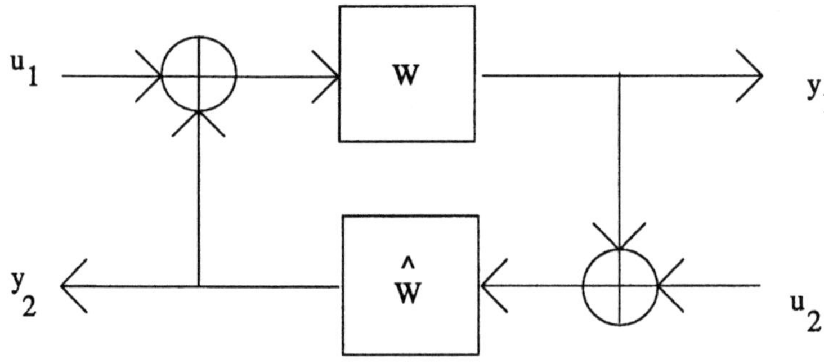

Figure 1. Interconnection of systems

Let $\{\mathcal{W}^\lambda(s)\}$ be a p by m family and let $\{\widehat{\mathcal{W}}^\lambda(s)\}$ be an m by p family; their *interconnection* is by definition the $p+m$ by $p+m$ family

$$\{\mathcal{W}^\lambda(s)\}.\{\widehat{\mathcal{W}}^\lambda(s)\} := \begin{pmatrix} I & -\mathcal{W}^\lambda(s) \\ -\widehat{\mathcal{W}}^\lambda(s) & I \end{pmatrix}^{-1} \begin{pmatrix} \mathcal{W}^\lambda(s) & 0 \\ 0 & \widehat{\mathcal{W}}^\lambda(s) \end{pmatrix}$$

(this is well-defined since the first matrix is invertible as a rational matrix, by the strict properness condition). This definition corresponds to the i/o behavior of the additive feedback connection in Figure 1.

The above and all other definitions for families apply also to single transfer matrices, simply by considering a one-element Λ.

A *state space system* (of dimension n) is a triple of matrices $\Sigma = (A, B, C)$, where $A \in \mathbb{R}^{n \times n}, B \in \mathbb{R}^{n \times m}, C \in \mathbb{R}^{p \times n}$. We write

$$W[\Sigma] := C(sI - A)^{-1}B$$

for its transfer matrix, and say that Σ *realizes* $W[\Sigma]$. The *McMillan degree* $\deg W$ of a transfer matrix $W(s)$ is the smallest possible dimension of a state space realization of it.

For a family $\{\mathcal{W}^\lambda(s)\}$ the degree is a function $\delta(\lambda)$; since this degree can be computed from the nonvanishing of determinants in a Hankel matrix, δ is lower semicontinuous; that is, if $\delta(\lambda_0) = \delta_0$ then in a neighborhood of λ_0 also $\delta(\lambda) \geq \delta_0$.

For each family $\{\mathcal{W}^\lambda(s)\}$ let δ_0 be the maximum possible value of $\delta(\lambda)$, and let Λ_0 be the set where this value is achieved. The requirement that the degree of the family be constant is equivalent to asking that $\Lambda_0 = \Lambda$; since Λ is connected and Λ_0 is open (because of semicontinuity), this is equivalent to Λ_0 being closed.

2.1 Parametric Stabilization

For any fixed $\gamma < 0$, we shall say that a family of transfer matrices is *γ-stable* if it can be written in such a way that for each λ and each i, j, all zeroes of $b_\lambda^{ij}(s)$ satisfy $\operatorname{Re} s \leq \gamma$.

A *γ-stabilizable* family is one for which there exists a $\{\widehat{\mathcal{W}}^\lambda\}$ so that the interconnection $\mathcal{W}^\lambda.\widehat{\mathcal{W}}^\lambda$ is *γ-stable*. A family is *stabilizable with arbitrary rates* if it is *γ-stabilizable* for all $\gamma < 0$.

The main result, to be proved in the next section, is as follows:

Theorem 1. *If a family is stabilizable with arbitrary rates then its McMillan degree is constant.*

Note that in the definition of stabilizable with arbitrary rates we allow the stabilizing family $\{\widehat{\mathcal{W}}^\lambda\}$ to be very different for each γ. It will follow from the Theorem, together with its already known converse ([Del]), that if a family is like this then it is also possible to build stabilizing families which depend continuously on γ in an appropriate sense. A family with constant degree is sometimes called "split". See [Kh] for properties of such families, and relations to the existence of coprime factorizations.

3 Proof of the Theorem

Let $\Sigma = (A, B, C)$ be an n-dimensional state space system, and pick any $\gamma < 0$. We shall say that Σ is *γ-stabilizable and detectable* if it holds that

$$\operatorname{rank}[sI - A, B] = n \quad \text{and} \quad \operatorname{rank}[sI - A', C'] = n$$

(prime indicates transpose) for each $s \in \mathbb{C}$ with $\operatorname{Re} s > \gamma$. A *canonical* Σ is one which is γ-stabilizable and detectable for all $\gamma < 0$, or equivalently, controllable and observable.

The main lemma that we need is as follows.

Lemma 2. *Let $\{\mathcal{W}^\lambda(s)\}$ be a family of transfer matrices. For each $\overline{\lambda}$ in the closure of Λ_0 there exists a realization $\Sigma^{\overline{\lambda}}$ of dimension δ_0 with the following property: if the family is γ-stabilizable for some γ then $\Sigma^{\overline{\lambda}}$ is γ-stabilizable and detectable.*

The Theorem follows from here, since arbitrary stabilizability then implies that for each such $\overline{\lambda}$ the realization $\Sigma^{\overline{\lambda}}$ is canonical, and hence of minimal dimension by standard realization theory, which in turn implies that $\overline{\lambda} \in \Lambda_0$, which is therefore closed.

In order to prove the Lemma, we first establish the following result, which will be used twice and can be interpreted in terms of the quotient topology on classes of systems under the natural action of $GL(n)$, and is closely related to a result given in [De2].

Proposition 3. *Let $\{\mathcal{W}^\lambda(s)\}$ be a family of transfer matrices. Assume that $\deg \mathcal{W}^{\lambda_j} \equiv q$ for some sequence $\{\lambda_j\}$ and some integer q, and that $\lambda_j \to \overline{\lambda}$ as $j \to \infty$. Then there exists some subsequence of the $\{\lambda_j\}$, which for notational simplicity we write again as $\{\lambda_j\}$, and for each element of this subsequence a system Σ^{λ_j} of dimension q, and there is a system $\Sigma^{\overline{\lambda}}$ of dim q, so that the following properties hold:*

1. $\Sigma^{\lambda_j} \to \Sigma^{\overline{\lambda}}$

2. $W[\Sigma^{\lambda_j}] = \mathcal{W}^{\lambda_j}$ *and* $W[\Sigma^{\overline{\lambda}}] = \mathcal{W}^{\overline{\lambda}}$

PROOF. First notice that, using a large enough ρ, it is possible to construct a continuously parameterized family of ρ-dimensional realizations $\{\Sigma(\lambda)\}$ of $\{\mathcal{W}^\lambda(s)\}$ each of whose members is observable. This follows easily from the "observable form" realization (see e.g. [Ba], formula 4.112, but instead of the "g" in the construction given there, use the product of all denominators). By continuity, and since $\{\lambda_j\}$ is convergent, the realizations Σ^{λ_j} are all in a bounded subset of $\mathbb{R}^{\rho\times\rho} \times \mathbb{R}^{\rho\times m} \times \mathbb{R}^{p\times\rho}$.

Minimal realizations of the transfer matrices in the family $\{\mathcal{W}^{\lambda}(s)\}$ can be obtained by restricting to the reachable sets of these observable realizations. Using orthogonal bases for the restricted maps, we conclude that the sequence $\mathcal{W}^{\lambda_j}$ can be realized by a bounded sequence of q-dimensional realizations. Thus there is a convergent subsequence of these realizations; call $\Sigma^{\overline{\lambda}}$ their limit. Continuity of transfer matrices on realizations gives the desired conclusions. Q.E.D.

Now we prove the Lemma. Let $\{\mathcal{W}^{\lambda}(s)\}$ and $\overline{\lambda}$ be as in the statement, and apply the Proposition, starting with any sequence of λ_j's in Λ_0 converging to $\overline{\lambda}$, to obtain (taking first a subsequence if necessary) a sequence $\Sigma^{\lambda_j} \to \Sigma^{\overline{\lambda}}$ of realizations of dimension δ_0 of $\mathcal{W}^{\lambda_j}, \mathcal{W}^{\overline{\lambda}}$ respectively. This gives the realization $\mathcal{W}^{\overline{\lambda}}$ in the Lemma. Note that the systems Σ^{λ_j} are all minimal, since they have dimension equal to the maximal degree δ_0, but that this is not necessary for the limit system $\Sigma^{\overline{\lambda}}$.

Assume that the family is γ-stabilizable for some γ. Let $\{\widehat{\mathcal{W}}^{\lambda}\}$ be as in the definition of γ-stabilizability. Since the McMillan degree of the family is bounded, may assume, taking if necessary a subsequence of $\{\lambda_j\}$, that $\deg \widehat{\mathcal{W}}^{\lambda_j} \equiv q$ for some q. By the Proposition applied to these, and again taking a subsequence if necessary, we obtain realizations

$$\widehat{\Sigma}^{\lambda_j} \to \widehat{\Sigma}^{\overline{\lambda}}$$

for the corresponding sequences of stabilizing transfer matrices. Note that each $\widehat{\Sigma}^{\lambda_j}$ is a minimal realization of its transfer matrix, hence canonical.

Given any two systems $\Sigma = (A, B, C)$ and $\widehat{\Sigma} = (\widehat{A}, \widehat{B}, \widehat{C})$ we can define the interconnection $\Sigma.\widehat{\Sigma}$ again according to Figure 1; this is the system

$$\begin{pmatrix} A & B\widehat{C} \\ \widehat{B}C & \widehat{A} \end{pmatrix}, \begin{pmatrix} B & 0 \\ 0 & \widehat{B} \end{pmatrix}, \begin{pmatrix} C & 0 \\ 0 & \widehat{C} \end{pmatrix}$$

and it satisfies that

$$W[\Sigma.\widehat{\Sigma}] = W[\Sigma].W[\widehat{\Sigma}] \ .$$

In particular, we may consider the interconnections of the respective sequence elements, $\Sigma^{\lambda_j}.\widehat{\Sigma}^{\lambda_j}$. Note that

$$\Sigma^{\lambda_j}.\widehat{\Sigma}^{\lambda_j} \to \Sigma^{\overline{\lambda}}.\widehat{\Sigma}^{\overline{\lambda}} \ .$$

Further, the composite systems $\Sigma^{\lambda_j}.\widehat{\Sigma}^{\lambda_j}$ are all minimal realizations of the respective transfer matrices, since this is true for the component systems. Thus γ-stability of the interconnection of transfer matrices implies that the composite systems $\Sigma^{\lambda_j}.\widehat{\Sigma}^{\lambda_j}$ are internally γ-stable (eigenvalues with real part at most γ). By continuity of eigenvalues, the same is true for the limit composite system $\Sigma^{\overline{\lambda}}.\widehat{\Sigma}^{\overline{\lambda}}$. This implies that $\widehat{\Sigma}^{\overline{\lambda}}$ is γ-stabilizable and detectable. The proof of the Lemma, and therefore of the Theorem, is then completed. Q.E.D.

4 Comments

It would be desirable to have a characterization of the property of γ-stabilizability (with a fixed γ). When the parameterization is for instance polynomial or even analytic, over the parameter spaces $\Lambda = \mathbb{R}$ or $\Lambda = \mathbb{R}^2$, there are parameterized realizations which are generically minimal (this follows from the results in [RS]). An argument as the one given above then proves that if the family is γ-stabilizable there must be a parameterized family of realizations consisting entirely of γ-stabilizable and detectable systems. This condition is "best possible" in the sense that it is also sufficient, and is related to stable coprime factorizations in the sense of [KS]. But the restriction to one- or two-parameter families is too strong, and much work needs to be done here.

REFERENCES

[Ba] Barnett, S., *Polynomials and Linear Control Systems*, Dekker, NY, 1983.

[De1] Delchamps, D.F., "Analytic stabilization and the algebraic Ricatti equation," *Proc. IEEE Conf. Dec. and Control*, 1983, pp. 1396–1401.

[De2] Delchamps, D.F., "Multivariable ARMAX identification: Some geometric questions," *Proc.17th Annual Conference on Information Sciences and Systems*, Johns Hopkins University, Baltimore, 1983, pp. 287–292.

[Kh] Khargonekar, P.P., "On matrix fraction representations for linear systems over commutative rings," *SIAM J. Control and Opt.* **20**(1982): 172–197.

[KS] Khargonekar,P.P. and E.D.Sontag, "On the relation between stable matrix fraction decompositions and regulable realizations of systems over rings," *IEEE Trans.Autom. Control* **27**(1982):627–638.

[RS] Rouchaleau,Y. and E.D.Sontag, "On the existence of minimal realizations of linear dynamical systems over Noetherian integral domains," *J. Comp. Syst. Sci.* **18**(1979):65–75.

[So1] Sontag, E.D., "Parametric stabilization is easy," *Systems and Control Letters* **4**(1984): 181–188.

[So2] Sontag, E.D., "An introduction to the stabilization problem for parametrized families of linear systems," in *Contemporary Mathematics, Vol.47, Linear Algebra and its Role in Systems Theory*, pp.369–400, AMS, Providence, RI, 1985.

Eduardo D. Sontag
SYCON - Rutgers Center for Systems and Control
Department of Mathematics, Rutgers University
New Brunswick, NJ 08903, USA
`sontag@fermat.rutgers.edu`
`sontag@pisces.bitnet`

Adaptive Control of Time-Varying and Nonlinear Systems Using a Multi-Model Approach

H. Unbehauen

This paper describes the application of a self-tuning controller to uncertain systems whose parameters are changing very rapidly, for example due to rapid changes of the operating conditions. Classical adaptive controllers are not appropriate for rapid parameter changes of the controlled process. The technique presented in this paper is based on the description of a nonlinear or time-varying system by a multi-model consisting of several linear submodels. Each submodel describes the system at one operating condition. The parameters of the submodels have to be estimated on-line. A statistical test method is applied for the fast detection of parameter changes. For the design of the self-tuning controller an LQG-approach has been applied.

1 Introduction

In classical adaptive control systems it is assumed that the plant parameters are changing only slowly with respect to the dominating time constant. In order to be more sensitive to fast parameter changes of the plant, classical adaptive control algorithms had been extended by the introduction of a weighting factor to provide forgetting of previous process data. Depending on the choice of the weighting factor the algorithm is more or less sensitive to parameter changes and also for stochastic disturbances. However, this situation is undesirable. To reduce this effect, adaptive forgetting factors had been introduced (Fortescue, 1981; Kulhavy, 1984; Bertin, 1986). The main idea of these methods is to choose a small forgetting factor if the parameters of the plant are changing, with the result that the covariance matrix assumes large values, and this will restart the parameter estimation. Another method had been proposed by Anderson (1985) who uses a time-varying parameter vector to describe the uncertain system. For the parameter estimation problem he applies the celebrated Kalman filter theory using a weighting sum of Gaussian distributed parameter vectors to describe a plant with abruptly changing parameters. The drawback of this technique, compared with simple parameter estimation, is the increased calculation time.

In this paper a new algorithm for systems with uncertain rapidly occurring parameter changes is presented. The basic idea of this algorithm is that parameter changes often are due to changing operating conditions of the controlled process. Therefore the process is described by a multi-model consisting of several submodels. (Diekmann and Unbehauen, 1985; Unbehauen and Jedner, 1986). Each submodel describes the system for one operating condition. These different operating conditions are detected and are defined by a statistical test procedure (Jedner and Unbehauen, 1988; Jedner, 1988). If a correct operating condition is determined and the corresponding actual parameter vector is estimated, the self-tuning controller will be calculated by a LQG-approach. The estimated parameters for every operating condition are stored such that they can be used again if the same operating condition occurs again. Due to the possibility of detecting previously defined and stored parameter sets for the multi-model the present "intelligent" control algorithm needs no adaptation if the process had already been working in a previously defined and estimated operating condition. Therefore, a new adaptation is necessary only if the system had never experienced the present operating condition before.

2 Identification of the multi-model structure

It is assumed that the process has L different operating conditions which are described by a multi-model consisting of L linear time-invariant submodels (Fig. 1). Each of these subsystems is characterized by an ARMAX-model (Unbehauen, 1988)

$$A_i(z^{-1})Y(z) = B_i(z^{-1})U(z) + C_i(z^{-1})E_i(z) \tag{1}$$

with the parameter vector

$$\mathbf{p}_i = [a_{i,1} \dots a_{i,n_a} \vdots b_{i,1} \dots b_{i,n_b} \vdots c_{i,1} \dots c_{i,n_c}]^T \tag{2}$$

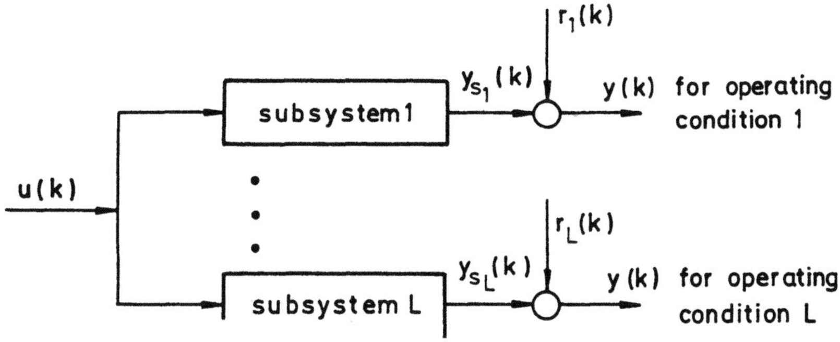

Figure 1. Description of the uncertain system by a multi-model

and the measurement vector

$$\mathbf{m}^T(k) =$$
$$[-y(k-1)\cdots-y(k-n_a)\vdots u(k-1)\cdots u(k-n_b)\vdots e_i(k-1)\cdots-e_i(k-n_c)] \tag{3}$$

for $i = 1, 2, \ldots, L$. Eqs.(1) to (3) have the corresponding representation in the time domain

$$y(k) = \mathbf{m}^T(k)p_i + e_i(k), \qquad i = 1, 2, \ldots, L, \tag{4}$$

where $y(k)$ represents the system output and $e_i(k)$ represents the equation error for operating conditions i. Let us assume for this section that the operating condition of the plant at time k would be already known and is denoted by $i = I^*(k)$ then the system can be described by the parameter vector

$$\mathbf{p}(k) = \mathbf{p}_i(k) \text{ for } i = I^*(k), \tag{5}$$

where $I^*(k)$ can be considered as an information variable. The procedure for determining $I^*(k)$ will be discussed in the next section. The aim of the identification of the multi-model is to find for all $i = 1, 2, \ldots, L$ the best submodel in the sense of the quadratic performance index

$$I(\mathbf{p}_i) = \sum_{k=1}^{N} \left[y(k) - \mathbf{m}^T(k)\mathbf{p}_i\right]^2 = \sum_{k=1}^{N} e_i^2(k) \stackrel{!}{=} \text{Min.} \tag{6}$$

This procedure is allowed, because for any $i = \ell(\ell \in \{1, 2, \ldots, L\})$ the minimum of $I(\mathbf{p}_\ell)$ depends only on $\mathbf{p}_\ell$ and is independent of the remaining parameter vectors $\mathbf{p}_i$, $i \neq \ell$ at all the other operating conditions. Therefore the optimal multi-model is obtained by minimizing the overall quadratic performance index

$$I(\mathbf{p}_1, \mathbf{p}_2, \ldots, \mathbf{p}_L) = \sum_{i=1}^{L} \sum_{k \in B_i} e_i^2(k) \stackrel{!}{=} \text{Min}, \tag{7}$$

where B_i $(i = 1, 2, \ldots, L)$ characterizes a set of time instants during which the process runs at the operating condition i. The advantage of this representation is that the equation error

$$e(k, \mathbf{p}_i) \equiv e(k) \equiv e_i(k) = y(k) - \mathbf{m}^T(k)\mathbf{p}_i \tag{8}$$

depends only on one parameter vector $\mathbf{p}_i$ at each time instant where i is the operating condition at time k. Thus the overall minimum of Eq.(7) is obtained by separate minimization of the performance indices of all L subsystems according to Eq.(6). Introducing the weighting factor

$$\mu_i(k) = \begin{cases} 1 & \text{if } i = I^*(k) \\ 0 & \text{else} \end{cases} \tag{9}$$

Eq.(6) can be written in the form

$$I(\mathbf{p}_i) = \sum_{k=1}^{N} \mu_i(k) e^2(k, \mathbf{p}_i). \tag{10}$$

The minimum of Eq.(10) can be found, e.g., by the well known consistent recursive least-squares (RLS) algorithm (Unbehauen, 1988). Because of the special choice of the weighting factors $i(k)$, it is not necessary to use simultaneously L algorithms at each sampling time. Regarding the special case of $i(k) = 0$, it can be shown that the covariance matrix within the RLS-estimation and the parameter vector do not change their values in this case. Therefore, the calculation time of the parameter procedure is not longer than for time-invariant systems. As the final result the following RLS-procedure is directly obtained to get the estimated parameter vectors of the multi-model:

$$\hat{\mathbf{p}}_i(k+1) = \hat{\mathbf{p}}_i(k) + \mathbf{q}_i(k+1)\hat{e}(k+1)\delta[i - I^*(k+1)], \tag{11a}$$

$$\mathbf{q}_i(k+1) = \mathbf{P}_i(k)\mathbf{m}(k+1)[1 + \mathbf{m}^T(k+1)\mathbf{P}_i(k)\mathbf{m}(k+1)]^{-1}, \tag{11b}$$

$$\mathbf{P}_i(k+1) = \mathbf{P}_i(k) - \left\{ \frac{\mathbf{P}_i(k)\mathbf{m}(k+1)\mathbf{m}^T(k+1)\mathbf{P}_i(k)}{1 + \mathbf{m}^T(k+1)\mathbf{P}_i(k)\mathbf{m}(k+1)} \right\} \delta[i - I^*(k+1)] \tag{11c}$$

$$\hat{e}(k+1) = y(k+1) - \mathbf{m}^T(k+1)\hat{\mathbf{p}}_i(k), \quad i = I^*(k+1) \tag{11d}$$

for $i = 1, 2, \ldots, L$. The value of the information variable $I^*(k+1)$ is i if the system is at sampling point $k+1$ in the operating condition i. The discrete δ-distribution in Eqs.(11a) and (11c) is defined through

$$\delta(\nu) = \begin{cases} 1 & \text{if } \nu = 0 \\ 0 & \text{else.} \end{cases} \tag{12}$$

The task of the information variable $I^*(k+1)$ is to inform the estimating algorithm, Eq.(11), at each sampling time about the operating condition.

The above described recursive algorithm can easily be initialized by

the parameter vectors

$$\mathbf{p}_i(0) = 0$$

and the covariance matrices

$$\mathbf{P}_I(0) = 10^3\mathbf{I},$$

where $\mathbf{I}$ is the unity matrix. An other possibility for the selection of $\mathbf{P}_I(0)$ would be to apply a weighted covariance matrix with a constant trace algorithm.

3 Detecting procedure for the operating condition

For the moment it is assumed that the error $e_i(k)$ in Eq.(4) for all L operating conditions would be Gaussian-distributed and would have the known variance

$$[\sigma_i]^2 = \text{Var}\{\hat{e}_i\} \quad i = 1, 2, \ldots, L. \tag{13}$$

The index i indicates that the variance of each operating condition may have different values. The value of the variance of the error signal according to Eq.(13) can be used for detecting changes in the operating conditions. This detection procedure delivers the information variable I^* which is needed in the estimation equations (11a,c,d).

In the following a procedure is derived which needs no measurement of an external signal for defining the information variable. This procedure uses a statistical test which takes into account the fact that the variance of the predicted equation error $\hat{e}(k)$ will be greater than the nominal value if the wrong subsystem is used for calculating the prediction error. For this purpose the test variable

$$t_i(k, M) = \frac{\sum_{m=1}^{M} e_i^2(k-1-m)}{[\sigma_i]^2} \quad i = 1, 2, \ldots, L \tag{14}$$

is introduced, where M is a freely selectable number $M \in N$. Under the condition that the equation error $e_i(k)$ is Gaussian-distributed and the system has been under operation condition i during the time instants $k, k+1, \ldots, k-M+1$, then the test variable $t_i(k,M)$ is Chi-square distributed.

To test whether the variance has changed a statistical hyperthesis test is applied.The hyperthesis H_0: "operating condition has not changed", is tested. This hyperthesis is rejected if the test variable $t_i(k,M)$ is greater than a critical value t_{crit}.The values of test are given in tables for several error probabilities a and for given M (Fisher, 1963). The selection of the value of a influences the probability of detecting a change of the operating condition. However, the value of a does not indicate the correctness of H_0. Therefore it is recommended to define a corresponding alternative hyperthesis H : "operating condition has changed such that $t_i(k,M)$ takes on its smallest value" (Jedner and Unbehauen, 1988).

The selection of M also influences the test variable $t_i(k,M)$ heavily. By increasing M the detection probability gets greater. But the disadvantage of a great value of M is a great dead time between the changing operating condition and its recognition. Therefore, a fixed M is not used for this test. The procedure starts with $M = 1$ values of the equation error. If H_0 is accepted and H is rejected, then the test is already finished, else the test is repeated with $M+1$ values of the equation error signal, until a decision can be made. It is obvious that thus M becomes dependent on the test variable $t_i(k,M)$. As $t_i(k,M)$ is a stochastical variable, M also becomes a stochastic variable. This M also depends on the signal/noise ratio of the measuring signals as well as on the selection of the submodels. In order not to be too sensitive to equation errors not being Gaussian-distributed, a minimal number M_{Min} of errors is used to calculate the test variable.

For the calculation of Eq.(14) it was above assumed that the variance $[\sigma_i]^2$ is already known. However, the detection algorithm can be extended by the recursive calculation of the variance in the form

$$[\sigma(k+1)]^2 = \{\rho_\nu[\sigma(k)]^2\Sigma(k) + e^2(k+1)\}/\Sigma(k+1) \tag{15a}$$

using

$$\Sigma(k+1) = \rho_\nu(k)\Sigma(k) + 1 \tag{15b}$$

and the weighting factor $\rho_\nu(k)$, where $\rho_\nu(k) < 1$ but for $k \to \infty$ converges to 1.

Now assuming the plant was at operating condition $i = i_0 = I^*(k-1)$ at time $k-1$ the following procedure can be applied to calculate the information variable and thus the operating condition of the plant at time k:
Step 1: Calculate

$$t_i(k,M) \text{ for } i = 1,2,\ldots,L \text{ and } M = M_{\min}\ldots M_{\max}$$

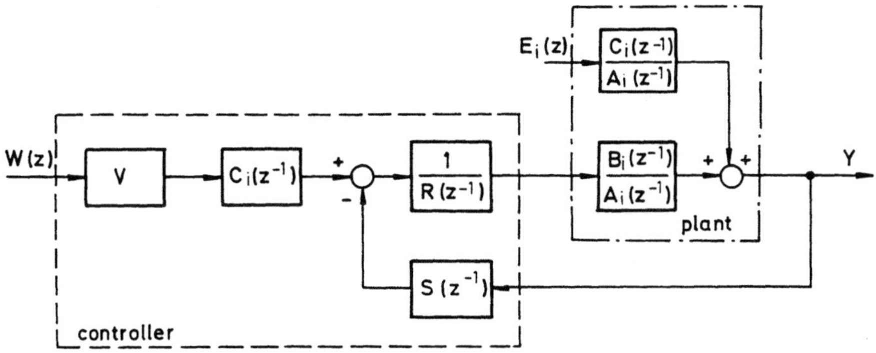

Figure 2. Basic control loop

Step 2: Determine

$$t_i(k) = \max_M \{t_i(k, M)/t_{\text{crit}}(M)\} \text{ for } i = 1, 2, \ldots, L$$

and

$$t_{i1}(k) = \min_i \{t_i(k, M)\} \text{ for } i_1 \neq i_0$$

Step 3: Find i_0 with

$$t_{i_0}(k) < 1$$

and

$$t_{i_1}(k) > 1 \text{ for } i \neq i_0$$

Step 4: If i_0 cannot be found then
a) Define a new operating condition if $t_i(k) > 1$ for $i = 1, 2 \ldots L$.
b) Cancel one parameter vector if more than one test value is less than 1.

4 The LQG-Self-tuning controller

Based on the above described estimation and detection algorithm any kind of self-tuning controller can be applied to the corresponding submodel described by Eq.(1). For the case presented here a LQG-self-tuning controller had been selected. The structure of the basic loop is given in Fig. 2, where $W(z)$ represents the reference signal and $S(z^{-1})$ and $R(z^{-1})$ are controller polynomials, whereas V is a constant factor to get a DC-amplification equal to one. The LQG-controller minimizes the criterion

$$I = E\left\{[w(k) - y(k)]^2 + \sigma_u[u(k) - \bar{u}]\right\} \stackrel{!}{=} \text{Min.} \tag{16}$$

with the weighting factor σ_u. The controller polynomials

$$R(z^{-1}) = r_0 + r_1 z^{-1} + \cdots + r_{n_r} z^{-n_r} \tag{17a}$$

$$S(z^{-1}) = s_0 + s_1 z^{-1} + \cdots + s_{n_s} z^{-n_s} \tag{17b}$$

can be calculated from (Åström, 1984)

$$P(z^{-1})P(z) = B_i(z^{-1})B_i(z) + \sigma_u A_i(z^{-1})A_i(z) \tag{18a}$$

$$C_i(z^{-1})P(z^{-1}) = A_i(z^{-1})R(z^{-1}) + B_i(z^{-1})S(z^{-1}), \tag{18b}$$

where $P(z^{-1})$ is a Hurwitz polynomial of order n_p, and $R(z^{-1})$ and $S(z^{-1})$ have the order

$$n_s = n_a - 1 \tag{19a}$$

$$n_r = \mathrm{Max}[n_b - 1, n_p + n_c - n_a]. \tag{19b}$$

For the spectral factorization of Eq.(18b) the algorithm of Kucera (1979) is used.

If steady-state deviations occur, e.g., due to deterministic disturbances, it would be easy to implement an integrator into the controller. In this case $A_i(z^{-1})$ in Fig. 2 has to be replaced by

$$A_i'(z^{-1}) = A_i(z^{-1})(1 - z^{-1}). \tag{20}$$

An other possibility would be to estimate the corresponding bias.

5 Application to a turbo-generator pilot plant

The considered pilot plant is a turbo-generator set which consists of an air pressure turbine driving a synchronous generator. Two output signals, the frequency f and the phase voltage u_A are the two controlled signals y_1 and y_2, whereas the two actuating signals u_1 and u_2 are given by the field current i_e of the synchronous generator and the pressure p of the turbine (see Fig. 3). The generator supplies a balanced load at unity power factor. The load can be switched between different operating points. The static characteristic between rotational speed n and field current i_e, presented in Fig. 4 shows the nonlinear behaviour of the plant. For the experiment described here the valve position was kept constant at 80%. Hence the plant was operating as a single input/single output system. The reference values for the field current i_e have been selected to be 0.2[A] in operating point 1 and 1.0[A] in operating point 2. In addition the reference signal was superimposed by a PRBS-signal of amplitude 0.1[A]. Because of the nonlinear characteristic the plant cannot be described by a single linear model in both operating points. Therefore a multi-model had been used

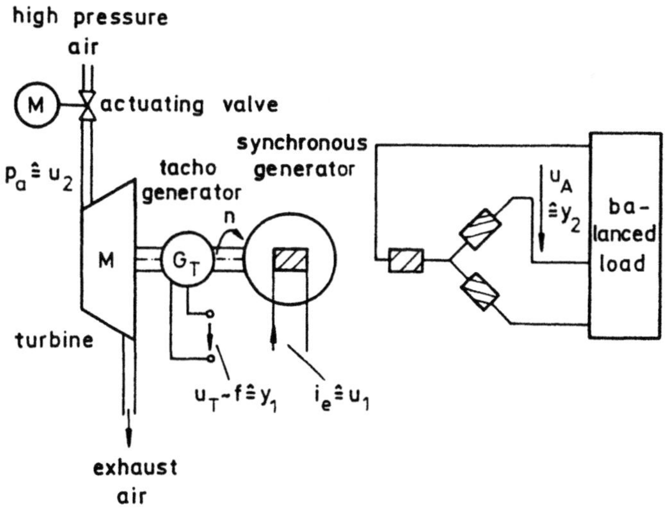

Figure 3. The pilot plant

by which both of the two operating points ($i = 1, 2$) are described by a separate linear submodel given by the difference equation of second order

$$y(k) = a_{1,i}y(k-1) + a_{2,i}y(k-2) + b_{1,i}u(k-1) + b_{2,i}u(k-2) + d_i. \quad (21)$$

The applied modified chi-square test is based on an error probability of $a = 0,1\%$ for the single tests with $M_{\min} = 5$ and $M_{\max} = 20$. At the beginning of the identification no knowledge either of the parameters of the submodels nor the number of linearization section had been available. The results of the on-line identification of the multimodel are given in Fig. 5 for two representative parameters. These results show that both linearization sections had been discovered by the statistical test procedure and hence a fast parameter adaptation is possible. It should be mentioned that changes in the operating conditions as well as in the parameters can only be detected if the plant is persistently excited.

Based on this identification of the operating condition and the corresponding parameter estimation an on-line realization of a LQG-self-tuning controller had been performed. (See Fig. 5.) This approach had been applied successfully also to other uncertain plants, working under different unknown operating conditions (Jedner 1988; Jedner and Unbehauen, 1989).

6 Conclusions

A new method for adaptive control of a class of uncertain systems, especially with uncertain time-varying or nonlinear characteristics had been

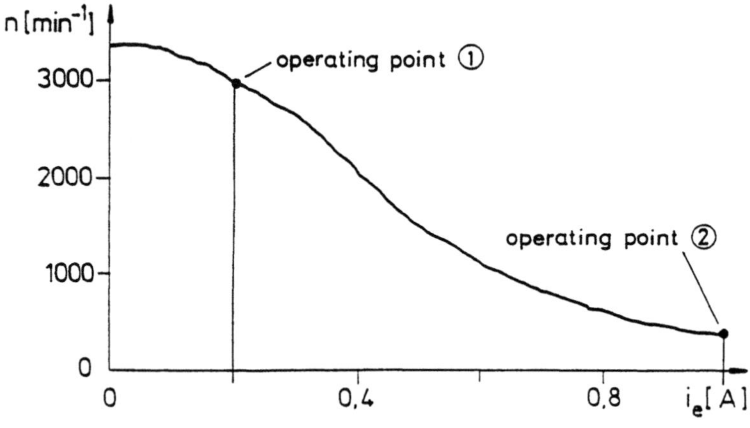

Figure 4. Nonlinear plant characteristic

proposed. The uncertain system was described by a linear multi-model. The parameter adaptation at a special operating point is necessary only if the system runs for the first time under the corresponding operating conditions. An algorithm for estimating the operating condition by a statistical test procedure was presented. The resulting LQG-self-tuning control algorithm had been applied to various uncertain plants with rapidly changing parameters and provided a considerably shorter adaptation time than the classical algorithms using a forgetting factor, which had been usually used in these situations.

REFERENCES

Anderson, P. (1985). Adaptive forgetting in recursive identification through multiple models. International Journal of Control 42, pp. 1175–1194.

Åström, K. and B. Wittenmark (1984). Computer controlled systems. Prentice-Hall, Englewood Cliffs.

Bertin, D. (1986). Tracking of nonstationary systems by means of different prediction-error directional forgetting techniques. Proceed. IFAC Workshop on Adaptive systems in Control and Signal Processing, Lund, Sweden pp. 91–96.

Diekmann K. and H. Unbehauen (1985). On-line parameter estimation in a class of nonlinear systems via modified least-squares. 7th IFAC-symposium on 'Identification and systems Parameter Estimation' York, pp. 149–153.

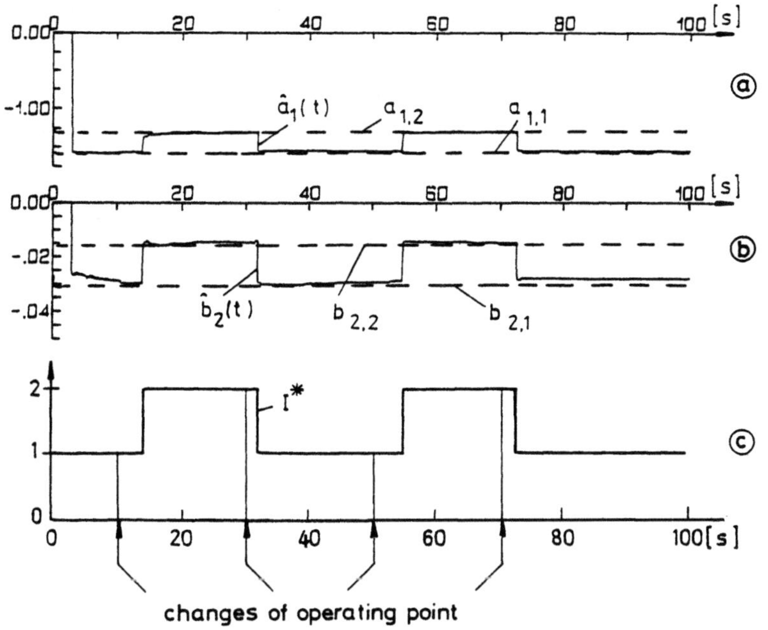

Figure 5. Estimated parameters $\hat{a}_1(t)$ and $\hat{b}_2(t)$ and detection of changes of the operating point

Fortescue T., L. Kershenbaum and B. Ydstie (1981). Implementation of self-tuning regulators with variable forgetting factors. Automatica 17 pp. 831–835.

Jedner U. (1988). Eine adaptive Regelstrategie für zeitvariante und nichtlineare Systeme. Dissertation Ruhr-Universität Bochum. Fortschritt-Berichte VDI-Reihe 8 No. 167 VDI-Verlag Düsseldorf.

Jedner, U. and H. Unbehauen (1988). Erkennung von Betriebszuständen einer Anlage mit Hilfe eines statistischen Testverfahrens. Automatisierungstechnik at 36, pp. 289–294.

Jedner U. and H. Unbehauen (1989). Intelligent adaptive control for a class of time-varying systems. In Borne P. (Ed.): Computing and computers for control systems. I. C. Baltzer AG, Scientific Publishing Co. (IMACs), pp. 123–126.

Kucera V. (1979). Discrete linear control. John Wiley & Sons Chichester, New York, Brisbane, Toronto.

Kulhavy, R. and M. Karny (1984). Tracking of slowly varying parameters by directional forgetting. Proc. 9th IFAC World Congress Budapest, Vol. 10, pp. 79–83.

Unbehauen H. and U. Jedner (1986). Multimodellansatz zur Identifikation von Regelstrecken mit schnellen Parameteränderungen. Zeitschrift für messen steuern, regeln (msr) 29, pp. 218–220.

Unbehauen, H. (1988). Regelungstechnik III (3. Auflage), Vieweg-Verlag Wiesbaden/Braunschweig.

H. Unbehauen
Ruhr-University Bochum
Dept. of Electrical Engineering
D-4630 Bochum 1, F.R.G.

Robust Stabilization and Disturbance Rejection for Uncertain Systems by Decentralized Control*

R.J. Veillette J.V. Medanić W.R. Perkins

This paper develops a decentralized control design which provides robust H_∞ disturbance rejection for a plant with structured uncertainty in a bounded admissible set. The design consists of an observer in each control channel, which includes estimates of the controls generated in the other channels and of the worst disturbance as determined by a state-feedback H_∞ solution. The observer gains are computed from a positive-definite solution of a Riccati-like algebraic equation. A convexity property of a matrix Riccati function is used to compute for the closed-loop system an H_∞-norm bound smaller than the predetermined bound, and to find an enlarged admissible set of plant uncertainities.

1 Introduction

Many past results in decentralized control are concerned with conditions for existence of stabilizing decentralized control laws. Davison reviews some of these results in [1]. One important necessary condition, introduced in [2], for the existence of any stabilizing time-invariant decentralized control law is that the plant have no unstable "fixed modes." The concept of a (decentralized) fixed mode is an extension of the concept of an uncontrollable or

*This work was supported by the Flight Dynamics Laboratory, Air Force Wright Aeronautical Laboratory (AFSC), USAF; by the Joint Services Electronics Program under Grant N00014-90-J-1270; and by Sundstrand Corporation.

unobservable mode in the centralized case. An algebraic characterization of fixed modes is given in [3]. Other results, for example [4] and [5], give plant connectivity conditions which guarantee that the whole plant can be made controllable and observable through a single control channel by some feedback applied in the other channels. Still other results, for example [6], [7], [8], and [9], on systems comprising several interconnected subsystems, give qualitative, quantitative, or topological restrictions on subsystem interconnections which guarantee that independent control laws applied to the subsystems will have a desired effect (such as stabilization) on the overall system. A design method in [10] consists of a hierarchical scheme where local controllers are used to stabilize the individual subsystems, and a global control is applied to attempt to restrict subsystem interactions. Some design methods are sequential in nature, as in [11] and [12], so that the design of the ith controller takes into account the dynamics introduced by the first $i-1$ controllers.

This paper presents a robust decentralized control design which guarantees stability and a predetermined bound on the H_∞ norm of the closed-loop system. (The H_∞ norm of a stable transfer function matrix $T(s)$ is defined as

$$\|T\|_\infty = \sup_{\omega \in \mathbb{R}} \sigma_{\max}\{T(j\omega)\}$$

where $\sigma_{\max}$ denotes the largest singular value.) The design does not explicitly depend on the connectivity or structure of the plant, and the controllers for the various channels are computed all at once, rather than sequentially. The decentralized design is an extension of state-feedback and centralized observer-based H_∞ control designs given in [13].

Consider the plant described by

$$\dot{x} = Ax + \sum_{i=1}^{q} B_i u_i + G_0 w_0, \qquad (1.1a)$$

$$y_i = C_i x + w_i, \quad i \in \{1, 2, \ldots, q\}, \qquad (1.1b)$$

$$z = \begin{pmatrix} H_0 x \\ u_1 \\ \vdots \\ u_q \end{pmatrix}, \qquad (1.1c)$$

where the y_i's are locally measured outputs, z is a vector signal to be regulated, the w_i's are square-integrable disturbances, and the u_i's are local control inputs. For convenience, introduce the notation

$$\sum_{i=1}^{q} B_i u_i \equiv Bu,$$

$$\begin{bmatrix} y_1 \\ y_2 \\ \vdots \\ y_q \end{bmatrix} = \begin{bmatrix} C_1 \\ C_2 \\ \vdots \\ C_q \end{bmatrix} x + \begin{bmatrix} w_1 \\ w_2 \\ \vdots \\ w_q \end{bmatrix} \equiv Cx + w,$$

and

$$w_e = \begin{bmatrix} w_0 \\ w \end{bmatrix}.$$

The A matrix of the plant has structured uncertainty described by

$$A = A_0 + \sum_{i=1}^{k} G_i L_i H_i, \tag{1.2}$$

where A_0 is known, the G_i's and H_i's give the structure of the uncertainty, and each unknown constant matrix L_i lies in some bounded set.

Given a particular bound on the largest singular value of each L_i, a decentralized control law is derived for plant (1.1) which guarantees that the closed-loop system will be stable and that the closed-loop transfer function matrix from w_e to z will have a predetermined H_∞-norm bound. The decentralized control law approximates a given robust state-feedback control law by means of q separate observers. Each observer assumes a particular worst disturbance, and estimates the controls generated by the other controllers.

All the main results in this paper rely on the following lemma, which is proved in [14]. Throughout this paper, matrix inequalities are meant in the sign-definite sense.

Lemma 1. *Let $T(s) = H(sI - F)^{-1}G$, with (F, H) a detectable pair. If there exist a real matrix $X \geq 0$ and positive scalars γ and δ such that*

$$F^T X + XF + \frac{1}{\gamma^2} XGG^T X + \frac{1}{\delta^2} H^T H \leq 0,$$

then F is Hurwitz, and

$$\|T\|_\infty \leq \gamma\delta.$$

In the following two sections, Lemma 1 is the main tool for deriving decentralized observer-based control laws for plant (1.1): First, the observer structure of the control law is fixed. This structure is meant to approximate a state-feedback H_∞ control law computed from the solution of an algebraic Riccati equation (ARE). Then, the observer parameters are chosen so that the matrices (F_e, G_{0e}, H_{0e}) describing the closed loop system, and some matrix $X_e \geq 0$, satisfy the inequality

$$F_e^T X_e + X_e F_e + \frac{1}{\alpha^2} X_e G_{0e} G_{0e}^T X_e + H_{0e}^T H_{0e} \leq 0. \tag{1.3}$$

To ensure that (1.3) is satisfied, observer gains are computed from the positive-definite solution of a particular Riccati-like algebraic equation. Finally, the pair (F_e, H_{0e}) is shown to be detectable. Lemma 1 then guarantees that the closed loop system is stable, and that $T(s) = H_{0e}(sI - F_e)^{-1}G_{0e}$ satisfies $\|T\|_\infty \leq \alpha$.

To obtain a control law robust with respect to the plant uncertainty (1.2), the same basic design procedure is applied to an augmented plant described by the known triple (A_0, G, H), where $G = [G_0 G_1 \ldots G_k]$, $H^T = [H_1^T H_2^T \ldots H_k^T]$. The control law guarantees stability and the H_∞ performance bound α for the known augmented closed-loop system; hence, provided each unknown matrix L_i satisfies $(I - \alpha^2 L_i^T L_i) > 0$, it guarantees stability and H_∞ performance robustly when applied to the actual (uncertain) plant. This is proved in the following way: The design computed for the augmented plant guarantees the existence of a matrix $X_e \geq 0$ satisfying the inequality

$$F_{0e}^T X_e + X_e F_{0e} + \frac{1}{\alpha^2} X_e G_e G_e^T X_e + H_e^T H_e \leq 0, \tag{1.4}$$

where (F_{0e}, G_e, H_e) describes the known augmented closed-loop system. By rearranging terms of (1.4), it is shown that the inequality (1.3) is satisfied for any admissible plant uncertainty described by (1.2).

Section 2 presents the design for the nominal plant without uncertainty. Section 3 presents a design which accounts for the structured plant uncertainty (1.2). Section 4 discusses the Riccati-like design equation. Section 5 shows that the design of Section 3 provides robust stability and disturbance rejection over a set of uncertainties larger than that assumed in Section 3, and with a smaller H_∞-norm bound. Section 6 contains an example illustrating the results of the control designs of Section 2 applied to a 3rd-order plant.

2 Control of the Nominal Plant

Consider plant (1.1), with $A = A_0$ known. For convenience, let $S_i = B_i B_i^T$ for each $i \in \{1, 2, \ldots, q\}$, and $S = BB^T = S_1 + S_2 + \ldots + S_q$. The problem is to design q controllers, where the ith controller uses the local measurement y_i to generate the local control u_i for plant (1.1). The developed decentralized control law stabilizes the nominal plant and provides an H_∞-norm bound for the closed-loop transfer function matrix from w_e to z.

The controllers are based on observers which form estimates of the state x for feedback, and are intended to approximate the state-feedback control law

$$u = -B^T X_0 x, \tag{2.1}$$

where $X_0 \geq 0$ satisfies the ARE

$$A^T X_0 + X_0 A + \frac{1}{\alpha^2} X_0 G_0 G_0^T X_0 - X_0 S X_0 + H_0^T H_0 = 0. \tag{2.2}$$

This state-feedback control provides a closed-loop system with H_∞ norm less than or equal to α. (See [13].) As an approximation to (2.1), the decentralized control law is given by

$$u_i = -B_i^T X_0 \xi_i, \quad i \in \{1, 2, \dots, q\}, \tag{2.3}$$

where $X_0 \geq 0$ solves (2.2), and ξ_i is the estimate of the state x computed by the ith observer. The ith observer has the form

$$\dot{\xi}_i = A\xi_i + \sum_{j=1}^{q} B_j \hat{u}_j + G_0 \hat{w}_0 + K_i(y_i - C_i \xi_i), \tag{2.4}$$

and so must form estimates of the disturbance w_0 and of the controls u_j, $j \neq i$. In a game setting where the state-feedback control u plays against the disturbance w_0 and initial conditions $x(0)$, the worst disturbance is $w_0 = \frac{1}{\alpha^2} G_0^T X_0 x$. (See [15].) Therefore, the ith observer estimates this worst disturbance by

$$\hat{w}_0 = \frac{1}{\alpha^2} G_0^T X_0 \xi_i, \tag{2.5}$$

and estimates all other controls by

$$\hat{u}_j = -B_j^T X_0 \xi_i. \tag{2.6}$$

Of course, $\hat{u}_i = u_i$ is given by (2.3). With (2.5) and (2.6), the controllers become

$$\dot{\xi}_i = \left(A + \frac{1}{\alpha^2} G_0 G_0^T X_0 - S X_0 - K_i C_i \right) \xi_i + K_i y_i \tag{2.7a}$$

$$u_i = -B_i^T X_0 \xi_i, \quad i \in \{1, 2, \dots, q\}, \tag{2.7b}$$

with observer gains $K_i, i \in \{1, 2, \dots, q\}$, yet to be determined.

Applying the q control laws (2.7) to the nominal plant (1.1) gives a closed-loop system of order $(q+1)n$. Transforming coordinates such that the last qn state variables are the errors $e_i = \xi_i - x, \ i \in \{1, 2, \dots, q\}$, gives

$$\begin{aligned} \begin{pmatrix} \dot{x} \\ \dot{e} \end{pmatrix} &= \begin{pmatrix} A - SX_0 & -BB_c^T X_{0c} \\ \alpha^{-2} G_{0c} G_0^T X_0, & A_{0c} - K_c C_c \end{pmatrix} \begin{pmatrix} x \\ e \end{pmatrix} \\ &\quad + \begin{pmatrix} G_0 & 0 \\ -G_{0c} & K_c \end{pmatrix} \begin{pmatrix} w_0 \\ w \end{pmatrix} \qquad (2.8a) \\ &\equiv F_{0e} x_e + G_{0e} w_e, \end{aligned}$$

$$z = \begin{pmatrix} H_0 & 0 \\ -B^T X_0 & -B_c^T X_{0c} \end{pmatrix} \begin{pmatrix} x \\ e \end{pmatrix} \equiv H_{0e} x_e. \tag{2.8b}$$

Here,

$$A_{0c} = \begin{pmatrix} A_{0\alpha} + S_1 X_0 & S_2 X_0 & \dots & S_q X_0 \\ S_1 X_0 & A_{0\alpha} + S_2 X_0 & \dots & S_q X_0 \\ \vdots & \vdots & & \vdots \\ S_1 X_0 & S_2 X_0 & \dots & A_{0\alpha} + S_q X_0 \end{pmatrix},$$

$$\begin{aligned} A_{0\alpha} &= A + \tfrac{1}{\alpha^2} G_0 G_0^T X_0 - S X_0, \\ B_c &= \text{Diag}\,(B_1, B_2, \dots, B_q), \\ C_c &= \text{Diag}\,(C_1, C_2, \dots, C_q), \\ G_{0c}^T &= [G_0^T \; G_0^T \dots G_0^T], \\ K_c &= \text{Diag}(K_1, K_2, \dots, K_q), \\ X_{0c} &= \text{Diag}(X_0, X_0, \dots, X_0). \end{aligned}$$

Define also

$$S_{0c} = X_{0c} B_c B_c^T X_{0c}.$$

In view of Lemma 1, the existence of a $(q+1)n \times (q+1)n$ matrix $X_{0e} \geq 0$ satisfying

$$F_{0e}^T X_{0e} + X_{0e} F_{0e} + \frac{1}{\alpha^2} X_{0e} G_{0e} G_{0e}^T X_{0e} + H_{0e}^T H_{0e} = 0 \tag{2.9}$$

will guarantee stability and an H_∞-norm bound for the closed-loop system (2.8). Look for X_{0e} of the form

$$X_{0e} = \begin{pmatrix} X_0 & 0 \\ 0 & X_1 \end{pmatrix}, \tag{2.10}$$

where X_1 is a $qn \times qn$ positive-definite matrix, as yet unknown.

First, write

$$F_{0e}^T X_{0e} + X_{0e} F_{0e} + \frac{1}{\alpha^2} X_{0e} G_{0e} G_{0e}^T X_{0e} + H_{0e}^T H_{0e} = \begin{pmatrix} U_{11} & U_{12} \\ U_{12}^T & U_{22} \end{pmatrix}. \tag{2.11}$$

The off-diagonal block U_{12} is identically zero, and (2.2) gives also $U_{11} = 0$. Hence, independent of K_c and X_1, (2.11) becomes

$$F_{0e}^T X_{0e} + X_{0e} F_{0e} + \frac{1}{\alpha^2} X_{0e} G_{0e} G_{0e}^T X_{0e} + H_{0e}^T H_{0e} = \begin{pmatrix} 0 & 0 \\ 0 & U_{22} \end{pmatrix}.$$

Here,

$$\begin{aligned} U_{22} &= (A_{0c} - K_cC_c)^T X_1 + X_1(A_{0c} - K_cC_c) \\ &\quad + \tfrac{1}{\alpha^2} X_1 (G_{0c}G_{0c}^T + K_cK_c^T) X_1 + S_{0c} \\ &= \tfrac{1}{\alpha^2} X_1 \{ Y A_{0c}^T + A_{0c} Y + \tfrac{1}{\alpha^2} Y S_{0c} Y - Y C_c^T C_c Y \\ &\quad + G_{0c}G_{0c}^T + (K_c - YC_c^T)(K_c^T - C_cY) \} X_1, \end{aligned} \tag{2.12}$$

with $Y = \alpha^2 X_1^{-1}$. It is now shown that it is possible to pick X_1 (or, equivalently, Y) and K_c such that $U_{22} = 0$. It is logical in view of Lemma 1 to try to eliminate the last term in (2.12); however, this is not generally possible, since in the decentralized problem K_c must be block-diagonal. The most that can be done in this regard is to eliminate the $n \times n$ main-diagonal blocks of $K_c - YC_c^T$: Divide Y into $n \times n$ blocks, as

$$Y = \begin{pmatrix} Y_{11} & Y_{12} & \dots & Y_{1q} \\ Y_{21} & Y_{22} & \dots & Y_{2q} \\ \vdots & \vdots & & \vdots \\ Y_{q1} & Y_{q2} & \dots & Y_{qq} \end{pmatrix},$$

and set

$$K_i = Y_{ii} C_i^T. \tag{2.13}$$

Then, (2.12) becomes

$$\begin{aligned} U_{22} &= \tfrac{1}{\alpha^2} X_1 \{ Y A_{0c}^T + A_{0c} Y + \tfrac{1}{\alpha^2} Y S_{0c} Y - Y C_c^T C_c Y \\ &\quad + G_{0c}G_{0c}^T + (Y - Y_D) C_c^T C_c (Y - Y_D) \} X_1, \end{aligned} \tag{2.14}$$

where $Y_D = \text{Diag}\,(Y_{11}, Y_{22}, \dots, Y_{qq})$. Therefore, if $Y > 0$ satisfies the Riccati-like algebraic equation

$$\begin{aligned} & Y A_{0c}^T + A_{0c} Y + \tfrac{1}{\alpha^2} Y S_{0c} Y - Y C_c^T C_c Y + G_{0c}G_{0c}^T \\ & + (Y - Y_D) C_c^T C_c (Y - Y_D) = 0, \end{aligned} \tag{2.15}$$

then $U_{22} = 0$, and (2.9) is satisfied. Since $Y > 0$ is assumed, $X_{0e} \geq 0$ holds automatically.

By Lemma 1, F_{0e} is Hurwitz and $T(s) = H_{0e}(sI - F_{0e})^{-1} G_{0e}$ satisfies $||T||_\infty \leq \alpha$ provided (F_{0e}, H_{0e}) is a detectable pair. To see this is the case, assume (A, H_0) is a detectable pair and $A_{0\alpha} \equiv A + \frac{1}{\alpha^2} G_0 G_0^T X_0 - S X_0$ is Hurwitz. (This condition guarantees the Hamiltonian matrix associated with the ARE (2.2) has no $j\omega$-axis eigenvalues.) In addition, assume $A_{0\alpha}+SX_0$ has no jw-axis eigenvalues. Now suppose λ is an eigenvalue of F_{0e} corresponding with an unobservable mode of (F_{0e}, H_{0e}); that is, some $v = \binom{v_1}{v_2} \neq 0$ satisfies

$$F_{0e}v = \begin{pmatrix} A - BB^T X_0 & -BB_c^T X_{0c} \\ \alpha^{-2} G_{0c} G_0^T X_0 & A_{0c} - K_c C_c \end{pmatrix} \begin{pmatrix} v_1 \\ v_2 \end{pmatrix} = \lambda v \tag{2.16}$$

and

$$H_{0e}v = \begin{pmatrix} H_0 & 0 \\ -B^T X_0 & -B_c^T X_{0c} \end{pmatrix} \begin{pmatrix} v_1 \\ v_2 \end{pmatrix} = 0. \tag{2.17}$$

The proof now consists of showing that $Re(\lambda) < 0$.

The lower block of (2.17) and the upper block of (2.16) combine to give $Av_1 = \lambda v_1$, while the upper block of (2.17) gives $H_0 v_1 = 0$. Since (A, H_0) is assumed a detectable pair, this implies that either $Re(\lambda) < 0$ or $v_1 = 0$. If $v_1 = 0$, then the lower block of (2.16) gives

$$(A_{0c} - K_c C_c) v_2 = \lambda v_2. \tag{2.18}$$

The bracketed expression in (2.12) is equal to zero; so, write

$$(A_{0c} - K_c C_c) Y + Y (A_{0c} - K_c C_c)^T + \frac{1}{\alpha^2} Y S_{0c} Y + G_{0c} G_{0c}^T + K_c K_c^T = 0. \tag{2.19}$$

Let η^* be a left-eigenvector of $A_{0c} - K_c C_c$ corresponding to the eigenvalue λ. Multiply (2.19) on the left by η^* and on the right by η to obtain

$$2Re(\lambda)\eta^* Y \eta + \frac{1}{\alpha^2} \eta^* Y S_{0c} Y \eta + \eta^* G_{0c} G_{0c}^T \eta + \eta^* K_c K_c^T \eta = 0. \tag{2.20}$$

Since every other term in (2.20) is non-negative, $Re(\lambda)\eta^* Y \eta \leq 0$ with $Y > 0$ assumed; therefore, $Re(\lambda) \leq 0$. Showing that $Re(\lambda) \neq 0$ will now complete the argument.

If $Re(\lambda) = 0$, then every term in (2.20) must be zero; so, $\eta^* K_c = 0$. Then λ is an eigenvalue of A_{0c}. But a similarity transformation on A_{0c} reveals that it can have no imaginary eigenvalues. Let

$$M = \begin{pmatrix} I & & & \\ I & I & & \\ \vdots & \vdots & \ddots & \\ I & 0 & \dots & I \end{pmatrix};$$

then,

$$M^{-1} A_{0c} M = \begin{pmatrix} A_{0\alpha} + SX_0 & S_2 X_0 & \dots & S_q X_0 \\ & A_{0\alpha} & \dots & 0 \\ & & \ddots & \vdots \\ 0 & & & A_{0\alpha} \end{pmatrix},$$

where $A_{0\alpha}$ is assumed Hurwitz, and $A_{0\alpha} + SX_0$ is assumed to have no imaginary eigenvalues.

Hence, by Lemma 1, F_{0e} is Hurwitz and the closed-loop transfer function matrix $T(s) = H_{0e}(sI - F_{0e})^{-1}G_{0e}$ from w_e to z satisfies $||T||_\infty \leq \alpha$.

The following theorem summarizes the result:

Theorem 2. *Let (A, H_0) be a detectable pair and α be a positive scalar. Suppose $X_0 \geq 0$ satisfies (2.2), $A_{0\alpha}$ is Hurwitz, and $A_{0\alpha} + SX_0$ has no jw-axis eigenvalues. Let $Y > 0$ satisfy the Riccati-like algebraic equation (2.15). If the observer gains K_i, $i \in \{1, 2, \ldots, q\}$, are given by (2.13), then the feedback control laws (2.7) stabilize the nominal plant (1.1), and the closed-loop transfer function matrix*

$$T(s) = H_{0e}(sI - F_{0e})^{-1}G_{0e}$$

from w_e to z (with F_{0e}, G_{0e}, and H_{0e}, defined in (2.8)) satisfies

$$||T||_\infty \leq \alpha.$$

3 Robust control of the uncertain plant

Consider the plant (1.1), now including the structured plant uncertainty (1.2). The design developed in this section accounts for this uncertainty, and gives a robust decentralized control law for the plant. The control law is obtained by computing an H_∞ disturbance-rejection control for an auxiliary plant. To construct the auxiliary plant, assume the actual dynamic matrix A is the nominal A_0, and replace the actual disturbance and output matrices G_0 and H_0 by the augmented matrices G and H defined by

$$G = [G_0 \; G_1 \ldots G_k], \; H^T = [H_0^T \; H_1^T \ldots \; H_k^T]. \tag{3.1}$$

In short, the auxiliary disturbance-rejection problem is formed by replacing the triple (A, G_0, H_0) of the actual plant (1.1) by the triple (A_0, G, H) of the design plant. This construction is similar in spirit to one given in [16], Theorem 2.1, for robust state-feedback designs. However, in [16], only robust stability is considered, and so G_0 and H_0 are omitted, and H_∞-norm bounds are not derived.

The control law of Section 2 applied to the design plant (A_0, G, H) is now shown to guarantee robust stability and H_∞ disturbance rejection for the actual plant (A, G_0, H_0) for all constant plant uncertainties in a given bounded set. Let $X \geq 0$ satisfying

$$A^T X + XA + \frac{1}{\alpha^2} XGG^T X - XSX + H^T H = 0 \tag{3.2}$$

characterize the state-feedback solution to the auxiliary problem. Then, the control laws corresponding to the auxiliary disturbance-rejection problem are

$$\dot{\xi}_i = (A_0 + \frac{1}{\alpha^2} GG^T X - SX - K_i C_i)\xi_i + K_i y_i \tag{3.3a}$$

$$u_i = -B_i^T X \xi_i, \tag{3.3b}$$

for $i \in \{1, 2, \dots, q\}$. The gains K_i are given by

$$K_i = Y_{ii} C_i^T, \tag{3.4}$$

where

$$Y = \begin{pmatrix} Y_{11} & Y_{12} & \dots & Y_{1q} \\ Y_{21} & Y_{22} & \dots & Y_{2q} \\ \vdots & \vdots & & \vdots \\ Y_{q1} & Y_{q2} & \dots & Y_{qq} \end{pmatrix} > 0$$

satisfies the Riccati-like algebraic equation

$$\begin{aligned} &YA_c^T + A_c Y + \tfrac{1}{\alpha^2} Y S_c Y - Y C_c^T C_c Y + G_c G_c^T \\ &+(Y - Y_D) C_c^T C_c (Y - Y_D) = 0 \end{aligned} \tag{3.5}$$

with

$$A_c = \begin{pmatrix} A_\alpha + S_1 X & S_2 X & \dots & S_q X \\ S_1 X & A_\alpha + S_2 X & \dots & S_q X \\ \vdots & \vdots & & \vdots \\ S_1 X & S_2 X & \dots & A_\alpha + S_q X \end{pmatrix},$$

$$A_\alpha = A_0 + \frac{1}{\alpha^2} GG^T X - SX,$$

$$B_c = \mathrm{Diag}\,(B_1, B_2, \dots, B_q),$$

$$C_c = \mathrm{Diag}\,(C_1, C_2, \dots, C_q),$$

$$G_c = \begin{pmatrix} G \\ G \\ \vdots \\ G \end{pmatrix} = (G_{0c}\, G_{1c} \dots G_{kc}), \quad G_{ic} = \begin{pmatrix} G_i \\ G_i \\ \vdots \\ G_i \end{pmatrix} \quad (i \in \{0, 1, \dots, k\}),$$

$$K_c = \mathrm{Diag}\,(K_1, K_2, \dots, K_q),$$

$$X_c = \mathrm{Diag}\,(X, X, \dots, X),$$

$$S_c = X_c B_c B_c^T X_c,$$

$$Y_D = \mathrm{Diag}\,(Y_{11}, Y_{22}, \dots, Y_{qq}).$$

The following theorem states the result.

Theorem 3. *Let (A_0, H_0) be a detectable pair, α be a positive scalar, and G and H be defined by (3.1). Suppose $X \geq 0$ satisfies (3.2), A_α is Hurwitz, and $A_\alpha + SX$ has no jw-axis eigenvalues. Suppose $Y > 0$ satisfies the Riccati-like algebraic equation (3.5), and the gains K_i are given by (3.4). Then, for any structured plant uncertainty (1.2) satisfying*

$$\sigma_{\max}\{L_i\} < \frac{1}{\alpha}, \quad i \in \{1, 2, \ldots, k\}, \tag{3.6}$$

the feedback control laws (3.3) stabilize the plant (1.1), and give a closed-loop transfer function matrix $T(s) = H_{0e}(sI - F_e)^{-1}G_{0e}$ from w_e to z satisfying

$$||T||_\infty \leq \alpha. \tag{3.7}$$

PROOF. By the development in Section 2, under the conditions of the theorem statement, the $(q+1)n \times (q+1)n$ matrix

$$X_e = \begin{pmatrix} X & 0 \\ 0 & \alpha^2 Y^{-1} \end{pmatrix} \geq 0$$

satisfies

$$F_{0e}^T X_e + X_e F_{0e} + \frac{1}{\alpha^2} X_e G_e G_e^T X_e + H_e^T H_e = 0, \tag{3.8}$$

where $H_e(sI - F_{0e})^{-1}G_e$ is the auxiliary closed-loop transfer function matrix, and (F_{0e}, H_e) is a detectable pair.

It is not the auxiliary transfer function matrix $H_e(sI - F_{0e})^{-1}G_e$, but the actual transfer function matrix $T(s) = H_{0e}(sI - F_e)^{-1}G_{0e}$, which is relevant. The following equations relate the actual system matrix F_e to the nominal system matrix F_{0e}, and the augmented disturbance and output matrices G_e and H_e to the actual disturbance and output matrices G_{0e} and H_{0e}:

$$F_e = F_{0e} + \sum_{i=1}^{k} \begin{pmatrix} G_i L_i H_i & 0 \\ -G_{ci} L_i H_i & 0 \end{pmatrix} \equiv F_{0e} + \sum_{i=1}^{k} G_{ei} L_i H_{ei}, \tag{3.9a}$$

$$\begin{aligned} G_e G_e^T &= G_{0e} G_{0e}^T + \textstyle\sum_{i=1}^{k} \begin{pmatrix} G_i \\ -G_{ci} \end{pmatrix} (G_i^T \; - G_{ci}^T) \\ &\equiv G_{0e} G_{0e}^T + \textstyle\sum_{i=1}^{k} G_{ei} G_{ei}^T, \end{aligned} \tag{3.9b}$$

and

$$H_e^T H_e = H_{0e}^T H_{0e} + \sum_{i=1}^{k} \begin{pmatrix} H_i^T H_i & 0 \\ 0 & 0 \end{pmatrix} \equiv H_{0e}^T H_{0e} + \sum_{i=1}^{k} H_{ei}^T H_{ei}. \tag{3.9c}$$

The definitions of G_{ei} and H_{ei} are clear from (3.9). Now, from (3.8) and (3.9),

$$\begin{aligned} &F_e^T X_e + X_e F_e + \tfrac{1}{\alpha^2} X_e G_{0e} G_{0e}^T X_e + H_{0e}^T H_{0e} \\ &= -\textstyle\sum_{i=1}^k \{H_{ei}^T (I - \alpha^2 L_i^T L_i) H_{ei} \\ &\quad + \left(\alpha H_{ei}^T L_i^T - \tfrac{1}{\alpha} X_e G_{ei}\right) \left(\alpha L_i H_{ei} - \tfrac{1}{\alpha} G_{ei}^T X_e\right)\}. \end{aligned}$$

By condition (3.6), $(I - \alpha^2 L_i^T L_i) > 0$; therefore,

$$\begin{aligned} &F_e^T X_e + X_e F_e + \tfrac{1}{\alpha^2} X_e G_{0e} G_{0e}^T X_e + H_{0e}^T H_{0e} \\ &\leq -\textstyle\sum_{i=1}^k H_{ei}^T (I - \alpha^2 L_i^T L_i) H_{ei} \leq 0. \end{aligned} \tag{3.10}$$

By Lemma 1, the theorem is proved once (F_e, H_{0e}) is shown to be a detectable pair. Suppose λ is an eigenvalue of F_e associated with an unobservable mode of (F_e, H_{0e}); that is, some $v \neq 0$ satisfies

$$F_e v = \left(F_{0e} + \sum_{i=1}^k G_{ei} L_i H_{ei}\right) v = \lambda v \tag{3.11}$$

$$H_{0e}^T H_{0e} v = \left(H_e^T H_e - \sum_{i=1}^k H_{ei}^T H_{ei}\right) v = 0 \tag{3.12}$$

Multiply (3.10) on the left by v^* and on the right by v to obtain

$$\begin{aligned} &2Re(\lambda) v^* X_e v + \tfrac{1}{\alpha^2} v^* X_e G_{0e} G_{0e}^T X_e v \\ &+ \textstyle\sum_{i=1}^k v^* H_{ei}^T (I - \alpha^2 L_i^T L_i) H_{ei} v \leq 0. \end{aligned} \tag{3.13}$$

All but the first term of (3.13) must be non-negative; hence,

$$Re(\lambda) v^* X_e v \leq 0. \tag{3.14}$$

If inequality holds in (3.14), then $v^* X_e v > 0$ and $Re(\lambda) < 0$. If equality holds, then all terms in (3.13) must be zero. Consequently, $H_{ei} v = 0$, since strict inequality in (3.6) implies $I - \alpha^2 L_i^T L_i$ is nonsingular. Therefore, (3.11) and (3.12) give $F_{0e} v = \lambda v$ and $H_e v = 0$. Since (F_{0e}, H_e) is a detectable pair, $Re(\lambda) < 0$. Therefore, (F_e, H_{0e}) is a detectable pair, and the theorem is proved. Q.E.D.

4 The Riccati-like algebraic equation

The Riccati-like algebraic equation (3.5) can be solved by the following iterative procedure:

(i) Let $Y^{(0)} = 0$.

(ii) For $i \in \{1, 2, 3, \ldots\}$, compute $Y^{(i)}$ as the positive-definite solution of the equation

$$\begin{aligned} &Y^{(i)}A_c^T + A_cY^{(i)} + \tfrac{1}{\alpha^2}Y^{(i)}S_cY^{(i)} - Y^{(i)}C_c^TC_cY^{(i)} + G_cG_c^T \\ &+(Y^{(i-1)} - Y_D^{(i-1)})C_c^TC_c(Y^{(i-1)} - Y_D^{(i-1)}) = 0. \end{aligned} \tag{4.1}$$

(iii) Compute the error $E = \sigma_{\max}\{M\}$, where

$$\begin{aligned} M \equiv\ & Y^{(i)}A_c^T + A_cY^{(i)} + \tfrac{1}{\alpha^2}Y^{(i)}S_cY^{(i)} - Y^{(i)}C_c^TC_cY^{(i)} + G_cG_c^T \\ &+(Y^{(i)} - Y_D^{(i)})C_c^TC_c(Y^{(i)} - Y_D^{(i)}) \\ =\ & (Y^{(i)} - Y_D^{(i)})C_c^TC_c(Y^{(i)} - Y_D^{(i)}) \\ &-(Y^{(i-1)} - Y_D^{(i-1)})C_c^TC_c(Y^{(i-1)} - Y_D^{(i-1)}). \end{aligned}$$

(iv) Repeat steps (ii) and (iii) until the error E is smaller than some desired tolerance.

Note that in (4.1), $Y^{(i-1)}$ and $Y_D^{(i-1)}$ are known; hence, (4.1) is simply an ARE, which can be solved without difficulty.

The given algorithm is not theoretically guaranteed to converge; however, computational experience shows it to be satisfactory in many cases. It has been used to solve the Riccati-like algebraic equation corresponding with plants having undesirable properties which might be expected to cause trouble. These properties include the presense of stable decentralized fixed modes near the imaginary axis, and the presence of unstable modes which are fixed with respect to each channel acting independently, but which are not decentralized fixed modes; in each case, the algorithm converges for α sufficiently large. Of course, the algorithm does not converge if the plant has any unstable fixed modes; in this case, the Riccati-like algebraic equation has no positive-definite solution.

For $\alpha = \infty$, the algorithm seems always to converge. As α is decreased, the algorithm converges more slowly, until finally, below some value $\bar{\alpha}$, the algorithm diverges. Whether this $\bar{\alpha}$ is the smallest value of α for which the Riccati-like algebraic equation has a positive-definite solution, or whether the solution algorithm simply becomes unstable for other reasons at that point, is under investigation.

Given the experience of computing decentralized control laws by the algorithm given above, we present the following conjectures:

Conjecture 1. *There exists a value of the parameter α for which the Riccati-like design equation (3.5) has a solution $Y > 0$ if and only if the corresponding plant has no unstable decentralized fixed modes.*

Conjecture 2. *If the Riccati-like design equation (3.5) has a positive-definite solution for some fixed parameter value $\alpha = \alpha_1$, then it has a positive-definite solution for every $\alpha > \alpha_1$.*

These two conjectures, if true, would be analogous to known results for the state-feedback and output-feedback cases, where the existence of appropriate solutions to design AREs is a necessary and sufficient condition for the existence of control laws providing the prescribed degree of H_∞ disturbance attenuation.

5 Enlarging the set of admissible uncertainties

Section 3 develops robust control laws for an uncertain plant. Provided the plant uncertainty satisfies (3.6), these control laws guarantee robust stabilization and disturbance rejection with H_∞-norm bound (3.7), where α is the parameter appearing in the design equations. This section gives results which, for the same designs, relax condition (3.6) to

$$\sigma_{\max}(L_i) < \frac{\sqrt{1+d^2}}{\alpha}, \quad i \in \{1, 2, \ldots, k\}, \tag{5.1}$$

and improve the bound (3.7) to

$$||T||_\infty \leq \frac{\alpha}{\sqrt{1+d^2}}, \tag{5.2}$$

where d is a non-negative scalar. These improvements rely on a convexity property of the matrix Riccati function R defined on the set of symmetric matrices by

$$R(X) = F^T X + XF + XGG^T X + H^T H. \tag{5.3}$$

This operator is studied in [17].

Theorem 4, which is proved in [17], identifies a class of easily computable matrices X for which $R(X) \leq 0$.

Theorem 4. *If $R(X_i) = 0$ for $i \in \{1, 2, \ldots, r\}$, and if scalars $\beta_i \geq 0$ satisfy $\sum_{i=1}^r \beta_i = 1$, then*

$$R\left\{\sum_{i=1}^{r} \beta_i X_i\right\} \leq 0. \tag{5.4}$$

Theorem 4 provides the means for improving the predetermined H_∞-norm bound (3.7) and uncertainty constraint (3.6) once a design from Section 3 is completed. Recall that, in each design, controller parameters are selected such that the closed-loop system satisfies the ARE

$$F_{0e}^T X_e + X_e F_{0e} + \frac{1}{\alpha^2} X_e G_e G_e^T X_e + H_e^T H_e = 0, \tag{5.5}$$

with $X_e \geq 0$. In addition to X_e, there are several positive semi-definite solutions X_i of (5.5); see [18]. Let a matrix $Z \geq 0$ be a convex combination of several solutions $X_i \geq 0$ of (5.5). Then, by Theorem 4,

$$F_{0e}^T Z + Z F_{0e} + \frac{1}{\alpha^2} Z G_e G_e^T Z + H_e^T H_e \leq 0. \tag{5.6}$$

If the range-space of the left-hand side of (5.6) includes the range-space of H_e, then for some scalar $d > 0$

$$F_{0e}^T Z + Z F_{0e} + \frac{1}{\alpha^2} Z G_e G_e^T Z + H_e^T H_e \leq -d^2 H_e^T H_e,$$

and so

$$F_{0e}^T Z + Z F_{0e} + \frac{1}{\alpha^2} Z G_e G_e^T Z + (1 + d^2) H_e^T H_e \leq 0. \tag{5.7}$$

As in the proof of Theorem 3, (5.7), (3.9), and (5.1) give

$$\begin{aligned} &F_e^T Z + Z F_e + \tfrac{1}{\alpha^2} Z G_{0e} G_{0e}^T Z + (1 + d^2) H_{0e}^T H_{0e} \\ &\leq -\textstyle\sum_{i=1}^k H_{ei}^T \left[(1 + d^2) I - \alpha^2 L_i^T L_i\right] H_{ei} \leq 0, \end{aligned} \tag{5.8}$$

which gives the desired H_∞-norm bound (5.2) via Lemma 1.

For designs where the parameter α is much larger than the minimum possible, the closed-loop system generally has an H_∞ norm much smaller than α. In this case, a convex combination $Z \geq 0$ of solutions of (5.5) can usually be found to satisfy (5.7) with $1 + d^2$ significantly greater than 1, so that (5.1) and (5.2) can represent a substantial improvement over (3.6) and (3.7).

6 Example

Consider the plant

$$\dot{x} = Ax + [B_1 \ B_2] \begin{pmatrix} u_1 \\ u_2 \end{pmatrix} + G_0 w_0 \tag{6.1a}$$

$$\begin{pmatrix} y_1 \\ y_2 \end{pmatrix} = \begin{pmatrix} C_1 \\ C_2 \end{pmatrix} x + \begin{pmatrix} w_1 \\ w_2 \end{pmatrix}, \quad z = \begin{pmatrix} H_0 x \\ u_1 \\ u_2 \end{pmatrix} \tag{6.1b}$$

where

$$A = \begin{pmatrix} -2 & 1 & 0 \\ 1 & -1 & -1 \\ 0 & -1 & -1 \end{pmatrix}, B_1 = \begin{pmatrix} 1 \\ 0 \\ 0 \end{pmatrix}, B_2 = \begin{pmatrix} 0 \\ 0 \\ 1 \end{pmatrix}, G_0 = \begin{pmatrix} 0 \\ 1 \\ 0 \end{pmatrix},$$

$$C_1 = (1\ 0\ 0), C_2 = (0\ 0\ 1), H_0 = (1\ 0\ -1).$$

The spectrum of A is {-2.80, -1.45, 0.25}; hence, the plant has an unstable mode. This section gives the results of H_∞-suboptimal control designs for this plant.

Each state-feedback control was computed from the solution $X_0 \geq 0$ of the ARE (2.2) for a given value of $\alpha > 0$, with $S = BB^T$ and $B = [B_1\ B_2]$. The LQ-optimal solution corresponds with $\alpha = \infty$. A nearly H_∞-optimal solution corresponds with $\alpha = 1.23$, the lowest value for which a solution $X_0 \geq 0$ of (2.2) was computed. For $\alpha = 1.22$, the Hamiltonian matrix associated with (2.2) has a pair of eigenvalues on the imaginary axis.

Each observer-based decentralized control law was computed for a given value of $\alpha > 0$ from matrices $X_0 \geq 0$ satisfying (2.2) and $Y > 0$ satisfying the Riccati-like equation (2.15). Solutions to (2.15) were computed by the simple iterative method given in Section 4. Computation was stopped when the matrix norm of the Riccati-like expression was less than 0.001. This algorithm converged for α ranging from ∞ to 2.7. For $\alpha = 2.6$ and below, the algorithm did not seem to converge. The lower α, the more iterations were required to satisfy this condition: For example, $\alpha = 10$ required only 6 iterations, while $\alpha = 2.7$ required 44 iterations.

Table 1 shows the controller gains for both the state-feedback and the decentralized control laws for various values of α. Note that the controller gains increase somewhat with decreasing α. Table 2 shows the closed-loop spectra and H_∞ norms. Some state-feedback poles are recognizable among the poles of the observer-based solutions.

7 Conclusion

From a positive-definite solution to a Riccati-like algebraic equation, a decentralized control law can be computed which guarantees robust stability and H_∞ disturbance rejection for a plant with structured uncertainty in a particular bounded set. Once the design is completed, a convexity property of the matrix Riccati function R can be used to show that robust stability will hold for an enlarged set of uncertainties, and to obtain an H_∞-norm bound smaller than the predetermined bound for the closed-loop system. How best to solve the Riccati-like algebraic equation remains an open research area; however, there is a simple and often effective means of finding a positive-definite solution, when one exists.

	State Feedback $K_f = -B^T X_0$	Decentralized Control $K_1 = Y_{11}C_1^T$	$K_2 = Y_{22}C_2^T$
$\alpha = \infty$	$\begin{pmatrix} -0.35 & -0.38 & 0.50 \\ 0.50 & 0.66 & -0.75 \end{pmatrix}$	$\begin{pmatrix} 0.38 \\ 0.84 \\ -0.44 \end{pmatrix}$	$\begin{pmatrix} -0.37 \\ -0.98 \\ 0.71 \end{pmatrix}$
$\alpha = 10$	$\begin{pmatrix} -0.35 & -0.38 & 0.50 \\ 0.50 & 0.66 & -0.75 \end{pmatrix}$	$\begin{pmatrix} 0.41 \\ 0.90 \\ -0.46 \end{pmatrix}$	$\begin{pmatrix} -0.39 \\ -1.03 \\ 0.76 \end{pmatrix}$
$\alpha = 5$	$\begin{pmatrix} -0.35 & -0.39 & 0.50 \\ 0.50 & 0.67 & -0.76 \end{pmatrix}$	$\begin{pmatrix} 0.51 \\ 1.13 \\ -0.57 \end{pmatrix}$	$\begin{pmatrix} -0.47 \\ -1.26 \\ 0.92 \end{pmatrix}$
$\alpha = 3$	$\begin{pmatrix} -0.36 & -0.41 & 0.52 \\ 0.52 & 0.71 & -0.79 \end{pmatrix}$	$\begin{pmatrix} 1.24 \\ 2.79 \\ -1.36 \end{pmatrix}$	$\begin{pmatrix} -1.05 \\ -2.79 \\ 1.97 \end{pmatrix}$
$\alpha = 2.7$	$\begin{pmatrix} -0.36 & -0.41 & 0.52 \\ 0.52 & 0.72 & -0.80 \end{pmatrix}$	$\begin{pmatrix} 3.11 \\ 7.31 \\ -3.96 \end{pmatrix}$	$\begin{pmatrix} -2.83 \\ -6.68 \\ 4.49 \end{pmatrix}$
$\alpha = 2$	$\begin{pmatrix} -0.38 & -0.46 & 0.55 \\ 0.55 & 0.80 & -0.86 \end{pmatrix}$	none	none
$\alpha = 1.23$	$\begin{pmatrix} -1.00 & -2.06 & 1.71 \\ 1.71 & 3.80 & -3.03 \end{pmatrix}$	none	none

Table 1. Controller gains for varying α.

REFERENCES

[1] E. J. Davison. *The decentralized control of large scale systems*, pages 61–91. Volume 1, JAI Press, 1984.

[2] S. H. Wang and E. J. Davison. On the stabilization of decentralized control systems. *IEEE Transactions on Automatic Control*, AC-18(5):473–478, October 1973.

[3] B. D. O. Anderson and D. J. Clements. Algebraic characterization of fixed modes in decentralized control. *Automatica*, 17(5):703–712, 1981.

[4] J. P. Corfmat and A. S. Morse. Control of linear systems through specified input channels. *SIAM J. Control and Optimization*, 14(1):163–175, January 1976.

[5] W. Yan and R. R. Bitmead. Decentralized control of multi-channel systems with direct control feedthrough. *Int. J. Control*, 49(6):2057–2075, August 1989.

	State Feedback		Decentralized Control		
	Spectrum	H_∞ norm	Spectrum		H_∞ norm
$\alpha=\infty$	-3.09 -1.47 -0.54	1.71	-3.21 -3.09 -2.76 -1.51	-1.46 $-1.20\pm j0.76$ -0.54 0.32	5.78
$\alpha=5$	-3.09 -1.47 -0.56	1.69	-3.25 -3.10 -2.74 -1.56	-1.47 $-1.25\pm j0.89$ $-0.49\pm j0.16$	4.42
$\alpha=2.7$	-3.09 -1.47 -0.60	1.62	-4.49 -3.28 -1.46 -1.33	$-3.06\pm j0.48$ $-2.15\pm j1.67$ -0.71	2.56
$\alpha=1.23$	-3.84 -2.74 -1.44	1.23	none		none

Table 2. Closed-loop spectra and H_∞ norms for varying α.

[6] E. J. Davison. The robust decentralized control of a general servomechanism problem. *IEEE Transactions on Automatic Control*, AC-21(1):14–24, February 1976.

[7] Ü. Özgüner and W. R. Perkins. Optimal control of multilevel large-scale systems. *Int. J. Control*, 28(6):967–980, December 1978.

[8] E. J. Davison. The robust decentralized control of a servomechanism problem with input-output connections. *IEEE Transactions on Automatic Control*, AC-23(2):325–327, April 1979.

[9] M. Ikeda and D. D. Šiljak. Generalized decompositions of dynamic systems and vector Lyapunov functions. *IEEE Transactions on Automatic Control*, AC-26(5):1118–1125, October 1981.

[10] D. D. Šiljak and M. B. Vukčević. Decentralization, stabilization, and estimation of large-scale linear systems. *IEEE Transactions on Automatic Control*, AC-21:363–366, June 1976.

[11] E. J. Davison and W. Gesing. Sequential stability and optimization of large scale decentralized systems. *Automatica*, 15(3):307–324, May 1979.

[12] D. S. Bernstein. Sequential design of decentralized dynamic compensators using the optimal projection equations. *Int. J. Control*, 46(5):1569–1577, November 1987.

[13] J. C. Doyle, K. Glover, P. P. Khargonekar, and B. Francis. State-space solutions to standard H_2 and H_∞ control problems. In *Proceedings of the American Control Conference*, pages 1691–1696, Atlanta, GA, 1988.

[14] R. J. Veillette, J. V. Medanić, and W. R. Perkins. Robust stabilization and disturbance rejection for systems with structured uncertainty. To appear in the *Proceedings of the 28th Conference on Decision and Control*, (Tampa, FL), December 1989.

[15] S. Weiland. Linear quadratic games, H_∞ and the Riccati equation. In *Proceedings of the Workshop on the Riccati Equation in Control, Systems, and Signals*, pages 156–159, Como, Italy, June 26-28, 1989.

[16] P. P. Khargonekar, I. R. Petersen, and K. Zhou. Feedback stabilization of uncertain systems. In *Proceedings of the 26th Allerton Conference on Communication, Control and Computation*, pages 88–95, Monticello, IL, 1988.

[17] R. J. Veillette and J. V. Medanić. An algebraic Riccati inequality and H_∞-norm bounds for stable systems. In *Proceedings of the Workshop on the Riccati Equation in Control, Systems, and Signals*, pages 63–68, Como, Italy, June 26-28, 1989.

[18] K. Mårtensson. On the matrix Riccati equation. *Information Sciences*, 3(1):17–50, 1971.

R. J. Veillette, J. V. Medanić, and W. R. Perkins
University of Illinois at Urbana-Champaign
Coordinated Science Laboratory
Urbana, IL
U.S.A.

Robust Stability of Linear State Space Models Via Bernstein Polynomials

A. Vicino M. Milanese

This paper presents recent results on robust stability of state space systems affected by parameteric variations. The case in which system characteristic polynomial coefficients are polynomial or rational functions of uncertain physical parameters is considered. A method is proposed which uses Bernstein polynomial expansions to test stability of families of matrices generated by parameters belonging to a hyperrectangular unceratinty set. The algorithm allows one to check positivity of a multivariate polynomial over a box. In particular, it includes a very easy test to check if the polynomial reaches its extremal value at one of the vertices of the box. It is also shown that the algorithm deriving from this approach can be combined with an algorithm presented in previous papers to compute efficiently the structured stability margin. Some numerical examples are given showing that computational improvements over the existing algorithm may be obtained for several problems.

1 Introduction

The problem of studying control systems stability robustness in the presence of structured perturbations has received much attention in the last years (see [1] for a comprehensive survey on robust control). The case in which physical parameters are present in the plant mathematical model whose values are not exactly known is one of the most investigated contexts. More specifically, the problem of studying invariance of the Hurwitz property of system characteristic polynomial when physical parameters belong to assigned uncertainty sets, frequently given as hyperrectangular domains, or the determination of the maximal stability radius in parameter space (called system *stability margin*) according to an assigned norm (common choices are l_∞ or l_2 norms), are problems of major interest. Many results are available for these problems for the case in which plant transfer function coefficients are linear functions of the physical parameters (see e.g., [2] - [8]). In this case nonconservative and computationally efficient solutions have been proposed in the literature. Considerably less results are available for the much more general case in which system characteristic polynomial coefficients are nonlinear functions of the parameters. For state space models subject to linearly correlated parameteric perturbations (this case corresponds to the situation in which system characteristic polynomial coefficients are polynomial functions of the parameters), the great majority of available results, in spite of their relative simplicity, provide only *sufficient conditions* for robust stability (see e.g., [9, 10]), whose degree of conservativeness is difficult to evaluate. Recently, solution algorithms have been proposed to compute the exact l_∞ stability margin [12, 13, 14, 15].

This paper proposes an algorithm based on Bernstein expansions of multivariate polynomials to test robust stability of a family of matrices whose entries are polynomial or rational functions of physical parameters. The key references [17, 18] give basic concepts on Bernstein expansions and on their use for constructing bounds on the range of univariate or bivariate polynomials on a hyperrectangle. Since the robust stability problem can be reduced to the problem of testing positivity of a multivariate polynomial on the given uncertainty domain, it can be solved by computing the minimum of the polynomial on that domain. Berstein expansion of a polynomial allows one to construct a sequence of upper and lower bounds on the minimum converging monotonically to it. The most interesting feature of the approach is that it provides naturally an easy test to check if the polynomial achieves the minimum on one of its vertices. This property allows one to detect if the vertices of a hyperrectangle in parameter space are the critical points of the set for stability invariance; in addition, if this is true, the answer to the robust stability problem is obtained without any additional computational effort. This attractive property has suggested to integrate the Bernstein expansion algorithm with the *signomial* algorithm proposed in [14, 15] to compute l_∞ stability margins, with the aim of improving computational efficiency, at least in all those problems where the vertices of the uncertainty sets are critical for stability.

The paper is organized as follows. Section 2 introduces notation and preliminar results. In section 3 Bernstein expansions of multivariate polynomials and two basic results are given; besides, application of these results for solving robust stability problems is illustrated. Section 4 reports several numerical examples showing the computational behavior of the algorithms presented.

2 Problem statement and preliminary results

We consider linear dynamical systems given in state equations

$$\dot{x}(t) = A(p)x(t) \tag{1}$$

where

$$A(p) = \sum_{i=1}^{l} f_i(p)A_i, \ p \in B(\rho) \ . \tag{2}$$

In (2) $B(\rho) \subset \mathbf{R}^q$ is the *uncertainty* set defined as

$$B(\rho) = \{p \in \mathbf{R}^q : \|p - p^o\|_\infty^w \leq \rho\} \tag{3}$$

where p^o, often referred to as *nominal* parameter vector, is such that $A(p^o)$ is (asymptotically) stable. The quantity ρ is a positive value and $\|\cdot\|_\infty^w$ represents a weighted l_∞ norm ($\|x\|_\infty^w = \max_i w_i^{-1}|x_i|, w_i > 0$), $A_i \in \mathbf{R}^{n,n}$ are given matrices and $f_i(\cdot) : B(\rho) \to \mathbf{R}$, $i = 1, \ldots, l$ are *polynomial* functions.

We say that $B(\rho)$ is (asymptotically) stable if each matrix of the set of matrices generated by parameters $p \in B(\rho)$ has eigenvalues in the open left half plane.

Robust stability analysis in the presence of parameteric perturbations in the form (2) is concerned with two related problems:

1. For a given uncertainty set $B(\rho)$, check if $B(\rho)$ is stable.

2. Given a nominal parameter vector p^o, compute the stability margin of system (1), defined as the maximal radius of the ball $B(\rho)$ whose interior points generate stable matrices $A(p)$.

To deal with the above problems, we consider the characteristic polynomial $\Delta(s,p)$ associated to system matrix $A(p)$ in (1)

$$\Delta(s,p) = \ \det\,[sI - A(p)] \ = \delta_0(p) + \delta_1(p)s + \cdots + \delta_{n-1}(p)s + s^n, \ p \in B(\rho) \tag{4}$$

where I is the identity matrix and $\delta_i(\cdot) : B(\rho) \to \mathbf{R}$, $i = 0, 1, \ldots, n-1$ are polynomial functions of p. We will use the well known Hurwitz stability criterion to characterize the domain of stable matrices Ω_s in parameter space. Let $H(p)$ be the Hurwitz matrix associated to the polynomial $\Delta(s,p)$

$$H(p) = \begin{pmatrix} \delta_{n-1}(p) & \delta_{n-3}(p) & \delta_{n-5}(p) & \cdots & 0 \\ 1 & \delta_{n-2}(p) & \delta_{n-4}(p) & \cdots & 0 \\ 0 & \delta_{n-1}(p) & \delta_{n-3}(p) & \cdots & 0 \\ 0 & 1 & \delta_{n-2}(p) & \cdots & 0 \\ \cdots & \cdots & \cdots & \cdots & \cdots \\ 0 & 0 & 0 & \cdots & \delta_0(p) \end{pmatrix} . \tag{5}$$

Let $H_i(p)$, $i = 1, \ldots, n$ denote the leading principal minors of $H(p)$ (called also *Hurwitz determinants* of $\Delta(s,p)$). The domain of stable matrices in parameter space Ω_s can be characterized by the Hurwitz necessary and sufficient conditions for stability of $\Delta(s,p)$

$$\Omega_s = \{p \in \mathbf{R}^q : \ H_i(p) > 0, \, i = 1, \ldots, n\} \,. \tag{6}$$

The assumption on existence of a stable (nominal) point p^o and the Orlando's formula (see e.g., [16]), allow one to conclude that knowledge of only the n-th order Hurwitz determinant (called occasionally 'critical' determinant) is sufficient for estabilishing stability of the entire set $B(\rho)$. This fact implies that invariance of the Hurwitz property of $\Delta(s,p)$ for $p \in B(\rho)$ is equivalent to strict positivity of $H_n(p)$. The critical surfaces bounding the maximal connected subset of the stability domain containing p^o are given by

$$\{p \in \mathbf{R}^q : H_n(p) = 0\} \,. \tag{7}$$

Notice that $H_n(p)$ is a polynomial in the parameters p_i. It is not difficult to verify that even if functions $f_i(\cdot)$ in (2) are allowed to be rational instead of polynomial in p, the problem of checking positivity of $H_n(p)$ is of the same nature as for the polynomial case.

The stability margin computation can be formulated as the following distance problem (see e.g. [15])

$$\begin{array}{rclcl}
\rho^* & = & \min_p \|p - p^o\|_\infty^w & = & \min_{p,\rho} \ \rho \\
 & & \text{subject to} & & \text{subject to} \\
 & & & & \left\{ \begin{array}{l} \rho \geq 0 \\ |p_i - p_i^o| \leq \rho w_i \quad i = 1, \ldots, q \\ H_n(p) \leq 0 \end{array} \right. \\
H_n(p) & = & 0 & &
\end{array} \tag{8}$$

Observe that (8) is a problem with polynomial constraints (and cost function). Optimization problems of this kind, where variables are restricted to be strictly positive, are called generalized geometric programming (or *signomial*) problems; they are in general nonconvex and may admit local extrema. In [15] an algorithm can be found for dealing with the specific problem (8). Roughly speaking, this algorithm constructs a sequence of convex problems which approximate the original nonconvex problem on the domain where the global extremum is assumed to lie. The extrema of these problems generate a sequence of lower bounds on the absolute minimum converging to it. The convergence speed of the algorithm is considerably affected by the extension of the domain where the global minimum is located. The Bernstein expansion algorithm presented in the next section allows one to reduce efficiently this domain in several circumstances. For this reason, a combined algorithm based on signomial programming and Bernstein expansions has been devised, with the aim of improving the computational efficiency of the signomial algorithm presented in [15].

3 Robust stability via Berstein expansion of multivariate polynomials

In this section we report two theorems which provide bounds on the range of a multivariate polynomial over a hyperrectangular domain. These results are constructive, in the sense that they generate an algorithm which allows one to compute sequences of convergent bounds for the extrema of the polynomial on the domain of interest. This algorithm, which is based on the expansion of a multivariate polynomial in the Bernstein form, has been originally developed in [17] for univariate polynomials and successively extended in a natural way to the bivariate case in [18]. A multivariate version of the algorithm can be readily applied to solve the robust stability problem, i.e. to check positivity invariance of $H_n(p)$ over a given uncertainty set $B(\rho)$.

In the following subsection, we define the Bernstein expansion form of a multivariate polynomial and give the two basic results which are easy generalizations of similar results in [18, 17].

3.1 Bernstein form of a multivariate polynomial

Let $Q(p)$ be a polynomial in the variables $p_i, \dots, p_q$

$$Q(p) = \sum_{i_1,\dots,i_q=0}^{r} \alpha_{i_1 i_2 \cdots i_q} \prod_{j=1}^{q} p_j^{i_j} \tag{9}$$

where r is the maximum degree of variables p_i and $\alpha_{i_1 i_2 \cdots i_q}$ are real coefficients. We want to construct bounds on the range of $Q(p)$ for $p \in B(\rho)$. Like in [18] we consider a 'unit' hypercube domain (called B_u) generated by q cartesian products of the closed interval $[0, 1]$ by itself in parameter space. In fact, any hyperrectangular domain can be mapped to B_u by a suitable linear change of coordinates. Our aim is that of constructing bounds on the extremal values of $Q(p)$, $p \in B_u$.

Define for $k \geq r$ the set of q-tuples

$$K = \{(i_1, \dots, i_q) : i_1 = 0, \dots, k, \ \dots, \ \dots, i_q = 0, \dots, k\} . \tag{10}$$

The Bernstein form of a multivariate polynomial on the domain B_u is defined as

$$Q(p) = \sum_{(i_1,\dots,i_q)\in K} \beta^{(k)}_{i_1\dots i_q} P^{(k)}_{i_1\dots i_q}(p) \tag{11}$$

where $P^{(k)}_{i_1\dots i_q}(p)$ are the *Bernstein polynomials* and $\beta^{(k)}_{i_1\dots i_q}$ are the *Bernstein coefficients.* The Bernstein polynomials are defined as

$$P^{(k)}_{i_1 \ldots i_q}(p) = \prod_{j=1}^{q} \binom{k}{i_j} p_j^{i_j} (1-p_j)^{k-i_j}, \quad p \in B_u \tag{12}$$

The corresponding coefficients of the expansion are

$$\beta^{(k)}_{i_1 \ldots i_q} = \sum_{l_1=0}^{i_1} \cdots \sum_{l_q=0}^{i_q} \left(\prod_{j=1}^{q} \binom{i_j}{l_j} \right) \rho^{(k)}_{l_1 \ldots l_q} \alpha_{l_1 \ldots l_q} \tag{13}$$

where

$$\rho^{(k)}_{l_1 \ldots l_q} = \left(\prod_{j=1}^{q} \binom{k}{l_j} \right)^{-1}, (l_1 \ldots l_q) \in K \ . \tag{14}$$

and $\alpha_{l_1 \ldots l_q} = 0$ if some l_j is such that $l_j > r$.

The following theorem gives bounds on the maximal and minimum of $Q(p)$, $p \in B_u$.

Theorem 1. *Let m and M be the minimum and maximum of the polynomial $Q(p)$ on the unit hypercube B_u and let*

$$\hat{m} = \min_{(i_1 \cdots i_q) \in K} \beta^{(k)}_{i_1 \ldots i_q}, \qquad \hat{M} = \max_{(i_1 \cdots i_q) \in K} \beta^{(k)}_{i_1 \ldots i_q} \ .$$

For any $k \geq r$ the following statements hold
i)

$$m \geq \hat{m} \quad \text{and} \quad M \leq \hat{M} \ . \tag{15}$$

ii) — Equality holds in the first (resp. second) inequality if and only if the minimal (resp. maximal) Bernstein coefficient $\beta^{(k)}_{i_1 \ldots i_q}$ is such that each i_j equals 0 or k, for $j = 1, \ldots, q$.

PROOF. The proof parallels that in [18].
i) — Inequalities (15) follow readily from the property of the Bernstein polynomials

$$\sum_{(i_1 \cdots i_q) \in K} P^{(k)}_{i_1 \cdots i_q}(p) = 1, \ \forall p \in B_u \ . \tag{16}$$

ii) —
Sufficiency.
Sufficiency follows from the fact that each value of $\beta^{(k)}_{i_1 \ldots i_q}$ for $i_j = 0$ or k coincides with the value of $Q(p)$ at a suitable vertex of B_u. Easy computations show that for $i_j = 0$ or k

$$\beta^{(k)}_{i_1 \ldots i_q} = Q(p_1^*, \ldots, p_q^*)$$

where

$$p_j^* = \begin{cases} 0, & \text{if } i_j = 0 \\ 1, & \text{if } i_j = k \end{cases} .$$

Necessity.

Necessity can be proven by contradiction. Let us consider the former inequality in (15) (analogous considerations apply for the latter). Assume that

$$m = \min_{(i_1 \cdots i_q) \in K} \beta^{(k)}_{i_1 \ldots i_q} = Q(p^*), \quad \text{for some } p^* \in B_u \tag{17}$$

and that

$$\min_{(i_1 \cdots i_q) \in K} \beta^{(k)}_{i_1 \ldots i_q} < \beta^{(k)}_{i_1 \ldots i_q}, \quad \text{for } i_j = 0 \text{ or } k \ \ (j = 1, \ldots, q) . \tag{18}$$

From the definition of the Bernstein form, property (16) and the definitions (11), (12), it follows that

$$Q(p^*) > \min_{(i_1 \cdots i_q) \in K} \beta^{(k)}_{i_1 \ldots i_q}, \ p^* \in B_u$$

which contradicts (17). Q.E.D.

The following Theorem 2, which represents the multivariate extension of the results for the univariate and bivariate cases [17, 18], gives uncertainty intervals for the minimum and the maximum of the polynomial on the unit hypercube. These intervals converge to the exact extremal values as k goes to ∞.

Let d and D be the approximation errors of the bounds given in theorem 1, i.e.

$$d = m - \hat{m} \quad \text{and} \quad D = \hat{M} - M \tag{19}$$

and ϵ the quantity

$$\epsilon = \sum_{i_1, \ldots, i_q = 0}^{r} \left(\sum_{j=1; i_j \neq 0}^{q} (i_j - 1)^2 \right) |\alpha_{i_1, \cdots, i_q}| \tag{20}$$

where $|\cdot|$ denotes absolute value.

Theorem 2. *For any $k \geq 2$ the following inequalities hold*

$$d \leq \epsilon(k-1)/k^2; \qquad D \leq \epsilon(k-1)/k^2 . \tag{21}$$

The proof of this theorem, which parallels closely those in [17, 18] for univariate and bivariate polynomials respectively, is omitted for space limitations.

3.2 Robust stability test for polynomials with nonlinear depending coefficient perturbations

Both theorems 1 and 2 can be used to check the sign of the n-th order Hurwitz determinant on a hyperrectangular domain. In particular, we notice that theorem 2 allows one to construct for each k an interval $I(k)$ containing the global minimum of the polynomial over a prescribed domain $B(\rho)$

$$I(k) = [\hat{m}, \hat{m} \; + \; \epsilon(k-1)/k^2] \, . \tag{22}$$

Since the error bound in (21), i.e., the width of intervals $I(k)$ around the global minimum, tends to 0 as $k \to \infty$, computation of the Bernstein coefficients allows one to construct a sequence of upper and lower bounds to the global minimum converging to the exact value. However, a quick examination of (10) – (11) shows that, although there may be ways of computing Bernstein coefficients avoiding the redundant computations implicit in (13) (see e.g. [18]), the computational burden may become excessive for increasing values of k, even for polynomials in few variables. In the light of these considerations, the results of the preceding subsection, if applied for a fixed value of k (say $k = r$), can be used to construct efficient, though only sufficient, tests for robust stability. The most attractive features of the basic results for practical applications are in two main points.

1. Condition ii) of theorem 1 allows one to compute the minimum of $H_n(p)$ very efficiently in all those problems in which the extremal is reached at one of the vertices of the hyperrectangular domain to be tested. More importantly, the same condition gives a systematic method to check if the extremum is actually achieved on a vertex of the domain.

2. If condition ii) of theorem 1 is not satisfied (for $k = r$), the result i) of theorem 1 and theorem 2 allows one to construct upper and lower bounds for the minimum of $H_n(p)$, without any additional computational effort. Positivity of the lower bound $\hat{m}$ is a sufficient condition for robust stability.

3.3 Stability margin computation

The tests referred to in points 1) and 2) of the preceding subsection can be used to improve performance of algorithms for computing l_∞ stability margins. In particular, in [14, 15] an algorithm for solving globally problem (8) has been proposed. This algorithm consists of a branch and bound procedure which generates a sequence of convex problems approximating with increasing accuracy the original nonconvex problem on domains containing the global extremum point. The algorithm generates sequences of upper and lower bounds on the global extremum converging monotonically to it. An important fact concerning its behavior is that the convergence speed is very sensitive to the extension of the domain of search. In our problem, this domain depends on the upper and lower bounds on the cost function ρ (the radius of the ball $B(\rho)$) computed at each step of the procedure. This consideration has been used in [15] to update iteratively the radius of the domain of search as long as the algorithm generates the tree of approximating problems. The introduction of this slight modification has caused remarkable convergence speed improvements, which have made it possible to use the algorithm successfully in a number of meaningful application examples. This fact has suggested the idea of using the Bernstein expansion test proposed in the preceding subsection as an additional tool for computing bounds on ρ, to be used during the construction of the branch and bound tree of the original signomial algorithm. A composite algorithm has been implemented, including both the signomial algorithm and the Bernstein expansion test. The resulting procedure retains the systematicity of the former algorithm and the excellent computational performance of the latter for problems in which vertices of the admissible uncertainty regions are the critical points for robust stability. Next section reports some application examples showing the performance obtained by the combined algorithm.

4 Examples

4.1 Numerical results

Several numerical examples have been worked out to compare the performance of the combined algorithm with that of each separate algorithm. The computing times reported in the following examples refer to codes implemented on a VAX 8800 computer.

Example 1. This example is taken from [19]. The control system characteristic polynomial coefficients are multilinear functions of a two dimensional parameter vector

$$\Delta(s,p) = s^3+(p_1+p_2+1)s^2+(p_1+p_2+3)s+(6p_1+6p_2+2p_1p_2+1.25)\,. \quad (23)$$

The nominal parameter vector is $p^o = [1.6, 0.3]^t$ and the perturbation weights are $w_1 = 0.15, w_2 = 0.05$.The n-th order Hurwitz determinant is

$$H_n(p) = \delta_0(p)H_{n-1}(p) = (6p_1+6p_2+2p_1p_2+1.25)(p_1^2+p_2^2-2p_1-2p_2+1.75)$$

The computed l_∞ stability margin is $\rho^* = 4$. The computing times to determine ρ^* with a tolerance of 0.5% were about 9 s for the signomial algorithm and 0.5 s for the combined algorithm. The Bernstein expansion based algorithm (with $k = r$) was not able to compute ρ^* with such a tolerance, due to the fact that condition ii) is not always satisfied and the fact that the bounds given in theorem 2 for $k = r$ are not sufficiently tight. □

Example 2. This example is taken from [12] and refers to the computation of the l_∞ stability margin for a closed loop control systems in which three uncertain parameters are present. The characteristic polynomial is

$$\Delta(s,p) = s^4+(10+p_2+p_3)s^3+(p_2p_3+10p_2+10p_3)s^2+(10p_2p_3+p_1)s+2p_1 \quad (24)$$

The nominal parameter vector is $p^o = [800, 4, , 6]^t$ and the perturbation weights are $w_1 = 800, w_2 = 2, w_3 = 3$. The n-th order Hurwitz determinant is

$$\begin{aligned} H_n(p) = \delta_0(p)H_{n-1}(p) = 2p_1(&10p_2^2p_3^3 + 10p_2^3p_3^2 + 200p_2^2p_3^2 + 100p_2^3p_3 + \\ &100p_2p_3^3 + p_1p_2p_3^2 + p_1p_2^2p_3 + 1000p_2p_3^2 + 8p_1p_3^2 + 1000p_2^2p_3 + \\ &8p_1p_2^2 + 6p_1p_2p_3 + 60p_1p_3 + 60p_1p_2 - p_1^2 - 200p_1) \end{aligned} \quad (25)$$

The stability margin obtained by application of the Bernstein algorithm, the signomial algorithm and the combined algorithm is $\rho^* = 0.3417$. The Bernstein algorithm is particularly efficient, because condition ii) of theorem 1 is always verified. The computing time to determine ρ^* with a tolerance of 0.5 % is 1 s. The corresponding computing time taken by the signomial algorithm is about 1.4 s, while the combined algorithm does the job in 0.9 s. □

Example 3. This example is taken from [13], where the l_∞ stability margin relative to the following class of perturbed polynomials is considered

$$\Delta(s,p) = s^4 + (p_1^3p_2)s^3 + (p_1^2p_2^2p_3)s^2 + (p_1p_2^3p_3^2)s + p_3^3 \quad (26)$$

with nominal parameter vector $p^o = [1.4, 1.5, .8]^t$ and perturbation weights

$$w_1 = 0.25;\ w_2 = w_3 = 0.20\ .$$

The Hurwitz determinant of order n is

$$H_n(p) = \delta_0(p)H_{n-1}(p) = p_3^3(p_1^6p_2^6p_3^3 - p_1^6p_2^2p_3^3 - p_1^2p_2^6p_3^4)\ . \quad (27)$$

The computed stability margin is $\rho^* = 1.09$. Computing times to reach this result with a tolerance of 0.5 % were about 2.5 s for the signomial algorithm, 20 s for the Bernstein expansion algorithm and about 2 s for the combined algorithm. □

Example 4. An inverse pendulum is considered. System state vector components are linear position and velocity of the carriage and angular position and velocity of the rod, i.e.

$$x = [y\ \dot{y}\ \theta\ \dot{\theta}]^t\ . \tag{28}$$

A horizontal force F applied to the carriage is considered as the command variable.

State equations of the system linearized around the origin in state space are

$$\dot{x} = Ax + Bu \tag{29}$$

where

$$A = \begin{bmatrix} 0 & 1 & 0 & 0 \\ 0 & \frac{-7b}{4m_a+7m} & \frac{3gm_a}{4m_a+7m} & 0 \\ 0 & 0 & 0 & 1 \\ 0 & \frac{-3b}{l(4m_a+7m)} & \frac{3g(m_a+m)}{l(4m_a+7m)} & 0 \end{bmatrix} \tag{30}$$

and

$$B = (1/(4m_a + 7m)) \begin{bmatrix} 0 \\ 7 \\ 0 \\ 3/l \end{bmatrix} . \tag{31}$$

In (30), (31) g is the gravity acceleration and parameters m, m_a, l, b, are defined as

m = mass of the carriage = 10 Kg;
m_a= mass of the rod = 1 Kg;
l = half length of the rod = 1 m
b = linear translation friction coefficient of the carriage = 0.2 Nm^{-1}s.

The parameter vector is $p = [m\ m_a\ l\ b]^t$. The nominal parameter vector is

$$p^o = [10\ 1\ 1\ 0.2]^t \tag{32}$$

and the corresponding weights are

$$w_1 = 1.;\ w_2 = w_3 = 0.1;\ w_4 = 0.01\ .$$

As it can be easily checked, the nominal open loop system is unstable. Thus, we assign a stabilizing state feedback $u = -H^t x$ on the nominal system and study the stability robustness guaranteed by this control law against possible variations of the four physical parameters involved.

Let us introduce the state feedback vector $H^t = [h_1\ h_2\ h_3\ h_4]$. We can compute state feedback placing the nominal system closed loop poles at the following values

$$s_1 = -0.5;\ s_2 = -0.7;\ s_3 = -1.0;\ s_4 = -2.\ . \tag{33}$$

The corresponding state feedback coefficients are

$$h_1 = -1.848;\ h_2 = -9.264;\ h_3 = 264.16;\ h_4 = 127.217\ .$$

The characteristic polynomial coefficients are

$$\begin{aligned}
\delta_3(p) &= (1/(l^2 m_a(4.m_a + 7.m))(7.bl^2 m_a - 64.918 l^2 m_a + \\
&\quad 380.067 l m_a) \\
\delta_2(p) &= (3/(l^2 m_a(4.m_a + 7.m))(-9.81 l m_a^2 - 9.81 l m m_a - \\
&\quad 4.312 l^2 m_a + 264.869 l m_a) \\
\delta_1(p) &= (1/(5 l^2 m_a(4.m_a + 7.m))(-147.15 b l m_a + 1364.67 l m_a) \\
\delta_0(p) &= 54.387/(l(4.m_a + 7.m))
\end{aligned}$$

The stability margin computed for this problem is

$$\rho^* = 6.26\ .$$

Computing times required to obtain this result with a tolerance of 0.5 % were about 30 s for the signomial algorithm, 1.5 s for the Bernstein expansion algorithm and about 0.9 s for the combined algorithm. □

4.2 Comments

Two main observations can be made on the basis of the numerical results obtained.

First, the Bernstein algorithm looses computational efficiency when the vertices of the uncertainty set are not the critical points for robust stability or when the multivariate polynomial contains high powers of the variables, i.e. is strongly nonlinear. This last consideration could be expected by observing that the number of Bernstein coefficients is combinatoric in the highest power of the variables in the polynomial.

Second, the combined algorithm works more efficiently than single algorithms in all the problems considered. The improvements are much more significant in two situations: 1) maximum degree of variables in the Hurwitz critical polynomial is large ($r > 6$ or 7) (in this case the Bernstein algorithm generally performs not satisfactorily); 2) the Hurwitz determinant has a large number of terms (> 90 or 100), but with moderate maximum degree (in this case, the signomial algorithm may loose efficiency).

5 Conclusions

In this paper we have been concerned with the problem of robust stability of linear dynamic systems with uncertain parameters entering polynomially in the state equation matrices. An method based on Bernstein expansions of multivariate polynomials has been presented for testing robust stability of families of matrices. The derived algorithm has been also used in conjunction with an existing algorithm based on geometric programming, to compute l_∞ stability margins, with the aim of improving computational efficiency of the preceding algorithm. The several numerical examples worked out show that computational efficiency of the combined algorithm is always better than that of the single algorithms and that in some cases this improvement may be remarkable (one order of magnitude difference in computing times).

Acknowledgement: This work was partially supported by funds of Ministero della Pubblica Istruzione.

REFERENCES

[1] D. D. Šiljak, "Parameter space methods for robust control design: a guided tour", *IEEE Trans. on Automat. Contr.*, vol. AC-34, pp. 674-688, 1989.

[2] S. P. Bhattacharyya, *Robust Stabilization Against Structured Perturbations.* Lecture Notes in Control and Information Sciences, 99, Springer Verlag, 1987.

[3] A. C. Bartlett, C. V. Hollot and H. Lin, " Root locations of an entire polytope of polynomials: it suffices to check the edges", *Proc. American Control Conference*, Minneapolis, 1987 and *Mathematics of Controls, Signals and Systems*,1, 1987.

[4] H. Chapellat and S. P. Bhattacharyya, "A generalization of Kharitonov's theorem: robust stability of interval plants", *IEEE Trans. on Automat. Contr.*, vol. AC-34, pp. 306-311, 1989.

[5] A. Tesi and A.Vicino, "Robustness analysis for uncertain dynamical systems with structured perturbations", *Proc. 27th CDC* , Austin (USA),pp. 519-525, 1988 (also to appear in *IEEE Trans. on Automat. Contr.*, 1990).

[6] D. Hinrichsen and A. J. Pritchard, "New robustness results for linear systems under real perturbations", *Proc. 27th CDC*, Austin (USA),pp. 1375-1379, 1988.

[7] B. Ross Barmish, "A generalization of Kharitonov's four polynomial concept for robust stability problems with linearly dependent coefficient perturbations", *IEEE Trans. on Automat. Contr.*, vol. AC-34, 1989.

[8] A. Vicino, " Maximal polytopic stability domains in parameter space for uncertain systems", *Int. J. Control*, vol.49, n.1, pp.351-361, 1989.

[9] R. K. Yedavalli, "Perturbation bounds for robust stability in linear state space models", *Int. J. Contr.*, vol. 42, pp. 1507-1517, 1985.

[10] K. Zhou and P. Khargonekar, "Stability robustness bounds for linear state-space models with structured uncertainty", *IEEE Trans. on Automat. Contr.*, vol. AC-32, pp. 621-623, 1987.

[11] R. Genesio and A. Tesi, "Results on the stability robustness of systems with state space perturbations", *Systems and Control Letters*, vol. 11, pp. 39-47, 1988.

[12] R. R. E. de Gaston and M. G. Safonov, "Exact calculation of the multiloop stability margin", *IEEE Trans. on Automat. Contr.*, vol. AC-33, pp. 156-171, 1988.

[13] A. Sideris and R. S. Sánchez Peña, "Fast computation of the multivariable stability margin for real interrelated uncertain parameters", *Proc. ACC*, Atlanta (USA), 1988.

[14] A. Vicino, A. Tesi and M. Milanese, "An algorithm for nonconservative stability bounds computation for systems with nonlinearly correlated parametric uncertainties", *Proc. of 27th IEEE Conference on Decision and Control*, pp. 1761-1766, Austin (USA), 1988.

[15] A. Vicino, A. Tesi and M. Milanese, "Computation of non conservative stability perturbation bounds for systems with nonlinearly correlated uncertainties", to appear in *IEEE Transactions on Automatic Control*, July 1990.

[16] A.T. Fuller, "Conditions for a matrix to have only characteristic roots with negative real parts", *J. Math. Anal. Applic.*, vol. 23, pp.71-98, 1968.

[17] T. J. Rivlin, "Bounds on a polynomial", *J. Res. Nat. Bur. Stand.*, vol. 74B, pp. 47-54, 1970.

[18] J. Garloff, "Convergent bounds for the range of multivariate polynomials", in *Lecture Notes in Computer science*, vol. 212, *Interval Mathematics 1985*, Springer-Verlag, pp. 37-56, 1985.

[19] J. Ackermann, H. Z. Hu, D. Kaesbauer, "Robustness analysis: a case study", *Proc. of 27th IEEE Conference on Decision and Control*, pp. 86-91, Austin (USA), 1988.

A. Vicino
Dipartimento di Sistemi e Informatica
Universitá di Firenze
Via di Santa Marta, 3 - 50139 Firenze (Italy)

M. Milanese
Dipartimento di Automatica e Informatica
Politecnico di Torino
Corso Duca degli Abruzzi, 24 - 10129 Torino (Italy)

Progress in Systems and Control Theory

Series Editor

Christopher I. Byrnes
Department of Systems Science and Mathematics
Washington University
Campus P.O. 1040
One Brookings Drive
St. Louis, MO 63130-4899
U.S.A.

Progress in Systems and Control Theory is designed for the publication of workshops and conference proceedings, sponsored by various research centers in all areas of systems and control theory, and lecture notes arising from ongoing research in theory and applications of control.

We encourage preparation of manuscripts in such forms as LATEX or AMS TEX for delivery in camera ready copy which leads to rapid publication, or in electronic form for interfacing with laser printers.

Proposals should be sent directly to the editor or to: Birkhäuser Boston, 675 Massachusetts Avenue, Suite 601, Cambridge, MA 02139, U.S.A.

PSCT1 Computation and Control: Proceedings of the Bozeman Conference, Bozeman, Montana, 1988
K. Bowers and J. Lund

PSCT2 Perspectives in Control Theory: Proceedings of the Sielpia Conference, Sielpia, Poland, 1988
B. Jakubczyk, K. Malanowski, and W. Respondek

PSCT3 Realization and Modelling in System Theory: Proceedings of the International Symposium MTNS-89, Volume I
M.A. Kaashoek, J.H. van Schuppen, and A.C.M. Ran

PSCT4 Robust Control of Linear Systems and Nonlinear Control: Proceedings of the International Symposium MTNS-89, Volume II
M.A. Kaashoek, J.H. van Schuppen, and A.C.M. Ran

PSCT5 Signal Processing, Scattering and Operator Theory, and Numerical Methods: Proceedings of the International Symposium MTNS-89, Volume III
M.A. Kaashoek, J.H. van Schuppen, and A.C.M. Ran

PSCT6 Control of Uncertain Systems: Proceedings of an International Workshop, Bremen, West Germany, June 1989
D. Hinrichsen and B. Mårtensson

GPSR Compliance

The European Union's (EU) General Product Safety Regulation (GPSR) is a set of rules that requires consumer products to be safe and our obligations to ensure this.

If you have any concerns about our products, you can contact us on ProductSafety@springernature.com

In case Publisher is established outside the EU, the EU authorized representative is:

Springer Nature Customer Service Center GmbH
Europaplatz 3
69115 Heidelberg, Germany

Batch number: 10577896

Printed by Printforce, the Netherlands